Engineering Solutions for Sustainability

Materials and Resources II

Engineering Solutions for Sustainability

Materials and Resources II

Edited by:
Jeffrey W. Fergus
Brajendra Mishra
Dayan Anderson
Emily Allyn Sarver
Neale R. Neelameggham

Organized by:
American Institute of Chemical Engineers
American Society for Civil Engineers
Association for Iron & Steel Technology
Society of Petroleum Engineers
Society for Mining, Metallurgy, and Exploration
The Minerals, Metals & Materials Society

Held at the TMS 2015 Annual Meeting,
Orlando, Florida, USA,
March 18-19, 2015

Funding support provided by:
American Institute of Mining,
Metallurgical, and Petroleum Engineers

Editors
Jeffrey W. Fergus
Brajendra Mishra
Dayan Anderson
Emily Allyn Sarver
Neale R. Neelameggham

ISBN 978-3-319-48613-0 ISBN 978-3-319-48138-8 (eBook)
DOI 10.1007/978-3-319-48138-8

Chemistry and Materials Science: Professional

Published by Springer International Publishers, Switzerland, 2016
Reprint of the original edition published by John Wiley & Sons, Inc., 2015, 978-1-119-17984-9

Printed on acid-free paper

This Springer imprint is published by Springer Nature
The registered company is Springer International Publishing AG
The registered company address is: Gewerbestrasse 11, 6330 Cham, Switzerland

Contents

Preface

This publication includes detailed capture of the content and a synthesis of the ideas presented at *Engineering Solutions for Sustainability: Materials and Resources II*, held in Orlando, Florida, March 18-19, 2015. The symposium was organized by The Minerals, Metals & Materials Society (TMS) and held in conjunction with their annual meeting. Funding support was generously provided by the American Institute of Mining, Metallurgical, and Petroleum Engineers (AIME).

This collection does not imply consensus of all attendees on the issues raised, nor does it represent an official position of any specific professional society, industry, organization, or agency represented. Rather, it reflects both the diverse perspectives and common themes that were generated throughout the event by conference participants with the guidance of the following organizers:

Brajendra Mishra, Colorado School of Mines
Iver Anderson, The Ames Laboratory
Brian Bliss, Association for Iron and Steel Technology (AIST)
Jeffrey Fergus, Auburn University
Ali M. Memari, The Pennsylvania State University
Jonathan Motherwell, Jonathan T. Motherwell and Associates, LLC
Carol Russell, Environmental Protection Agency
Emily Allyn Sarver, Virginia Tech
Darlene Schuster, AIChE's Institute for Sustainability
Deborah Shields, Colorado State University

I

Plenary Synthesis

Engineering Solutions for Sustainability: Materials and Resources II
Edited by: Jeffrey W. Fergus, Brajendra Mishra, Dayan Anderson, Emily Allyn Sarver, and Neale R. Neelameggham
TMS (The Minerals, Metals & Materials Society), 2015

ENGINEERING AND PRODUCING A SUSTAINABLE MATERIAL SUPPLY: CHALLENGES AND OPPORTUNITIES

Dayan Anderson, Brajendra Mishra, Jeffrey Fergus

Introduction

"Think not of what you see, but what it took to produce what you see."
-Benoit Mandelbrot

With impending and burgeoning societal issues affecting both developed and emerging nations, the global engineering community has a responsibility and an opportunity to truly make a difference and contribute to a sustainable future. A view of sustainability from the perspective of materials and resources encompasses two interdependent aspects: the production of those materials and resources and the designs of the products and applications that will use them. This symposium, a follow-on to the 2009 *Engineering Solutions for Sustainability: Materials and Resources* workshop hosted by the same groups in Lausanne, Switzerland, focused on materials and resources integral to meeting basic societal sustainability needs in critical areas of energy transportation, housing, and recycling. The following plenary speakers shared valuable perspectives from their respective sectors; these insights and the resulting symposium discussions are reflected in this synthesis.

- *Global Materials Resource Challenges (Opportunities) for the 21st Century* - Diran Apelian, Worcestor Polytechnic Institute
- *Sustainability: A Business Imperative, Not a Moral Sacrifice* - Behrooz Fattahi, TheEnerTrain Institute.
- *Sustainability Using Biotechnology for the Chemical Industries* - June Wispelwey, AIChE
- *A Healthy Home Is a Fractal Home* - Matthew Grocoff, THRIVE Net Zero Energy Collaborative
- *Sustainable Development Practices in the Minerals Industry* - Jessica Elzea Kogel, Imerys
- *Sustainable Policy from Washington and the States: A Role for the Engineer* - Mark Burtschi, ArcelorMittal USA

Enabling Change for a Sustainable Future

<u>The Role of the Engineer</u>

"The engineering profession must lead the way and be seen to lead the way towards a more sustainable future."
-World Federation of Engineering Organizations

As defined by the Word Federation of Engineering Organizations (WFEO), engineers can advocate for and contribute to a positive future: through cleaner technologies, by substituting high impacts with lower impacts, by designing systems to mitigate environmental problems, by improving efficiencies and through innovation. However, engineers are typically evaluated against short-term financial goals, which may often be at odds with long-term sustainable development mission and goals, as well as being at odds with the perspectives and values of the local community. In this challenging environment, engineers are expected to (1) show sensitivity to local communities, (2) provide robust information to decision-makers, (3) assume leadership, (4) understand risk and (5) seek innovative solutions. To help engineers navigate these challenges and be effective as they can be, the WFEO has established a Model Code of Practice that embraces 10 principles summarized in Table 1.

Table 1: WFEO Code of Practice for Sustainable Development

WFEO Code of Practice for Sustainable Development
1. **Knowledge & Competency:** Maintain and continuously improve awareness and understanding of environmental stewardship, sustainability principles and issues related to your field of practice.
2. **Limits to Competency:** Use expertise of others in the areas where your own knowledge is not adequate to address environmental and sustainability issues.
3. **Social Impacts:** Incorporate global, regional and local societal values applicable to your work, including local and community concerns, quality of life and other social concerns related to environmental impact along with traditional and cultural values.
4. **Sustainability Outcomes:** Implement sustainability outcomes at the earliest possible stage employing applicable standards and criteria related to sustainability and the environment.
5. **Costing and Economics:** Assess the costs and benefits of environmental protection, eco-system components, and sustainability in evaluating the economic viability of the work, with proper consideration of climate change and extreme events.
6. **Planning and Management:** Integrate environmental stewardship and sustainability planning into the life-cycle planning and management of activities that impact the environment, and implement efficient, sustainable solutions.
7. **Innovation:** Seek innovations that achieve a balance between environmental, social and economic factors while contributing to healthy surroundings in both the built and natural environment.
8. **Communication and Consultation:** Develop locally appropriate engagement processes for stakeholders, both external and internal, to solicit their input in an open and transparent manner, and respond to all concerns – economic, social and environmental in a timely fashion in ways that are consistent with the scope of your assignment. Disclose information necessary to protect public safety to the appropriate authorities.
9. **Regulatory and Legal Requirements:** Ensure that projects comply with regulatory and legal requirements and endeavor to exceed or better them by

the application of best available, economically viable technologies and procedures.
10. **Risk Mitigation:** Where there are threats of serious or irreversible damage but a lack of scientific certainty, implement risk mitigation measures in time to minimize environmental degradation.

The Need for Sustainable Policy

"Usually the first problems you solve with the new paradigm are the ones that were unsolvable with the old paradigm"
-Joel A. Barker

Society is at a crossroads with numerous paths to choose from and distinguishing those that will best enable our transition to a sustainable future from those that will not grows increasingly difficult in the complex and volatile regulatory landscape today. Our ability to develop and commercialize the diversity of sustainable technologies such a future will require is as dependent on the implementation of *sustainable policy decisions* as the raw materials and energy inputs needed to manufacture and bring them to market. Regulatory uncertainty in both the developed and developing world creates a high-risk environment for investment. Firms are increasingly reluctant to invest in research, new product development or raw material production when (a) a statute or regulation may only have the lifespan of a single election cycle, (b) each decision made by an agency is subject to potential challenges in the courts, (c) investments run the risk of being nationalized, or (d) projects are stalled by injunctions, lawsuits or other administrative delays to such an extent the investments are eventually abandoned.

In this context, policy must mature if it is to be sustainable and if it is to succeed in enabling a sustainable future. In many respects, the 'low hanging fruits' are gone and a new integrated approach informed by accurate information is paramount to success. More sophisticated policy that recognizes the technical nuances, needs and limitations across different sectors and disciplines will be necessary if it is to be effective. Among the many key observations made by the delegates of the *Engineering Solutions for Sustainability: Materials and Resources* workshop in 2009 was the necessity for all disciplines to escape a "silo mentality" in order to unlock our collective potential to solve the global challenges we face. This need is not limited to the development of technical engineering and design solutions. The regulatory world is much broader than ever before and thus the implications of a decision by a single agency, regional jurisdiction or nation must be considered with much greater scrutiny in terms of its potential impact on the greater whole.

Engineers have an opportunity to make a profound difference by informing more sophisticated policy from the local to the international level. This can be achieved by engaging with industry coalitions, horizontal associations and online forums as well as through direct engagement with their own elected representatives. Engineers have the ability to translate complex issues into simple, manageable concepts. In the workplace, they are often called upon to mediate among various disciplines and departments and are well-suited to then facilitate and integrate different ideas and options to reach a feasible solution. Engineers, through their training, possess traits and develop skills such as persistence, organization, critical analysis and ethics which are essential ingredients to becoming an effective communicator with policymakers. However, other skills not typically taught in engineering schools include: relationship building, sales, negotiation, and gauging an audience. These are skills that can be developed through professional development training programs as well as through exposure to lessons learned through the biographies and experiences of leaders in industry, politics, civil service and different professions.

Table 2: The Game and Art of Influence – Rules for Success

Influencing 101 Trust	• Never lie or mislead. • If you make a mistake in your communication, call back immediately to resolve and clarify.
Influencing 201 Flexibility	• The game of influence requires a willingness to refine your ideas until it becomes acceptable or palatable. • Never close the conversation; seek the opportunity to find a compromise. • Remain vigilant, persistent and "Live to fight another day."
Influencing 301 Expertise	• Engineers can influence the conversation with policy makers: (1) as the individual subject matter expert or (2) as a representative of a group of experts. • Provide concrete examples of how decision(s) will affect you, your company, and association. • Come prepared to offer solutions to the problem the policymaker needs to solve.
Influencing 401 Professionalism	• Proper framing of the issue is required. • Engineers are ambassadors of their trade. • Apply your engineering skills (persistence, organization, critical analysis, etc.). • It's not about being *right*, it's about *how* the problem can be solved.

Changing Our Design Paradigm

"Bottomless wonders spring from simple rules which are repeated without end."
-Benoit Mandelbrot

As evidenced by the emergence of new standards such as the NetZero Energy Building Certification℠ and the Living Building Challenge™[1] (Figure 1 and Figure 2), a new paradigm for design that embraces principles of *biomimicry* is beginning to "take root" (Table 3). To extend

[1] http://living-future.org

this metaphor, designs should strive for the complexity found in Nature. Designers should strive to "design like an old growth forest, not like a tree farm." The former is a self-sustaining system because biodiversity and complexity gives it the resilience it needs to adapt to change and survive. In contrast, a tree farm is based on a more linear process and simplified system; when the artificial inputs and management cease, the system is more susceptible to collapse.

Complexity is not the same as complication and complexity in Nature is in fact based on very simple rules. Nature self-assembles with similarity at every scale. There is an observed "performance code" for Nature and biomimicry attempts to apply this code to the materials, products and systems we create. In Nature, complex patterns are indicators of health and simple patterns are indicators of decline. Radically simple things in nature fail. By looking to nature to guide design, the products and systems we create can provide services to society for a longer period of time. Notable examples of biomimicry in design are offered below:

- By 'freezing' a whirlpool of water and creating a mold from that shape, a device known as the Lily Impeller is capable of mixing 10 million gallons of water using only 200 watts over a period of 30 seconds.[2]
- Bones which provide incredible structural strength are surprisingly porous and lightweight. Algorithms are being adopted in structural designs for automobiles to reduce material intensity without compromising strength and safety.
- The leaves of aquatic plants such as the lotus are extremely rough at the microscopic level, creating hydrophobic surfaces that form water beads which remove dust particles as they roll off due to gravity. These same 'self-cleaning' principles are being applied to paints, glass, textiles and other applications.
- The adaptive strategy of owls, a nocturnal predator, depends on "silent" flight. Research is underway to replicate the shape and texture of owls wings which could have profound implications in technologies seeking to reduce noise and vibrations (wind turbines, aircraft underwater naval vessels, etc.).
- Conventional water and power delivery systems are centralized with linear and perpendicular distribution networks that require considerable energy to transmit and transport. However highly efficient systems are found in nature. Over 60,000 miles of plumbing exists in the human circulatory system and it is run by a pump that averages 1.3 watts. Trees and the human circulatory system are composed of repeating patterns of larger diameter conduits branching into narrower spurs. Natural watersheds exhibit this same fractal pattern.
- A biodegradable alternative to plastic known as shrilk (made from discarded shrimp shells and proteins derived from silk) was inspired by the fractal layers of an insect's cuticle.[3]

---

[2] http://www.paxwater.com/biomimicry

[3] http://news.harvard.edu/gazette/story/2014/05/promising-solution-to-plastic-pollution/

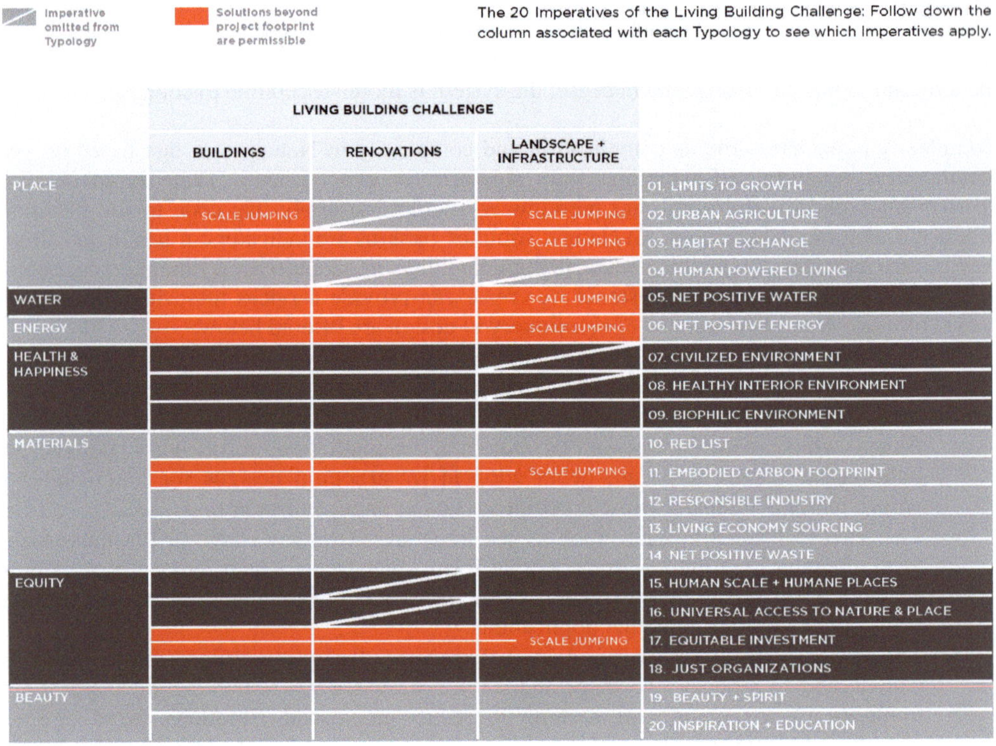

Figure 1. Living building challenge. (Source: International Living Future Institute. 2013. *Living Building Challenge 3.0: A Visionary Path to a Regenerative Future*. https://living-future.org/sites/default/files/reports/FINAL%20LBC%203_0_WebOptimized_low.pdf)

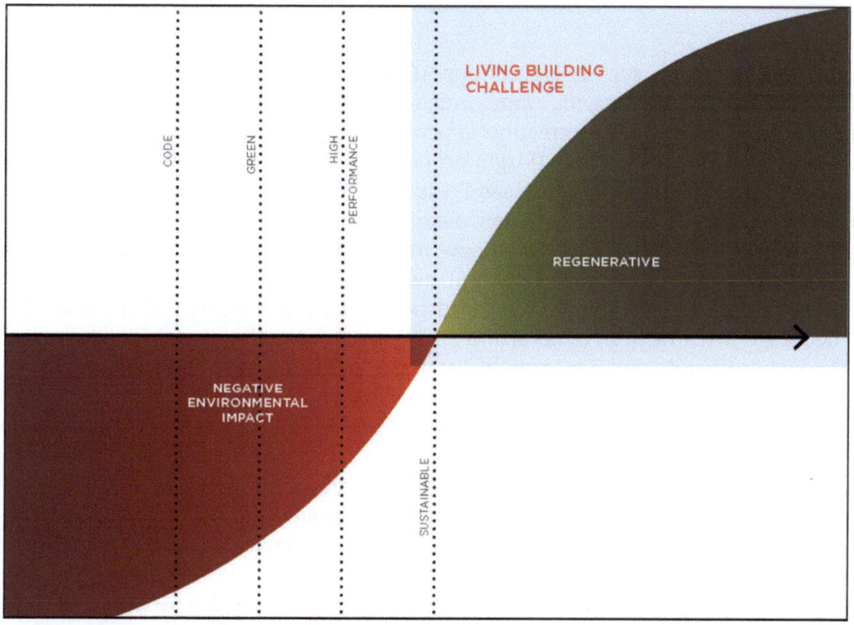

Figure 2. Living building challenge. (Source: International Living Future Institute. 2013. *Living Building Challenge 3.0: A Visionary Path to a Regenerative Future*. https://living-future.org/sites/default/files/reports/FINAL%20LBC%203_0_WebOptimized_low.pdf)

Table 3: Principles of Biomimicry[4]

Principles of Biomimicry
Nature creates conditions conducive to life
Nature fits form and function
Nature uses only the energy it needs
Nature recycles everything
Nature rewards cooperation
Nature banks on diversity
Nature demands local expertise
Nature curbs excesses from within

Innovations and Policies for Sustainable Development

The global supply of minerals and other non-renewable resources is finite. However, if we revolutionize and innovate in how we produce and manage them throughout the product life cycle, the services these non-renewable resources provide in the form of materials, metals, chemicals and energy can be long-lasting to the benefit of both present and future generations.

[4] Benyus, Janine M. 2002. Biomimicry: innovation inspired by nature. New York: Perennial.

Innovations within each sector create opportunities to enable and reinforce innovations elsewhere; collectively enabling the transformations a sustainable future will require. The most fundamental aspect of this transformation is achieving a sustainable raw materials supply mix, defined as a combination of primary (virgin ore) and secondary (recycled material) resources that together maximize net benefits of mineral supply across generations.[5] To achieve this end, numerous efforts must be pursued simultaneously:

1) We need to discover and develop new mineral resources, and all nations endowed with mineral and energy resource potential must be equally committed to contributing to the global raw materials supply. We need political environments and frameworks that will encourage investment in mineral and energy exploration.[6]
2) Through research and innovation we must strive to develop functionally equivalent renewable materials for minerals (e.g., plastics, composites, etc.,) to the extent that such substitution is feasible and for which the related land-use, water and energy demands are weighed carefully against conventional choices.
3) As with substitution and wherever feasible, the recycling of metals and other construction materials must be pursued. A significant transformation in manufacturing processes must take place and reducing the water and energy intensity of recycling must remain a research priority.
4) For the majority of applications for which minerals are and will remain the only viable material choice, we must strive for more efficient use of these non-renewable resources through improved mining and processing methods that optimize recovery and generate the maximum value.
5) The correct pricing of all resources is required that properly accounts for scarcity as well as the environmental and social costs of production.

The following discussions will highlight various perspectives on sustainability, innovation and policy challenges within the mining, energy, chemical and recycling sectors.

Mining & Minerals

"The World Needs Mining. Mining Needs to Change."
-Kellogg Innovation Network Catalyst Development Partner Working Group

To have a serious dialogue about mining, minerals and sustainability, the entire life cycle of the products and services that support our technology-driven society today must be considered. However, such a discussion should begin with a brief sojourn into mining history to put the transformative opportunities the industry and society faces into proper perspective. Humans throughout their history have had a very intimate relationship with minerals and mining as evidenced by the iron oxides used in the Paleolithic cave paintings in Spain to the Neolithic flint mines of Belgium. These early miners used rudimentary hand-tools and were mining for local markets. Similar tools are still used today in many developing nations where artisanal and small-scale mining (ASM) accounts for 80% of global sapphire production, 20% of gold production

[5] Engineering Solutions for Sustainability: Materials and Resources, 2009, Lausanne Switzerland.
[6] See, for example, Canada Mining Innovation Council Footprints Project: http://cmic-footprints.ca/.

and 20% of the world's diamond mining[7]. With the advent of the industrial revolution, innovation was driven by the ever-increasing demand for minerals leading to the more mechanized and large-scale mining operations we see today. While these large-scale industrial operations directly employ approximately 7 million people worldwide, ASM operations provide livelihoods for some of the most impoverished and isolated populations estimated at 100 million. While each sector faces unique social, environmental and governance challenges, both are necessary to maintain a sustainable raw materials supply and with proper management can achieve positive social outcomes. These are important distinctions to make because the ever-increasing shift from local to global markets for *all* resources has profound implications on what the *vision of our future* will look like and how it will take shape.

An increasing body of research and experiences in the field are serving to shape what 'responsible mining' looks like which includes doing business with transparency, making decisions that consider the social and environmental impacts that can be experienced locally as well as globally, and taking due consideration for future generations into account.

The global challenge the mining industry faces today is underpinned by three external pressures: (1) the global supply of minerals is finite, (2) demand for minerals is expected to increase and (3) mining is shifting from developed to developing nations. This presents a very complex and challenging situation compounded further by many internal pressures: on average ore deposits are deeper, they are trending towards lower grades, and mining is now taking place in regions of the world that are more environmentally sensitive. Pressures to mine these more sensitive areas (e.g., tropical rainforests, areas of high biodiversity, deserts with low water availability, higher elevations, polar regions, the sea floor, etc.,) are magnified when well-developed countries elect not to develop their own known resources and/or discourage investment in mineral exploration within their own borders. These pressures are amplified further still when national research priorities are directed away from developing mining and processing technologies that are less water- and energy-intensive, produce fewer emissions and minimize or even eliminate health and safety risks.

For nearly 25 years the mining industry has examined the contributions it can make to sustainable development which has led to (a) the establishment of industry wide codes, principles and commitments, (b) the development of individual corporate visions, policies, programs, reports and (c) individual professionals developing performance measures and operations-based metrics. An effective and credible industry-led path towards sustainable development first requires industry-wide commitment to publicly stated principles and declarations. The International Council on Mining and Metals (ICMM) members are required to adopt 10 principles and six supporting position statements and are evaluated against these principles annually. Professional societies and trade associations focused on different commodities, sectors or regions within the minerals industry have since developed guiding principles and codes of practice of their own. Together, these organizations have commissioned extensive research and collaborative initiatives to develop guidance on what are collectively being referred to as "sustainable mining practices."

---

[7] https://www.worldbank.org/en/topic/extractiveindustries/brief/artisanal-and-small-scale-mining.

Concurrent to these efforts at the corporate level, minerals professionals representing 12 leading technical professional organizations from across the globe came together to write and sign the Milos Declaration & Milos +10. This declaration clearly states how the minerals professional community will contribute to sustainable development and further demonstrates a universal commitment to SD across the entire minerals community. In order for industry to progress on its trajectory to fully adopted and integrated sustainable development practices, individual companies must commit to transparent reporting verified through rigorous third-party certifications based on tangible metrics. They have done so through several different initiatives, including but not limited to:

- The GRI guidelines, now in their fourth generation (G4), are the leading global standards in sustainability reporting. They are applicable to organizations of any size, type, sector or geographic region, and have been used by thousands of organizations worldwide. GRI reporting promotes transparency and accountability in sustainability performance, as well as continuous improvement against a universal framework.
- Although no single all-encompassing SD certification exists in the ISO system per se, there are ISO certifications that address two of the three pillars of SD: environment and social responsibility (ISO 14001 and ISO 26000 respectively). By obtaining these certifications companies demonstrate they have credible SD programs in place that meet clearly defined standards.
- Engineers and professionals often are not aware of what is happening at the corporate or industry level and corporations face the challenge of finding ways to drive these principles down to the operational level. At each stage in the mine life cycle, operations as well as academia continue to develop and propose metrics and performance indicators that track social, environmental and economic impacts.

Mining, oil and gas development are unique to other industries in their ability to generate wealth on a large scale. When direct and indirect revenues are taken into account, the mining, quarrying and petroleum sectors combined drive more than 45% of the world's measured economic activity[8]. Mining provides the building blocks for human development and serves as a socio-economic engine that can be harnessed to make significant positive contributions on a global scale, but arriving at this end point requires leadership, adherence to best practices and a strong vision for a new way of doing business.

This new *positive footprint* paradigm is emerging as thought leaders in the minerals industry have joined "dreamers" from other sectors, nonprofits, religious organizations, academia and government to collaboratively "reinvent mining." Convened by the Kellogg Innovation Network (KIN), this diverse coalition created a new vision and mindset for the industry called the Development Partner Framework (DPF) to inspire a shift from "individual company-focused wealth creation to collective investments with shared purpose and vision."[9] Under this new paradigm, the mining industry of the future has an opportunity to create shared purpose, flourishing ecosystems and global competitiveness wherever it operates through four key mechanisms referred to as enablers (Table 4).

[8] Reinventing Mining: Creating Sustainable Value, KIN Catalyst Mining Company of the Future Initiative

[9] Kellogg Innovation Network

Table 4: Enablers of the KIN Catalyst Development Partner Framework

Co-Inspire	By engaging a diversity of experts (geopolitical, government, civil society) an environment of co-inspiration will "support leaders through this brave new process"
Co-Collaborate	Miners will need to integrate multiple perspectives (e.g., ecology, economists and systems thinkers) into their designs. New policies, laws and regulatory environments will be needed to co-create the architecture and implement this new model of doing business.
Co-Innovate	The face of mining could rapidly change if new products and services were invented and reinvented with the same pace and momentum of other industries with robust networks of schools, start-ups and funders.
Co-Educate	The next generation of miners will need to be systemic designers and not one-off problem solvers that embrace the ethos "Do right to the land that is entrusted to you."

(Source: Adapted from "Reinventing Mining: Creating Sustainable Value" by the KIN Catalyst)

This new framework is still in its infancy and will require a system innovation approach to bring out the scale of transformation needed in the industry. To inform and shape its implementation plan, the working group examined the "Six Steps to Significant Change" model developed by the Forum for the Future[10]:

1. Experience the need for change
2. Diagnose the system
3. Create pioneering practices
4. Enable the tipping
5. Sustain the transition
6. Set the rules of the new mainstream

Energy

"The frog does not drink up the pond in which it lives."
-Native American Proverb

In recent years, sustainability in energy production and consumption has emerged as perhaps the most critical imperative of this century and is being touted as the grand challenge. As our lives are being altered by advancing technology in ways that we never imagined just a few short years ago, the awareness and understanding of the complexities in utilization of the nature and the impact of the footprint that we leave behind must lead us to a fundamental shift in the way we manage our individual and professional lives.

As the chart in Figure 3 shows, fossil derived fuel is the main supplier to energy demand in all parts of the world and will be not only in the near future but, perhaps, for the major part of this century. Gradually, nuclear energy and renewable energy capabilities will have to be built to sustain and fill the energy demand and supply gap that will ensue in coming years. With

[10] https://www.forumforthefuture.org/blog/introducing-forum%E2%80%99s-six-steps-significant-change

available fossil fuel supply, a gas-nuclear-wind energy combination will meet the needs of most nations providing a balance of energy and environmental sustainability.

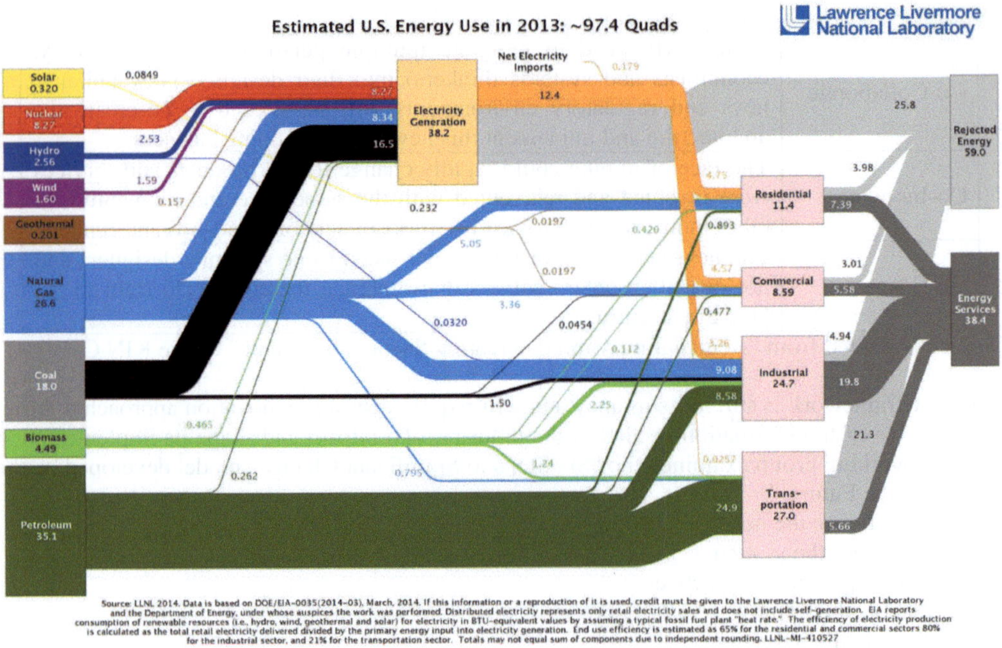

Figure 3. Estimated U.S. Energy Use in 2013. (Source: Lawrence Livermore National Laboratory, 2014. Data is based on DOE/EIA-0035/2014-03)

The oil and gas industry faces a very challenging future. Energy demand continues to grow, while the mature fields are declining. Out of the 1.2 trillion barrels of proven oil reserves, roughly two thirds are extremely challenging to develop and produce due to a variety of complicating problems such as depth, tight and thin formations, development and application of leading edge technologies, and a multitude of environmental considerations, and regulatory issues. The task is to extend the life of the industry through improving the success of finding new discoveries in frontier and remote basins, increasing the ultimate recovery from existing fields, as well as the need to find ways to tap into unconventional resources. At the same time, we must intensify the focus on sustainability, which calls for meeting humanity's needs without harming future generations.

Technology has always been the enabler for the industry to meet the demand for the last 150 years. But now, more than ever, the need for quick development and rapid deployment of technologies in the field is paramount. And, more importantly, we are realizing that there is only a small window of opportunity for us to act. However, these are not single industry challenges. Through the synergies, we can then transform from the current incremental pace into a fundamental shift in the ways we incorporate sustainability into the operational framework.

At first glance, the notion of sustainability may seem intimidating to many. The petroleum industry, however, has been focusing on the concept or its components for a long time but under different descriptive terms such as optimizing production, maximizing reserves, reducing cost, cutting waste, increasing efficiency, optimizing processes, minimizing footprint, maximizing safety, reducing environmental impact, and increasing corporate social responsibility. Considerable efforts have been invested in time and capital or human resources, in achieving these goals. Thus, for the industry, the gap between what we have been doing and what we need to do is much narrower than what it is perceived to be. It is only in recent years that the concept of sustainability has moved from the background into limelight. Sustainability has become the cornerstone of the operation, from the upstream to downstream. Environmental impact, social responsibility, sustainable solutions, and best practices are incorporated across the wide span of exploration and production life cycle from seismic acquisition to construction and drilling and onto production and abandonment.

Current incremental changes towards sustainability are not sufficient — we need a fundamental shift in the way we make business decisions. Sustainability must become the principle moral and business imperative in the lives. We must fully understand the complex interdependencies of nature, societal obligations, and the way we conduct the business.

The concept of prudent development of the natural gas and oil resources means development, operations, transport and delivery systems that achieve a widely acceptable balance of economic growth, environmental stewardship and sustainability, energy security, and human health and safety. This of course involves tradeoffs among these factors. The petroleum industry in concert with the rest of the world is finding sustainability presenting both risk and opportunities for its business, thus the need for incorporating sustainability as a business imperative and not a moral sacrifice.

Chemicals

The increase in petroleum pricing in the past decade, in combination with finding more sustainable feedstocks, has focused research on using bio-based feedstocks for chemical production for a sustainable supply. However, with the recent increase in production and availability of shale gas and the impact of the reduction in price for petroleum during the past six months, many are questioning the future of bio-based feedstocks. This review of current and potential methods for chemical production provides a perspective on the applications of biotechnology.

Most chemicals and polymers are built upon five basic chemicals – methane (i.e. natural gas), ethane, propane, butane, and aromatics (Figure 4).

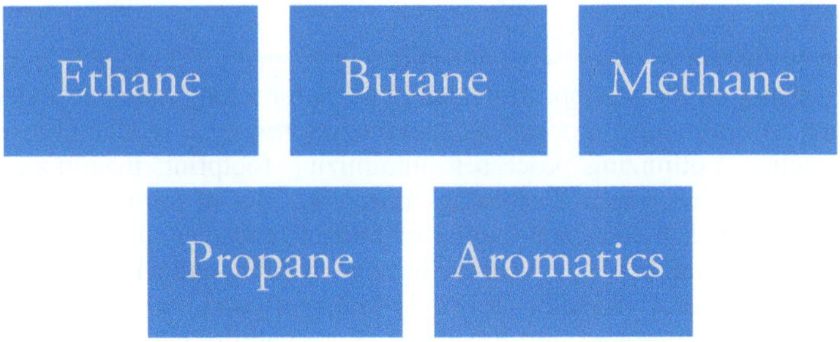

Figure 4. Basic building blocks of chemicals and polymers.

Only 2% of methane is used to produce chemicals, such as ammonia and methanol. Methane is also used to make synthesis gas, a mixture of carbon monoxide and hydrogen that makes other basic chemicals. Ethane is consumed for producing ethylene in steam crackers. Propane is mostly used for home or commercial heating, with a small amount going to chemicals. About two-thirds of butanes are used as liquefied petroleum gases (LPGs) and the remainder is consumed in other chemical production. Aromatics, consisting of benzene, toluene and xylene (BTX), are used for chemical production and in fuels as octane enhancers. The chemicals include polystyrene, phenolic resins, polycarbonate, nylon, polyurethane, polyesters and phthalic anhydride.

As the first three building blocks can be produced from low-cost natural gas and natural gas liquids, there are few opportunities for biotechnology to replace these routes, usually unless incentives are offered. The well-known ethanol incentive is a prime example of a situation where the legislation drove the commercial practice of fermenting corn to ethanol for consumption in gasoline to achieve emission standards. A second example that does not require incentives is Brazil's sugarcane-based ethanol production, where the energy to yield ratio is high enough for commercial production.

The last two building blocks are typically produced from crude oil, which has fluctuated with supply, demand, and geopolitical factors. Oil plummeted from $100/barrel in July to $50/barrel in December, 2014. It is with these last two building blocks that the greatest opportunity for biotechnology to produce certain chemicals economically.

Many opportunities for aromatic production methods are currently being pursued using bio-based chemistries, as the following examples suggest. Genomatica is developing a bio-based route for three key nylon intermediates (hexamethylene diamine, caprolactone and adipic acid). These intermediates have a total market of $18 billion/year and are used to make nylon-6 and nylon-6,6. Braskem and Amyris are developing a process to biologically produce isoprene. Braskem has manufactured bio-based polyethylene at its Trifo, Brazil complex since December 2013. Coca Cola recently invested in the development of Virent's bio-based p-xylene to make 100% renewable polyethylene terephthalate bottles.

There are also a number of bio-based polymers that have been produced for years due to their unique properties. NatureWorks has produced polylactic acid for over a decade, and it is one of the highest volume bioplastics. DuPont and Tate and Lyle produce 1,3 propanediol which is then used for engineering polymers with unique properties. Cargill and others produce bio-based polyols from soy for incorporation in flexible foam used in upholstered furniture, bedding, carpet backing and automotive seating. The opportunities for sustainable, economical processes in the chemical industry that use biotechnology are greatest when targeting the C4 and BTX building blocks. In this targeted area, producers are having the most success, with more processes and plants under development. Thus, biochemical industry has been consistently moving towards sustainable use of bio-resources that allows it to decouple its growth from a non-renewable to a renewable form of material.

Recycling

The global population is estimated to exceed ten billion by the middle of this century placing materials and resources into severe constraint. From the energy (all forms) and environmental (land, water and air) sustainability perspective, exploitation of resources to generate new materials will be immensely difficult and costly. Three factors contribute to this difficulty and, in many cases, unmanageable sustainability. (a) Natural resources, both in quality and quantity are dwindling and, therefore, enhancing production to meet the demands will be energetically and economically non-viable (see Figures 5a and 5b). (b) In the case of many materials, the demand will simply outstrip the supply that can come from the primary production, such as the critical materials. (c) The impact on environment from the natural resource exploration not only will cause irreparable damage but also generate massive amounts of process wastes that can have hazardous consequences.

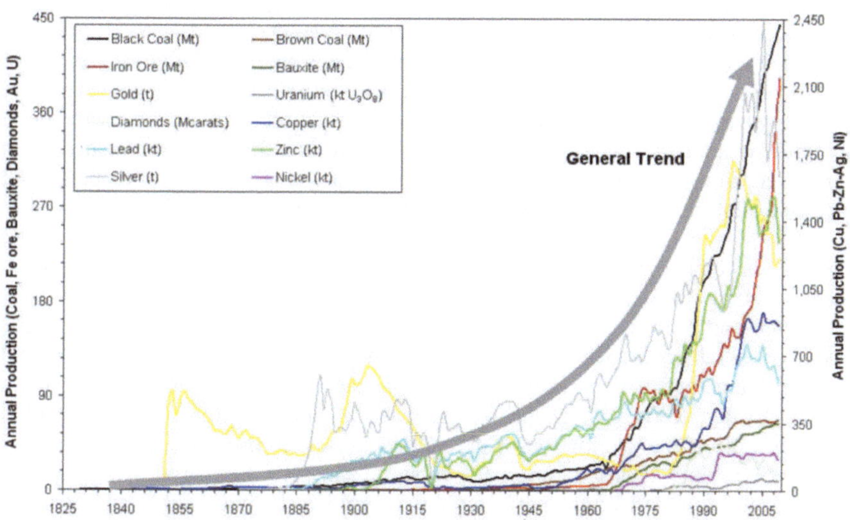

Figure 5a: Global trends in metal production. (Source: *The Sustainability of Mining in Australia – Key Production Trends and Their Environmental Implications for the Future*, Executive Summary, http://users.monash.edu.au/~gmudd/sustymining.html. Used with permission.)

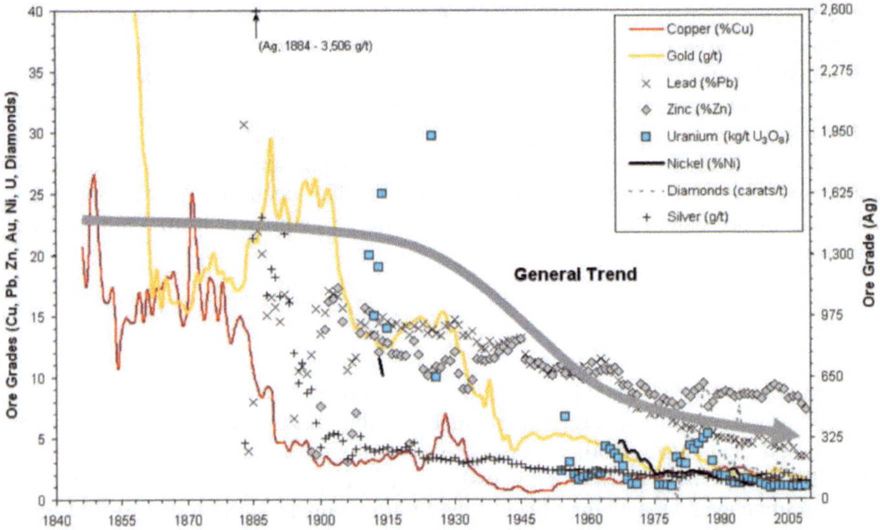

Figure 5b: Decline in natural resource quality. (Source: *The Sustainability of Mining in Australia – Key Production Trends and Their Environmental Implications for the Future*, Executive Summary, http://users.monash.edu.au/~gmudd/sustymining.html. Used with permission.)

Thus, increasingly, the U.S. government, academia, domestic industry, and the public acknowledge the imperative that we need to conserve energy and natural resources while exercising judicious stewardship of the environment. The issue of sustainability is and should be paramount in how we design, manufacture, use, and retire the many products we consume throughout the world. Inorganic materials are not renewable; the need exists for the development of technologies to address materials recovery and recycling. Engineering research supporting materials recovery and recyclability is inherently multidisciplinary and must respond to the needs of a multiplicity of commercial stakeholders from throughout the materials supply chain.

When does the waste become a recycling opportunity? When the economic value from recycling waste and scraps can compete economically with the manufacture of new products from virgin feedstocks – then and only then will recycling flourish as a business. Despite growing efforts to recycle metals, we fail to recover half of the domestic post-consumer metal scrap reclaimable from retired products, and we continue to rely on primary metals production to fulfill two thirds of our manufacturing needs. Use of primary metals, in lieu of scrap, increases global energy consumption as well as the production of greenhouse gases. In order to augment recycling rates the materials community needs to upgrade recovery and recycling processing technologies to maximize the capture of post-consumer scrap and minimize the quantity of manufacturing scrap (see Figure 6).

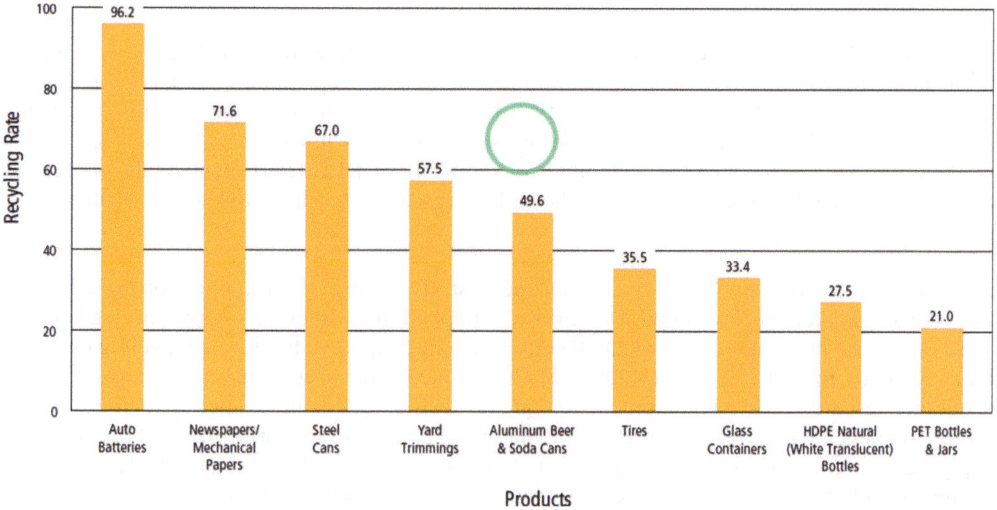

Figure 6: 2013 Recycling rates of various commodities. (Source: EPA)

The US Department of Energy published an analysis of the criticality of selected rare metals, the most critical elements were identified to be dysprosium, neodymium, terbium, yttrium and europium - what are also known as the heavy rare earth elements. Based on the demand and supply position of common rare-earths, the prices of common metals like Ce, Nd, Sm, La and Y, went up by as much as 150% to 700% in a short period of six months in 2010 and the global forecast shows each of these metals continues to be in short supply.

Until now, recycling of rare earths has not been implemented on a large industrial scale. It has been reported that recycling of rare earth based materials would have a stabilizing effect on price, supply, and quality. In addition, an infrastructure does not exist for the recycling of rare earth based materials. Higher volume application of rare earth based materials seems eminent, and therefore, the time is right to develop both the technology and infrastructure. Researchers have shown that aqueous processes, as well as molten slag electrorefining techniques, are viable methods for returning high purity metals, but have limitations in their ability to be selective and cannot handle all kinds of wastes, such as high density permanent magnet swarf. Liquid-liquid extraction using metallic solvents presents interesting opportunities that overcome the shortcomings of the other methods. However, more research is required to develop technologies for commercial use.

Several constraints on recycling of rare earth based materials were reported in an analysis by Okie-Institute AV, such as i) need for an efficient collection system, ii) need for sufficiently high prices for primary and secondary rare earth compounds, iii) losses of post-consumer goods by exports to developing countries and iv) the long life time of products such as electric motors and wind turbines. It has been suggested that 40-45% REE magnets are scrapped during manufacturing stage. Other researchers have suggested various pyrometallurgical and hydrometallurgical routes to recover REE from these scrapped magnets. Efforts have also been made to recover REE from used Ni-MH batteries. During pyrometallurgical treatment of these batteries the REEs report to the slag. Various hydrometallurgical routes have been investigated to recover these elements. Recycling of rare earths from phosphors, as discussed above, provides an efficient way to recover high value heavy rare earth elements. Not much work has been done on recycling of rare earths from catalysts. Catalysts primarily contain low value light rare earths like lanthanum and cerium which might be one of the reasons why not much effort has been put in this direction. However, once the economics of recycling of REE from spent catalysts becomes favorable, one would expect to recover the light rare earths feasibly and return them back to manufacturing.

A number of extraction processes have been successfully evaluated for application but not many have been commercially developed. However, the impending problem of supply shortage and the soaring prices of rare earths have made the environment conducive to build a recycling economy of these metals to address the problems. Such a strategy, if successfully implemented, would encourage research and development of green technologies and other critical areas by minimizing dependence on the unpredictable nature of overseas rare earth supply. This change in supply scenario, of course, will depend on the specific type of metal and material and its demand. In addition, some of the 'exotic' metals, such as molybdenum, rhenium, ruthenium, tungsten, PGMs, for which demand is predicted to stay high and the indigenous resources low, recycling and recovery from spent secondary resources will be the only viable option for sustainable growth. The sustainability solutions in the recovery and recycling area, thus, is embedded in the life-cycle management of all engineering commodities where all vested groups take responsibility (Figure 7).

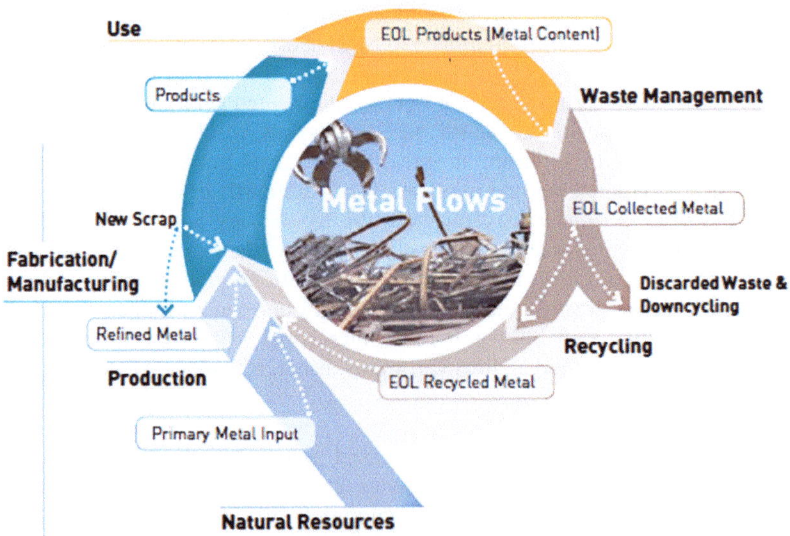

Figure 7: Life-cycle management of engineered materials. (Source: UNEP, *Assessing Mineral Resources in Society: Metal Stocks & Recycling Rates*, http://www.unep.org/pdf/Metals_Recycling_Rates_Summary.pdf. Used with permission.)

Summary of Proceedings Articles

The individual articles in this proceedings volume that were contributed by speakers provide additional insight and details on the sourcing, processing and selection of materials as well as on the supply of energy to support these and other activities.

Materials Resources

The first article [Kogel] provides an overview of needs for improving the sustainable development performance of mining and emphasizes that collaboration among the various stakeholders (e.g. mining companies, government, and communities) is needed to effectively address the issues throughout the mine-life cycle. This is followed by an article on the supply and demand of rare-earth metals [Mishra]. The growing need for such critical materials places pressure on the mining industry to meet the growing demand and requires that improvements in mining operations be complemented with other approaches, such as expanded recycling.

There are several articles on specific approaches to address challenges in the mining industry. These include reducing the sulfur emissions in copper smelters [Jie] as well as using sulfur to reduce CO_2 emissions and energy consumption [Neelamaggham]. There are also articles on a process for removal of phosphorus from iron ores [Kim] and for capturing chlorine in the metallurgical processing of zinc [Guo]. Process optimization can be used to both reduce the raw materials used as well as to reduce CO_2 and other emissions [Cavaliere]. Such optimizations can be improved using thermodynamic calculations especially in gas phase processes such as vacuum distillation [Kong]. Energy reductions can be achieved using alternate processesor through the modification of existing process, such as the recrystallization of brass alloys [Zhao].

Process improvements can also be achieved by addressing the materials used. For example, using renewable biological resources, rather than fossil fuels, is a more sustainable approach to producing polymeric materials. In addition, waste polymers can be used as the carbon source in steel making, which also reduces the use of fossil fuels [Hong]. The waste materials from mining operations can be used for engineering products, which both decreases the negative impact of the mining operation (i.e. less waste emitted) and reduces the needed resources. For example, tailings from mining have been used in ceramic composites for roof tiles and other building materials [Pérez and Badillo] and granite waste materials have been used in structural ceramics [Diaz].

Materials Selection

The selection of materials for engineering applications is important both in matching materials demand to the available supply as well as to reduce the environmental impact during use of the product. One article describes a software package that can be used to quantify the environmental impacts associated with the selection of particular materials and thus to understand the implications of materials selection [Fredriksson].

For example, the development of biodegradable polymers reduces the environmental impact of discarded or landfilled plastics [Akpan]. Protective treatments, such as corrosion inhibitors for rebar, can have negative environmental impact, so modification of these inhibitors is an important improvement [Okenihi]. Selection of materials for building can reduce energy use in heating or cooling the building and positively impact other sustainability metrics [Solnosky].

Energy use is particularly important in mobile applications where the energy needed to move the vehicle increases with the weight of the vehicle. This is particularly important in aerospace applications. One of the articles describes the use of a silver coating to provide lightning protection, so that a polymer composite can be used in an aircraft [Rajesh]. Lightweight materials are also important for automotive applications [Bushi] and there is interest in increasing the use of magnesium alloys, which requires improvements in the techniques for alloy processing [Su]. Lighweight materials are also key for use in orthopedia and dental applications [Gheisari].

Power Production

Several articles address issues associated with energy conversion and storage. This includes the use of biomass rather than fossil fuels for briquettes [Ilochonwu]. Another article discusses the use of high temperature fuel cells, which have excellent fuel flexibility and thus enable the efficient conversion of fossil fuels as well as fuels derived from renewable sources [Fergus]. There is an article on the use of superconducting transmission cables to reduce energy losses in the power grid [Sheehy].

Energy storage is important both for mobile applications, such as electronics and electric vehicles, but is also critical for enabling renewable energy conversion technologies such as solar and wind. There is an article on the use of hydrogen for energy storage to reduce the use of carbon [D'Errico], as well as an article on electrochemical storage using magnesium batteries [Shailaja].

II

Materials Resources

II

Engineering Solutions for Sustainability: Materials and Resources II
Edited by: Jeffrey W. Fergus, Brajendra Mishra, Dayan Anderson, Emily Allyn Sarver, and Neale R. Neelameggham
TMS (The Minerals, Metals & Materials Society), 2015

SUSTAINABLE DEVELOPMENT AND THE MINERALS INDUSTRY

Jessica Elzea Kogel

GeoIntellus; 2301 Laurel Lane; Augusta, GA, 30904, USA

Keywords: minerals, mining, contribution to sustainable development, resource efficiency

Abstract

The consumption of non-renewable mineral and metal resources is driven by society's need for the products and services that are derived from these essential raw materials. However, the global supply of minerals is finite and demand is expected to increase as a greater proportion of the world's growing population becomes more affluent. To meet this demand, mineral resource development will likely accelerate and mining will continue to shift from developed to developing nations where the vast majority of untapped mineral deposits are located. Mining projects carry a potentially significant social, economic and environmental footprint that, depending on how the project is managed, can have a positive or negative impact on current as well as future generations. The mining industry recognizes this challenge and has adopted business practices that address the problem of how to produce minerals in a sustainable and responsible way.

Introduction

For much of history humans have relied on minerals which, until the mid-19th century, were mostly extracted by hand from small mines. Large-scale mechanized mining operations began in Europe during the Industrial Revolution and since 1900 the rate of mineral consumption has grown almost exponentially [1]. Today minerals provide the basic raw materials for a vast array of manufactured consumer goods and services including innovative products that directly address some of modern society's most difficult environmental challenges. Examples include solar panels, high performance batteries, wind mill turbines, recyclable packaging, and energy efficient light bulbs. Mining also generates significant wealth and impacts economies on a global scale [2]. By some estimates mining's total direct and indirect contribution to global GDP is more than 45%.

Because minerals are critical for sustaining the built environment and modern economies, demand for them is expected to grow as global population grows. The UN predicts that the world's population will reach over 9 billion by the year 2050 and the highest rates of population growth are expected to take place in developing nations. At the same time the population demographic is expected to shift towards a more urban and middle class distribution [3] which will accelerate the demand for minerals and metals as rising affluence drives consumption of manufactured goods and services.

Regions where rapid population growth is expected to occur also host the vast majority of earth's remaining untapped mineral resources [4]. These regions include Africa, Afghanistan and parts of SE Asia. Many of these mineral rich areas are where geopolitical risk is high, technology is limited and infrastructure is poor. The confluence of these factors has the potential for creating significant and long lasting negative social and environmental consequences. The Democratic Republic of Congo is a prime example of such a region.

To insure an adequate supply of minerals for the present as well as for the future, steps must be taken to extend the life of earth's remaining mineral reserves. At the same time economic benefits must be maximized and social and environmental impacts minimized on multiple temporal and spatial scales. Achieving this outcome is immensely challenging and complex. It requires a new business philosophy that integrates social, economic and environmental considerations into core business strategies and in day-to-day operations. Elements of this business approach include adopting strategies that focus beyond the bottom line and the short-term quarterly or annual business cycle; efficient use of resources; cradle to grave material flow analysis to understand the entire manufacturing life cycle from mine to the post-consumer terminal point for manufactured goods; and designing processes and products to deliver sustainable social, environmental and economic benefits.

Sustainable Development and Mining

One of the most broadly accepted definitions of sustainable development is "... development that meets the social, economic and environmental needs of the present without compromising the ability of future generations to meet their needs" [5]. This definition states that for development to be sustainable it must generate wealth, advance social justice, and remain within the limits imposed by ecosystem and resource availability. It also states that development today must not compromise future generations. Our challenge is to apply this concept to mining in a practical and meaningful way.

Many argue that mining is not sustainable because it extracts non-renewable mineral resources and negatively impacts the environment. This is a simplified and one dimensional view of sustainability that is more relevant for renewable than non-renewable resources. There are other overriding factors that must be considered when applying SD concepts to non-renewable resources. Although single mineral deposits are finite, an uninterrupted supply of minerals is required to sustain today's technology-based society and to generate capital that fuels the growth of the world economy. Mining converts non-renewable mineral resources into capital that sustains social, economic and environmental activities that support human development. Mining, therefore, is unsustainable when the interaction between mineral resource development and economic growth, social development, and the environment is ignored [6].

It is important to acknowledge that the mining activity or operation itself is not sustainable but that mining can be managed in such a way as to contribute to positive environmental, social and economic outcomes. This is often referred to as the Triple bottom line, or the 3 pillars of sustainability which asks business to look past short term profitability towards the longer term objectives of environmental protection and resource conservation on a local and global scale; social equity and well-being for employees and affected communities; and economic prosperity for the business, its shareholders and its stakeholders [7]. Examples of how mining meets these triple bottom line objectives include: producing the raw materials that support human civilization; generating wealth through jobs, taxes, royalties, and upstream value that contributes to GDP; and contributing to the environment by adopting land use practices that improve biodiversity, protect fragile ecosystems and reduce water use [8].

In addition to the 3 pillars there are two more pillars that are unique to mining [9]. These are resource efficiency and health and safety (Figure 1). Resource efficiency is critically important in the case of mining because minerals have a finite life economically and physically. Therefore they must be extracted, processed and used as efficiently as possible. The goal is to extend the life of these irreplaceable natural assets and to extract their maximum value for human benefit without compromising current and future generations to meet their needs. This requires a sustainable mineral supply which can be accomplished by discovering and developing new resources, substitution of functionally equivalent renewable materials (i.e. plastics, composites) for minerals, more efficient use of the resource through improved mining methods and improved processing methods, and recycling. In 2011 approximately 40% of the U.S. metal supply was derived from recycling [10]. Mining can be a high risk occupation so worker health and safety is also core to a sustainable mining industry.

Figure 1: The five pillars of sustainability for mining after Laurence [9].

Commitment to Sustainable Development in the Mining Industry

The mining industry began the process of adopting sustainable development (SD) as a paradigm for driving business value and strategy in the late 1990's. Initially the focus was at the executive level and on crafting an industry-wide commitment to SD. The next phase of implementation involved individual companies developing corporate strategies with underlying visions and goals. The final stage, and perhaps most challenging phase, was to implement SD principles at the operations level where the daily pressures of running a mining operation may be in conflict with longer time line SD programs.

Not all mining companies or mining industry sectors have moved along this trajectory at the same rate so there is a wide range in acceptance and implementation of SD practices across the industry. However, there are some general trends that describe the industry's overall progress towards sustainability. For example, most large multinational metals mining companies have embraced SD and recognize that SD not only creates shareholder value but preserves the social license to mine. Other mining industry sectors such as industrial minerals and coal have also adopted SD and corporate social responsibility as core values and within these sectors larger companies tend to be the leaders. Small scale and artisanal miners, on the other hand, have not adopted SD to the same extent. In fact many artisanal miners fall on the opposite end of the spectrum: they often extract and process minerals using practices that are not sustainable, are harmful to the environment and hazardous to human health [11].

Industry Commitment to SD

About 10 years ago the International Council on Mining and Metals (ICMM), a consortium of 22 mining and metals companies and 34 global mining associations, started down the path of developing an industry-led commitment to SD. Through this process ICMM members voluntarily adopted publically stated principles and declarations, transparent reporting following standards set out by GRI G4 guidelines, and rigorous third party certifications such as ISO 14,000. ICMM conducts an annual assessment of the progress that each member company is making against these performance commitments including the integration of 10 principles and six supporting position statements into corporate policy [12]

At about the same time that ICMM was developing the 10 principles minerals professionals representing 12 leading technical professional organizations from across the globe came together to write and sign the Milos Declaration [13]. This declaration clearly states how the minerals professional community will contribute to sustainable development and further demonstrates the universal commitment to SD across the entire minerals community.

Individual Professional Practice Guidelines

To be effective the concepts of sustainability and sustainable development must be integrated into strategic decision-making processes as well as day-to-day operations. Operations level acceptance is critical because publically disclosed position statements, codes and principles declaring industry-wide commitment to SD are only as credible as the performance of individual operations. It is up to the individual mining professional who is responsible for day-to-day operations to insure that a plant or mine is operated sustainably or that a project is designed to contribute to sustainable development.

In many companies daily operations are managed against short term financial metrics and not longer timeline SD metrics. In fact it is not uncommon for plant and mine operators to have little understanding of how their daily decisions play into these goals. This gap can be bridged through clearly stated and shared operations specific SD metrics, by including SD related performance measures in individual performance goals and by giving individuals more accountability and responsibility for SD performance. The mining industry's success in improving health and safety outcomes demonstrates the importance of individual worker commitment to a safe mining operation. The same holds true for SD.

Mining Life Cycle Framework for Sustainable Development

The life of a mine consists of discrete phases or activities that together comprise the mining lifecycle (Figure 2). This lifecycle applies to all mined materials including coal, industrial minerals, metals, and precious metals. It provides a framework for both understanding and managing the environmental, social and economic impacts of mining at every stage of the mining project (Table 1)

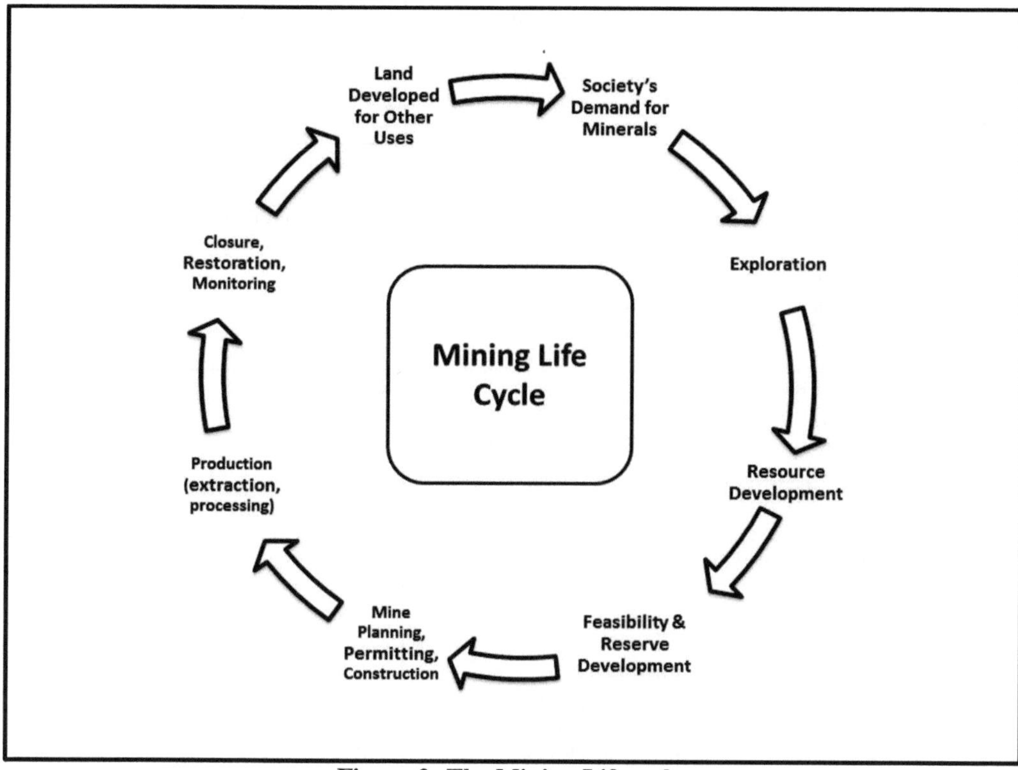

Figure 2: The Mining Lifecycle

<u>Exploration</u>

Exploration focuses on the discovery of new mineral deposits and typically takes place in two phases. The first phase involves searching for a deposit on the surface. It may begin with a geologist identifying promising areas and then searching for signs of mineralization. Prospectors rely on outcrops, stream beds and road cuts for clues to subsurface geology. Stratigraphic position, topographic elevation or geomorphologic features are also commonly employed prospecting tools. In addition to these simple tools, more sophisticated approaches such as geophysical surveys, remote sensing, and geochemical analysis may be used to gather information about potential mineral deposits. Once a deposit has been located, the second phase of exploration usually involves drilling into the subsurface to recover samples for testing. Typically a small number of test holes are drilled on a random or widely spaced grid pattern to confirm the discovery.

At this point the social and environmental impacts of exploration drilling such as noise, water contamination, soil erosion, disturbance of wildlife habitat, disruption of land use for other activities (i.e. farming, hunting, timber, community recreation) and disturbance of culturally significant sites must be considered. These impacts are important because, if mismanaged, the

Table 1: Mining lifecycle stages and impacts at each stage.

	Activities	Environmental Impacts	Social Impacts	Economic Impacts
Mineral Resouce Development	Drilling, Feasibility, Mine Planning, Mine Design, Permitting	Land Disturbance, Vegetation Disturbance, Noise, Dust, Water Consumption, Energy Consumption,	Worker H&S, Community H&S	Jobs
Mine Development & Operation	Infrastructure Development, Overburden Removal or Underground Mine Construction, Ore Extraction	Land Disturbance, Noise, Dust, Asthetics, Water Consumption, Energy Consumption, Water Discharge, Air Emissions, Biodiveristy Protection , Resource Efficiency	Worker H&S, Community H&S, Capacity Building, Skills Development, Enhanced Community Services,	Jobs, Royalties, Taxes, Capital Investment
Ore Handling	Transport ROM to Plant (truck, pipeline, conveyor, other), Storage	Land Disturbance, Vegetation Disturbance, Noise, Dust, Water Consumption, Energy Consumption,	Worker H&S, Community H&S	Jobs
Processing	Crushing, Screening, Grinding, Separation, Concentration, Particle Size Fractionation, Physical or Chemical Removal of Contaminants, Drying	Waste Generation, Recovery and Reuse; Water Consumption, Contamination, Discharge and Reuse; Energy Consumption and Recovery; Air Emissions, Resource Efficiency; Noise	Worker H&S, Community H&S, Capacity Building, Skills Development, Enhanced Community Services (water, power, roads, schools, hospitals)	Jobs, Royalties, Taxes, Capital Investment
Transportation	Product to customer by rail, truck, ship.	Air emissions, Noise, Dust	Worker H&S, Community H&S	Jobs
Waste Disposal	Process and mine waste to, impounds, backfill or deep well injection	Land Surface Distrubance, Water Contamination, Asthetics	Worker H&S, Community H&S	Jobs
Closure	Backfilling, Removal of Equipment, Removal of Buildings & Pipelines, Monitoring, revegetation	Waste Disposal, Biodiversity Protection, Water Discharge, Aesthetics	Worker H&S, Community H&S, Future Land Use	Jobs, Revenues from post mining land use (timber, agriculture etc)

company risks losing its reputation and social license to operate. This license is not a physical document but refers to acceptance of the mining company and its projects by the local community. The social license to operate is earned by developing and maintaining good relationships with all stakeholders and is based on open dialogue, transparency, timely responses to community concerns, and ethical behavior.

Resource Development

The next step towards bringing a deposit into production involves drilling the deposit on an increasingly dense grid pattern. Closely spaced drilling is required for estimating the size of a mineral resource which is defined as a mineral deposit of sufficient size and quality to have "reasonable prospects for economic extraction." According to this definition, which is used by most codes governing resource and reserve estimation, there are three pieces of information needed to evaluate the resource potential of a mineral deposit. These are tonnage and grade, and economic viability.

Methods used to estimate tonnage and grade include calculations based on maps, cross-sections and spreadsheets as well as calculations based on 3-D computer-generated geostatistical modeling techniques. Economic viability is determined by deposit depth, deposit continuity, distance to the processing plant, distance to market, access to transportation, availability of basic infrastructure (roads, power), and other cost factors. The economic assessment completed at this stage is very preliminary and serves to determine the feasibility of the project and the risk associated with continued investment.

Environmental and social considerations are also examined at this stage of the mining project. Often this is done through the formalized process of environmental and social impact assessment (EIA and SIA). EIA focuses on the project's physical and biological impacts on the natural environment while SIA focuses on the project's social impact on the local community.

Feasibility Studies and Reserve Development

Feasibility studies are detailed engineering and economic analyses that include mine design, cash-flow analysis (capital cost, operational cost, and revenue), mineral processing flow-sheets, closure plans, reclamation plans and plant design. Feasibility studies are used to delineate the reserve which is the portion of the measured or indicated resource that is economically minable. There may be additional drilling at this stage as well as continued environmental work including wetland delineation, submission of permit applications to appropriate local and federal agencies as well as baseline monitoring.

Mine Design, Construction and Production

Mine design and planning encompasses a broad range of activities that are mainly concerned with determining the size of the mine, mine layout, mining method, production requirements, and equipment needs. The mining process begins with designing a mine that takes into account the geotechnical aspects of the site. In the case of an open pit operation, the mining engineer designs the highwalls and ramps to reduce the risk of wall failure and ensure safe working conditions. The same considerations guide the design of underground operations.

The next step is to construct the mine. All required local, state or federal permits must be secured before this stage can begin.

Overburden is removed using scrapers, excavators or loaders. Once overburden is removed ore is extracted using a range of methods including loaders, dredges, drag lines, blasting or hydraulic mining. For open pit mines cut and fill methods are used whereby overburden removed from the first pit is stockpiled and then the overburden from each subsequent pit is placed in the previously mined out pit. The in-filled area is then sloped and graded in preparation for re-vegetation and final reclamation.

Construction of underground mines involves digging shafts and underground excavations to access the ore. The geometry, depth, and orientation of the ore body determine the mining method used for ore extraction. At this stage structures including headframes, hoisting-machine buildings, machine repair shops, and other administration buildings are constructed on the surface however most of the mine's footprint is underground.

Closure and Reclamation

After the ore has been removed, the mine is closed and the land is graded, re-vegetated and developed for a variety of post mining uses that, if managed appropriately, will bring long term value to the local community. Some common examples of land use after mining include wildlife habitat creation, community recreation areas, timber plantations, farming, hunting, agriculture and fishing. Community engagement is important throughout this process because communities are left with the land after it has been mined and the best outcomes occur when communities have a role in deciding how to repurpose the land.

Companies focus significant resources on this critical phase of the mining life cycle. In addition to re-vegetating the land, constructing ponds or creating wetlands they establish environmental monitoring programs of the site to ensure that there are no long term environmental issues after the mining project has finished. Mismanagement at this stage puts a company's reputation at risk.

A Vision for the Future

William McDonough and Michael Braungart in their 2013 book, *The Upcycle: Beyond Sustainability*, introduce the concept of constantly improving and moving from being "less bad" to becoming "more good" [14]. They call this the upcycle. Conventional eco-efficient approaches are aimed at reducing or minimizing damage with the goal of shrinking negative footprint. This is where many companies focus their efforts today. They do this because becoming more efficient generates positive results to the bottom line. However, even better results may be realized by going beyond efficiency and focusing on sustainable value generation through innovation that creates a "positive footprint" [15].

So how do we go from "less bad" to "more good"? How do we harness the value that mining generates and use it to not only solve real social, environmental and economic problems today but to create a better future for the next generation? The answer is not simple but there are 4 areas where the industry needs to shift or sharpen focus [16].

- People affected by mining must be treated fairly and with respect.
- Mining companies must create a culture of transparency by being more inclusive in engaging with stakeholders including NGOs.

- The industry must step up efforts to support social and economic development especially in underdeveloped countries where wealth tends to be concentrated in natural capital (minerals, oil, gas) while wealth in advanced economies tends to be concentrated in physical and human capital thus creating disparity between the two.
- As grades decline and demand for minerals grows, the industry must change the way it operates through innovation, new technologies and new processes.

It isn't up to industry alone to make changes. Other stakeholders must participate. Local and national government agencies must enact policies that support this commitment; the public must engage with open minds and with a willingness to participate in the process; mining professionals in both the private and public sector must innovate and seek new solutions to technical problems that are barriers to sustainability; non-government organizations and the international community must be at the table as well. Through collaborative problem solving the mining industry will meet the SD challenge.

References

1. Lorie A. Wagner, "Materials in the Economy- Material Flows, Scarcity, and the Environment," 2002, *U.S. Geological Survey Circular 1221*

2. Paul Collier, *The Bottom Billion: Why the Poorest Countries are Failing and What Can Be Done About It* (Oxford University Press, 2007), 224.

3. United Nations Department of Economic and Social Affairs Population Division: http://esa.un.org/wpp/

4. Jonathan Clayton, "Mapping Mineral Deposits and Mining Around the Globe." 11 April 2013, Available at http://raconteur.net/business/mapping-mineral-deposits-and-mining-around-the-globe

5. United Nations (1987)."Report of the World Commission on Environment and Development." General Assembly Resolution 42/187, 11 December 1987. Available at http://www.un-documents.net/k-001303.htm (Accessed 28 December 2011)

6. George A, Nooten, "Sustainable Development and Nonrenewable Resources—A Multilateral Perspective," 2007, (Proceedings for a Workshop on Deposit Modeling, Mineral Resource Assessment, and Their Role in Sustainable Development, Edited by Joseph A Briskey and Klaus J. Schulz, USGS Circular 1294).

7. J. Elkington, *Cannibals with Forks: The Triple Bottom Line of 21st Century Business* (Oxford: Capstone Publishing Limited, 1997), 69-97.

8. Michelle E. Jarvie-Eggart, ed. *Responsible Mining* (Society for Mining, Metallurgy & Exploration, 2015), 804.

9. Laurence, D.C., "Establishing a Sustainable Mining Operation – An Overview," *Journal of Cleaner Production*, 2011, Vol. 19, Issue 2-3

10. Sibley, S.F., 2011, "Overview of flow studies for recycling metal commodities in the United States, chap. AA of Sibley, S.F., ed.," *Flow studies for recycling metal commodities in the United States: U.S. Geological Survey Circular 1196*, p. AA1–AA25, Available at http://pubs.usgs.gov/circ/circ1196-AA/.

11. "Working together – how large-scale miners can engage with artisanal and small-scale miners," Available from: http://www.icmm.com/page/17638/new-publication-on-engaging-with-artisanal-and-small-scale-miners.

12. http://www.icmm.com/our-work/sustainable-development-framework/10-principles

13. http://www.sdimi.org/sdimi_milos.htm

14. William McDonough and Michael Braungart, *The Upcycle: Beyond Sustainability--Designing for Abundance*, (North Point Press, 2013), 227.

15. Stuart L. Hart and Mark B. Milstein, "Creating Sustainable Value," *Academy of Management Executive*, 2003, Vol. 17, No 2, 56-69

16. Mark Cutifani, "The need to meet Challenges has never been Greater," Available from: http://www.csr21.org/content/icmm-need-meet-challenges-has-never-been-greater

Engineering Solutions for Sustainability: Materials and Resources II
Edited by: Jeffrey W. Fergus, Brajendra Mishra, Dayan Anderson, Emily Allyn Sarver, and Neale R. Neelameggham
TMS (The Minerals, Metals & Materials Society), 2015

GLOBAL SUSTAINABILITY OF RARE EARTH METALS AND COMPOUNDS

Brajendra Mishra

Colorado School of Mines
Metallurgical & Materials Engineering
1500 Illinois Street, Golden, CO 80401, USA

Keywords: Rare earths, Sustainability, Energy, Supply and Demand

Abstract

Rare earths metals, including yttrium and scandium, are being increasingly used in clean energy technologies, colored phosphors, lasers and high intensity magnets. The commitment to clean energy technologies by governments and the projected growth in power and transportation sectors across the globe ensure that the demand for rare earth metals and compounds would continue to escalate. This demand implies that, to ensure unhindered technological innovation, it is essential to possess secure supply chains for rare earth elements. In order to ensure secure rare earth supply and attenuate supply-demand imbalance post 2014, it is of utmost importance to look at opportunities to process intelligently, recycle and reuse Rare Earth Elements from secondary sources, such as post-consumer and manufacturing process wastes. It is anticipated that China's control on rare-earth market will drop from over 90% to below 70% by 2017.

Introduction

As a non-renewable resource, metals and materials supply to the industry for use in manufacturing commercial commodities will come under significant pressure at some point in the near future. The supply from nature will get curtailed, as recovery from earth will become progressively difficult. The sources are likely to get leaner and their exploitation exceedingly more non-viable. This change in supply scenario, of course, will depend on the specific type of metal and material and its demand. In addition, from the US perspective, excessive dependence on foreign imports for some metals will make the availability under higher constraint due to socio-political and economic reasons. Some of the 'exotic' metals, such as rare-earths, molybdenum, rhenium, ruthenium, tungsten, tellurium, lithium, PGMs, etc, for which demand is predicted to stay high and the indigenous resources low, recycling and recovery from spent secondary resources will be only viable option for sustainable growth.

The energy technologies that underpin America's economy and quality of life rely on materials with highly specialized properties. These properties of materials enable the advanced engineered products and systems that propel our dynamically evolving energy economy, and support our national energy security. The unique luminescent and magnetic properties of the rare earths, for example, are essential to efficient lighting and electric vehicles, driving global 'green' energy solutions and accelerating national investment in clean energy technologies [1].

Sometimes only small quantities of such materials, with their transformational properties, are necessary to power the innovation behind the technology. Compact fluorescent lamps reduce the energy requirement for lighting by 80%, but would not exist without minute amounts of yttrium, terbium and europium. Hybrid vehicles and some advanced wind turbine designs will not work without the 'super power' properties of larger, but still relatively small, amounts of neodymium and dysprosium that are used to make higher power, lighter weight magnets for electric motors and generators. But such unique properties, even if they only require small amounts of material, are simply not always available. While industries understand that the rare earths and their unique properties hold the key to accelerating many modern innovations for clean energy technologies, we now recognize that the economic viability and stable supply of such materials has to be assured. Some of the rare earths, and a few other elements, form a group of materials used in modern technologies, whose properties are the 'holy grail' to America's energy future. When there are global supply and access risks and no easy substitutes, these become 'critical materials'[1].

DOE conducted [2] detailed studies of many 'critical materials' essential for America's transition to clean energy technologies and identified five rare earths – neodymium, europium, terbium, dysprosium, and yttrium – as 'high risk.' While Nd and Dy are indispensable at present for advanced permanent magnets, Eu, Tb, and Y are vital in phosphors for energy efficient lighting. Economic projections indicate that these elements could experience worldwide supply deficits of up to 30% by 2016, stalling domestic manufacturing and preventing the adoption of emerging clean energy products. DOE also identified other elements such as lithium and tellurium as near-critical. Lithium plays a crucial role in emerging energy storage and battery technologies, particularly important for use in hybrid and electric vehicles; tellurium is essential in certain photovoltaic thin-films, but its geochemical scarcity and production supply risks threaten future U.S. leadership in photovoltaic manufacturing. It should be noted that with innovative research, careful stewardship of resources and efficiency improvement, metals and materials can come and go from this list of 'at risk' elements.

Rare Earth Elements and Selected Applications [3]

The prominent high-tech applications of rare-earth elements used in energy harnessing, storage, transport and usage devices are listed below:

- **Sc**andium **Optical fibers**, light aluminum-scandium alloy for aerospace components, additive in mercury-vapor lamps, ceramics, phosphors

- **Y**ttrium **Microwave filters**, yttrium is used as host for the red fluorescent lamp phosphor Y2O3:Eu3+; yttrium is also important for ceramics: yttria-stabilized zirconia

- **La**nthanum **NiMH battery**, high refractive index glass, flint, hydrogen storage, battery electrodes, camera lenses, fluid catalytic cracking catalyst for oil refineries

- Cerium **Polishing powders**, chemical oxidizing agent, yellow colors in glass and ceramics, catalyst for self-cleaning ovens, fluid catalytic cracking catalyst for oil refineries, ferrocerium flints for lighters

- Praseodymium **Rare-earth magnets,** including for hard disk drives lasers, core material for carbon arc lighting, colorant in glasses and enamels, additive in didymium glass used in welding goggles, ferrocerium, firesteel (flint) products

- Neodymium **Rare-earth magnets** including for hard disk drives, ceramic capacitors, lasers, violet colors in glass and ceramics Promethium Nuclear batteries

- Samarium **Rare-earth magnets, electro-mechanical relays**, lasers, neutron capture, masers

- Europium Red and blue phosphors, lasers, mercury-vapor lamps, NMR shift reagent

- Gadolinium **Rare-earth magnets**, **computer memories**, high refractive index glass or garnets, lasers, X-ray tubes, neutron capture, MRI contrast agent

- Terbium **Optical fiber, ceramics**, green phosphors, lasers, fluorescent lamps

- Dysprosium **Rare-earth magnets** including for hard disk drives, **lasers**

- Holmium **Lasers**

- Erbium **Optical amplifiers**, vanadium steel, lasers

- Thulium **Portable X-ray machines**

- Ytterbium **Optical amplifiers, infrared lasers**, chemical reducing agent

- Lutetium **Optical fiber**, PET scan detectors, high refractive index glass

The processing step is where the rare earth oxides (REO) are refined and separated into the individual elements. The fabrication and manufacturing stage is comprised of the transformation of the group of elements into intermediate (capacitors, phosphors, permanent magnets, etc.) and final (catalysts, wind turbines, computers, batteries, etc.) products. A few of the unique fabrication and manufacturing stages are indicated on this representational figure, as are a few of the final products. The use stage is where final products are employed. After use, the products are discarded into the waste management and recycling system. [4]. Figure 1 shows the life cycle of rare-earth elements from its extraction from ores through, processing and product manufacturing leading to recovery, recycling and disposal. Also, included in the product development stage is the concept of developing substitutes that help enhance resource productivity. Higher is the resource productivity, lower is the demand on resource recovery and urban mining.

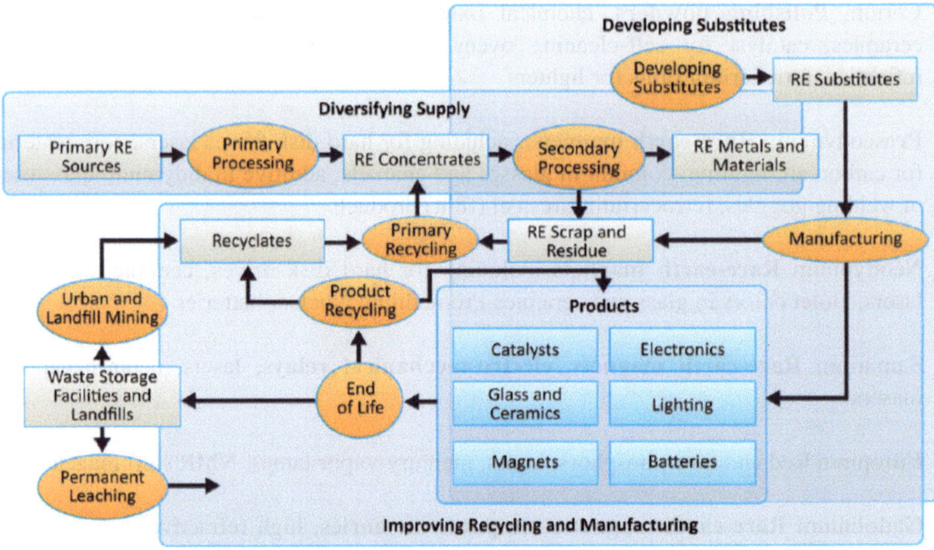

Figure 1: Life cycle of the rare earths at the global level [1].

<u>The Supply Chain:</u>

Increasingly, the U.S. government, academia, domestic industry, and the public acknowledge the imperative that we need to conserve energy and natural resources while exercising judicious stewardship of the environment. The issue of sustainability is and should be paramount in how we design, manufacture, use, and retire the many products we consume throughout the world. Inorganic materials are not renewable; the need exists for the development of technologies to address materials recovery and recycling. Research supporting materials recovery and recyclability is inherently multidisciplinary and must respond to the needs of a multiplicity of commercial stakeholders from throughout the materials supply chain [5].

Despite growing efforts to recycle metals, we fail to recover half of the domestic post-consumer metal scrap reclaimable from retired products, and we continue to rely on primary metals production to fulfill two thirds of our manufacturing needs. Use of primary metals in lieu of scrap increases global energy consumption as well as the production of greenhouse gases. In order to augment recycling rates the materials community needs to upgrade recovery and recycling processing technologies to maximize the capture of post-consumer scrap and minimize the quantity of manufacturing scrap.

Rare earth elements are a group of 17 elements, which include 15 Lanthanides, Scandium and Yttrium. In spite of what the name suggests these elements are not "rare". In fact, Rare earths like cerium are more abundant in Earth's crust than some other common metals such as Copper. However, in recent years rare earth elements have become strategically critical for developed and developing economies around the world which is primarily due to the shortage of discovered minable resources [6]. Before 1948, the placer deposits of Brazil and India were the chief sources of rare earth metals for the rest of the world (Figure 2). With increasing demand newer supply sources were needed and for a while the monazite deposits in South Africa played an important role before the production was dominated by Bastnasite reserves in Mountain Pass and China [7].

Rare earths are being increasingly used in clean energy technologies, colored phosphors, lasers and high intensity magnets. There are important defense applications such as fighter jet engines, missile guidance systems and space based satellite and communication systems based on these metals. Additionally, rare-Earth metals are important alloying addition to steels. The commitment to clean energy technologies by governments and the projected growth in power and transportation sectors across the globe ensure that the demand for rare earth metals and compounds would continue to escalate. This demand implies that, to ensure unhindered technological innovation, it is essential to possess secure supply chains for rare earth elements. Currently, China dominates as the producer of over 95 percent of world output of rare earth minerals and having 36 percent of proven reserves [8]. The United States continues to be one of the largest consumers and importer of rare earths and the trend is expected to continue as the demand increases. According to a forecast done by IMCOA, the world rare earth demand is projected to rise to 220,000 tons by 2015 and the Chinese production is expected to be around 160,000 tons [9]. In addition, the demand of rare earths in China itself has increased by 380% between 2000 and 2009, which is believed to be the primary reason behind the export cuts on rare earth [8].

Between 2005 and 2010 the Chinese government had cut down rare earth supply by 39.6 percent, and the trend is expected to continue. These developments have made rare earth elements a strategically important material as evident by the Rare Earths and Critical Material Revitalization Act of 2010 approved on September 29, 2010 which aims to establish an R&D program within the DOE to assure long term supply of rare earth materials [10]. In December 2010 the US Department of Energy published an analysis of the criticality of selected rare metals, the most critical elements were identified to be Dysprosium, Neodymium, Terbium, Yttrium and Europium - what are also known as the heavy rare earth elements [11]. In order to ensure secure rare earth supply and attenuate supply-demand imbalance post 2014, it is not only necessary to encourage and support exploration of newer REE reserves, build a rare earth stockpile, challenge China on its export policy, etc., but it is also of utmost importance to look at opportunities to recycle and reuse Rare Earth Elements from secondary sources. Table 2 and Table 3 show the forecasted global rare earth demand and supply based on applications as well as individual metals. Based on the demand and supply position of common rare-earths, the prices of common metals like Ce, Nd, Sm, La and Y, went up by 150% to 700% in a short period of six months between January and August 2010.

<u>Potential for New Rare Earth Supply:</u>
Current rare earth mines production is listed below, country-wise:

1. China: 100,000 MT
(Production and export: 93,800 MT and 31,000 MT, respectively).

2. United States: 4,000 MT
3. India: 2,900 MT
4. Russia: 2,400 MT

5. Australia: 2,000 MT
6. Vietnam: 220 MT
7. Brazil: 140 MT
8. Malaysia: 100 MT

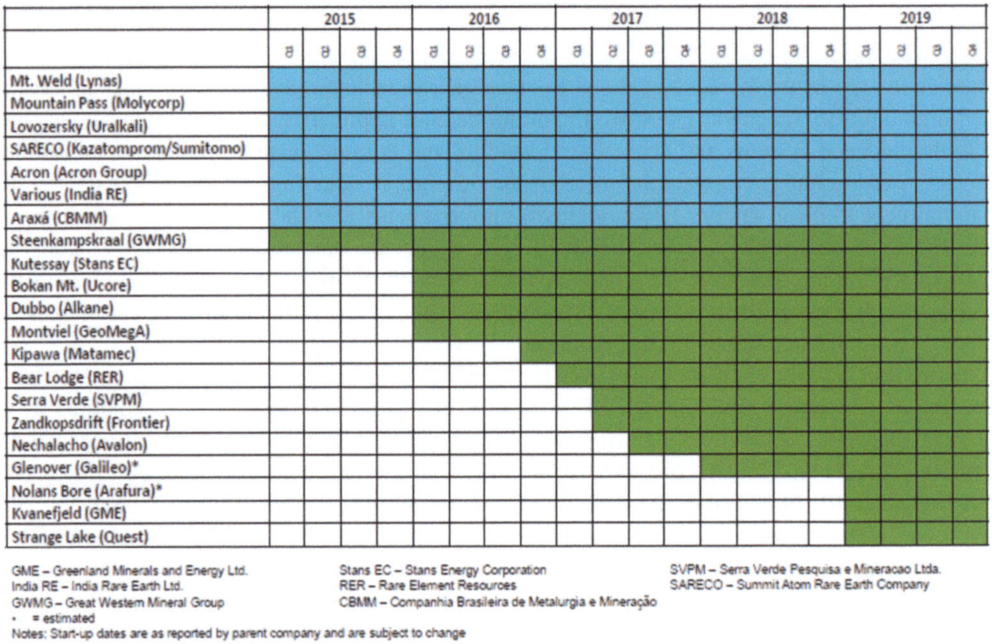

Figure 2: Potential global sources and their expected production dates: [12]

Table 2: Forecast Global Rare Earths Demand and Supply in 2017 (t REO ±20%) (Source: IMCOA and Rare Earths Industry Stakeholders) [9]

Application	China	Japan & NE Asia	**USA**	Others	Total
Catalysts	**9,500**	**2,500**	**13,000**	**1,500**	**26,500**
Glass	**6,000**	**1,000**	**1,000**	**1,000**	**9,000**
Polishing	**11,000**	**2,500**	**3,500**	**1,000**	**18,000**
Metal Alloys	**23,500**	**3,500**	**2,500**	**1,000**	**30,500**
Magnets	**26,000**	**5,000**	**3,000**	**1,000**	**35,000**
Phosphors	**4,000**	**2,000**	**1,000**	**500**	**7,500**
Ceramics	**3,000**	**2,000**	**2,000**	**1,000**	**8,000**
Other	**3,500**	**2,000**	**4,000**	**1,000**	**10,500**
Total	**86,500**	**20,500**	**30,000**	**8,000**	**145,000**
Market Share	**60%**	**14%**	**21%**	**5%**	**100%**

Table 3: Forecast Global Demand and Supply for Individual Rare Earths in 2017 [9]

Rare Earth Oxide	Demand		Supply/Production	
	REO Tonnes	Per Cent	REO Tonnes	Per Cent
Lanthanum	40,800	28.1%	47,800	27.4%
Cerium	51,295	35.4%	75,500	43.2%
Praseodymium	8,400	5.8%	8,500	4.8%
Neodymium	28,925	19.9%	27,400	15.6%
Samarium	1,400	1%	3,175	1.8%
Europium	375	0.3%	450	0.2%
Gadolinium	2,125	1.5%	2,000	1.2%
Terbium	475	0.3%	250	0.2%
Dysprosium	900	0.6%	1,000	0.6%
Erbium	1,065	0.7%	525	0.3%
Yttrium	8,975	6.2%	7,250	4.1%
Ho-Tm-Yb-Lu	265	0.2%	1,150	0.6%
Total	145,000	100.0%	175,000	100.0%

References:

1. Critical Materials Institute: Accelerating Energy Innovations, Proposal submitted to US Dept. of Energy for the creation of DOE Energy Hub, [2012].
2. D. Bauer, D. Diamond, J. Li, M. McKittrick, D. Sandalow, P. Telleen, "Critical Materials Strategy," U.S. Department of Energy, December, 2011. Available at: http://energy.gov/sites/prod/files/DOE_CMS2011_FINAL_Full.pdf
3. International Electronics Manufacturing Initiative, Rare Earth Metals: Current Status & Future Outlook, Advanced Manufacturing Technology Report, [2014].
4. X. Du and T.E. Graedel, "Global In-use Stocks of the Rare-earth Elements", Environmental Science & Technology, Vol. 45, pp. 4096-4101, [2011].
5. The economic Benefits of the North American Rare Earth Industry, "Rare Earth Technology Alliance Report", American Chemistry Council, April 2014.
6. United States Congressional Research Service, Rare earth elements: The Global Supply Chain, March Humphries, 2010.
7. D. Sandlow, "Keynote Address"- Technology and Rare Earth Metals Conference 2010, Washington D.C., 17 March, 2010.
8. United States Geological Survey, China's Rare-Earth Industry, Pui-Kwan Tse, 2011.
9. Kingsnorth, D., IMCOA, "Rare Earths: Facing New Challenges in the New Decade", presented by Clinton Cox SME Annual Meeting 2010, 28 Feb - 03 March 2010, Phoenix, Arizona.
10. Rare Earths and Critical Material Revitalization Act of 2010. H.R. 6160. 22 Sep 2010.
11. U.S. Department of Energy. "Critical Materials Strategy". December, 2010.
12. D. Merriman, "The European REE market and its place in the global industry", Roskill Report, ERES 2014, Milos, Greece, Sept. 2014.

Engineering Solutions for Sustainability: Materials and Resources II
Edited by: Jeffrey W. Fergus, Brajendra Mishra, Dayan Anderson, Emily Allyn Sarver, and Neale R. Neelameggham
TMS (The Minerals, Metals & Materials Society), 2015

DEVELOPMENT ON COPPER SMELTERS IN CHINA TODAY

Yan Jie
China ENFI Engineering Corporation
NO.12 Fuxing Avenue, Haidian District,
Beijing, China 100038
yanj@enfi.com.cn

Abstract

This paper introduces and studies systematically the Cu production of Chinese Copper smelters in recent years and in future years, as well as the technology development and CAPEX of Chinese new Copper smelters. In the paper, treatment trend on copper scraps, waste copper materials and electronic wastes are also analyzed.

Introduction on Chinese and World's Cu Production in Recent Years

The world's Cu production has increased by 26% in 13 years from 14.82 million tons in 2000 to 18.77 (21.37) million tons in 2013. The Cu production in China has increased 5 times from 1.37 million tons in 2000 to 6.84 million tons in 2013, making China one of the biggest Cu production countries in the world.

Lots of Copper smelters have been built in recent years with capacity from 100,000 t/a to 400,000 t/a. and process for Copper smelters is more and more advanced with better environmental performance, which represents development trends in Cu smelting today. Yet cathode copper production from Cu concentrate in China is 1,596,000 t/a, and Cu concentrate for Cu smelting in China depends mainly on imports.

Chile has world's largest Cu resources. In Chile, total Cu in Cu concentrates is 5.24mt/a, with cathode copper production 3.09mt/a in 2011; copper in Cu concentrate is 5.786mt/a in 2013, making Chile the largest Cu concentrate producer in the world. Yet, in terms of cathode copper production, Chile ranks the second in the world with 2.75mt/a, only next to China. Chile's Cu concentrates are mainly exported to China. Cu production in Japan, with little change in last decade, is 1.33 million tons in 2011 and

1.47 million tons in 2013, remaining the third in the world. The Cu concentrate depends mainly on imports. The USA, once the world's largest Cu production country, is now the fourth, and its Cu production has decreased to 1.273 (1.01) million tons in 2013. The reason of reduction is resource exhaustion in USA, and smelting from imported concentrate is unprofitable. Some smelters are shut down. One typical example is Hildalgo Copper smelter of Dodge Company, which was once the largest Copper smelter with flash furnace and was shut down due to resource exhaustion. Canada, similar to USA, has seen a cut of about 50% on cathode copper production in recent years. The famous Kidd Creek Copper smelter in Canada, once the second Copper smelter adopting Mitsubishi continuous smelting process and one of the inventors of stainless steel permanent cathode electrolyzing process, was shut down in 2013 because of resource exhaustion. Cu production in other traditional Cu producing countries like Russia, Germany, Poland, Australia and South Korea shows little change in decades. Some Copper smelters are newly built in emerging countries like Zambia, Peru, Congo, Iran, Kazakhstan, and Indonesia by using rich Cu resources in these countries, with substantial increase in cathode copper production.

The year of 2014 witnesses an ample supply of Cu concentrate. Chinese Copper smelters have continuous imports of Cu concentrates as TC/RC（91USD/9.1USD）rises continuously. The TC/RC will rises to 107 USD/10.7USD in 2015 year. Customs data show that imports of Cu concentrate in China are 1.29 million tons in September 2014, up by 26.64% compared with last September and hitting a record high. Copper smelters have also remained high operation rate. Based on data of Bureau of Statistics, Chinese cathode copper production in September 2014 reaches 714,000 tons, an increase of 33,000 tons, up by 4.77% compared with August, and by 12.05% compared with last September, thus reaching a single month production record. Total production from January to November 2014 is 7,170,000 tons, up by 12.0% compared with the same period of last year. Based on current capacity of Chinese smelters, production capacity for Chinese Copper smelters in 2014 is predicted to reach 8.00 million tons, which will hit a record high.

Technology Level and CAPEX for New Copper Smelter in China

From 2006, Chinese government requires that production capacity for Copper smelter is no less than 100kt/a cathode copper per single series. Twenty years ago, the world's largest Copper smelter produces 200kt/a cathode copper per single series, and now the largest is 400kt/a cathode copper per single series. Three Copper smelters with 400 kt/a per single series have been built and commissioned in succession in recent years in China.

It is also required by Chinese government that new Copper smelter must adopt advanced and mature smelting technologies of the world, which include: Oxygen-enriched Flash Smelting Process, TSL (ISA & Ausmelt) Process, Side-blowing Smelting Process and Oxygen Bottom-Blowing Smelting Process (SKS Process) invented independently by ENFI and Shuikoushan. In recent years, three 400kt/a single-series Copper smelters are newly built in China, which adopt flash smelting and flash converting process (double flash process) with CAPEX about RMB 6~7 billion. A decade ago, TSL process was adopted for old copper smelter modification as well as 100~300kt/a greenfield Copper smelters. CAPEX for new 100kt/a Copper smelters with TSL process from Cu concentrate to cathode copper is about RMB 1.9~2.1 billion. In recent years, new 100~300kt/a Copper smelters prefer SKS process independently invented by ENFI and Shuikoushan. CAPEX for new Copper smelters with SKS process from Cu concentrate to cathode copper is about RMB 1.8~1.9 billion.

Requirements on environmental performance and energy consumption for Copper smelters by countries in the world are tougher, especially in China. Recent required indicators are: S recovery in offgas$>$97.5%, S collection in H_2SO_4 and slag $>$99%, SO_2 concentration in vented offgas $<$400mg/Nm^3, particle concentration in vented offgas$<$80mg/Nm^3, and comprehensive energy consumption per ton blister copper $<$ 180kgce/t. All the indicators have reached the world's leading level. To improve environmental performance, Chinese government has set particular emissions limit for new Copper smelter in 47 cities, that is: SO_2 concentration in vented offgas $<$ $100mg/Nm^3$, particle concentration in vented offgas $<10mg/Nm^3$. It actually means that no new Copper smelter is allowed in those cities.

In recent years, besides flash smelting process and TSL process which have achieved great progress after absorption and innovation, brand new oxygen bottom-blowing Cu smelting process has also witnessed rapid development.

Oxygen bottom-blowing smelting process, i.e. Shui Kou Shan Copper Smelting Process (SKS Process) for short, is developed from the 1990s in Shuikoushan Mining Administration (subordinate to Minmetals) and is commercialized from 2008. Now the technology is jointly owned by China ENFI Engineering Corporation (hereinafter referred to as ENFI), Shuikoushan Nonferrous Metals Co. (hereinafter referred to as Shuikoushan), and Dongying Fangyuan Nonferrous Metals Co., Ltd (hereinafter referred to as Dongying Fangyuan).

Ever since Dongying Fangyuan's first 100kt/a single series with SKS process was commissioned, the process has been adopted successfully in five Copper smelters of similar capacity with good indicators. Now two smelters separately processing 1mt/a Cu concentrate and 1.5mt/a Au & Cu concentrate are under construction.

The SKS process is especially fit for processing complex copper concentrates, or mixture of complex Cu concentrate and complex Au concentrate to recover associated precious metals.

The main equipment for SKS smelting process is a horizontal rotary bath smelting furnace. Enriched air with 75% oxygen is blown via sleeve-type furnace lances from furnace bottom into bath to produce copper matte. Melt is settled and separated at one furnace end, and then high grade copper matte and furnace slag are tapped separately. Cu can be recovered from furnace slag after flotation with Cu <0.3% in discard slag. There is no need of water-cooling for furnace, so heat loss is reduced. Wet concentrate can be directly added into furnace without carbonaceous fuel, thus realizing self-heating.

Furnace lining life can be more than 3 years.

Besides, ENFI has developed oxygen-enriched bottom-blowing continuous converting technology in corporation with Yuguang Gold and Lead Corporation (hereinafter referred to as Yuguang Company) and Dongying Fangyuan. Based on National 863 Projects "Key Technology and Device Research on SKS Continuous Cu Smelting Process", pilot test, full-scale test and industrial trial project on SKS process are successfully completed and put into operation. This new technology, with low carbon smelting and cleaner production, is very promising.

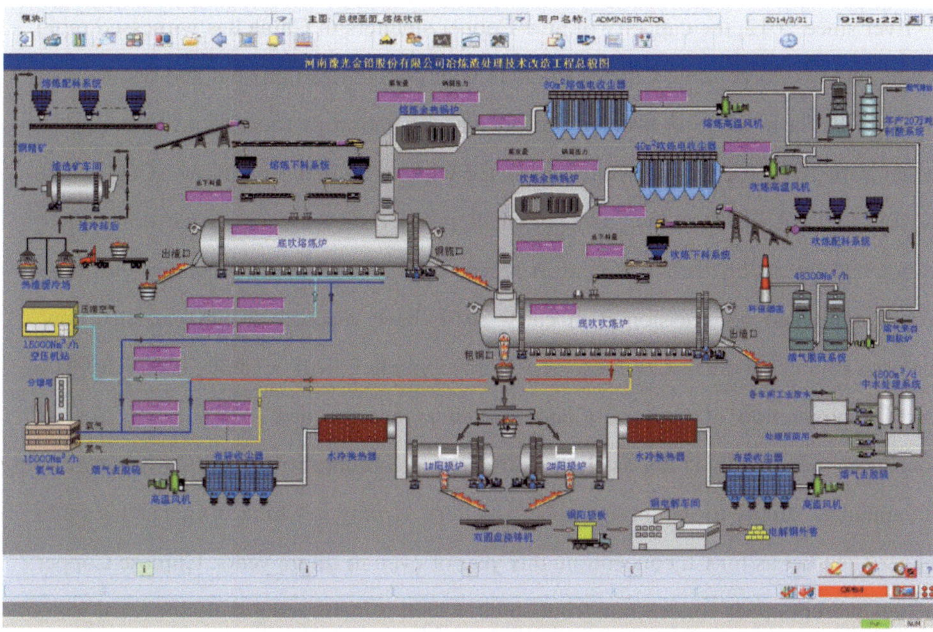

Figure 1--Double SKS Process Adopted in Medium-scale Chinese Cu Smelters Commissoned in Recent Years

Main processes like flash process, TSL process and SKS process have witnessed great improvements. Besides, new technologies are adopted in Chinese Copper smelters for energy saving and emission reduction, such as pure oxygen combustion, purging brick, acid-making from high concentration SO_2 offgas, offgas desulfurization, dry method As recovery process, parallel flow electrolysis, cyclone electrowinning for Cu removal, which are all applied in Chinese Copper smelters. With these technologies, project benefits can be improved and CAPEX will be usually up by 5-20%.

Future Copper Smelter Capacity in China

In recent years, under more and stricter environment requirements in China, most of old Copper smelters with traditional blast furnace, reverberatory furnace and electric furnace have been transformed. The rest small-scale backward Copper smelters will be shut down before 2017, with total Cu production of 900~1200kt/a.

Ever since 2012, the capacity of new Copper smelters in China has witnessed rapid growth ,and two large-scale Copper smelters with double flash process and one Copper smelter with flash smelting process are put into operation in succession. One Copper smelter with TSL process has been constructed without commission. Now TC/TC value has risen to (110USD/11USD), Copper smelters, once shut down or reduced in output, are put into operation with full load in recent days. Two Copper smelters with SKS process, built and commissioned in early 2014, have reached design capacity now. Increased copper production of these new Copper smelters will be about 1600~1800kt/a. Now Cu production of Chinese Copper smelters has reached a new peak. Yet Cu concentrate supply now is not as short as in 2012. According to survey, Cu concentrate is in abundant supply for Chinese Copper smelters, turning in China from a seller's into a buyer's market as for Cu concentrate this year or even in future years. Chinese Copper smelters prefer copper concentrate from china, which is increased greatly in recent years, because: (1) discount coefficient for Chinese Cu concentrate decreases to 84%; and (2) more precious metals in Chinese Cu concentrate makes Copper smelters more profitable by comprehensive recovery of metals.

In future 3-5 years, there will be 4-5 new and expanded Copper smelters being commissioned in succession, with SKS smelting technology and SKS continuous converting technology as the first choice, to meet requirements of new energy consumption standard (blister <180kgce/t) and environmental standard (base exhaust volume for 21000m^3/t Cu). PS converter may never be adopted in Chinese Copper smelters in the future. Of course, China is an open country, which can be demonstrated by various processes adopted in Chinese Copper smelters now. Therefore, the double side-blowing process newly invented will be promoted in the future, as well as flash converting process and Mitsubishi continuous converting process. That's why Chinese Copper smelter is said to be the museum of World Cu smelting. Nowadays, China is leading the world in Cu smelting technology development, which is the pride of Chinese metallurgical workers.

In a word, increased copper production from new and expanded smelters will be about 80~90kt/a in the future, which will nearly meet the requirements of Chinese government in weeding out backward Copper smelters. In future five years, the copper production in China will remain the status quo. Capacity of newly built Copper smelters maintains balance with that of eliminated backward smelters. The buyers' market pattern will be remained in Cu concentrate supply.

Besides, sulfuric acid from Copper smelters in China is unmarketable now, which will be one of the key factors in restraining Chinese copper smelters' production capacity in future years.

Figure 2--Future to build Copper smelter in China

Development on Chinese and World's Copper Scraps Recovery Technology

The best smelter specialized in processing copper scraps is Kayser Smelter (belonging to Nord Deutsche Affinerie) in Germany, with multiple middle-low grade copper scraps (copper content 20%-80%) as feed, and blister as capturing agent to recover precious metals like Pt and Pd in waste electronic components.

The smelter is transformed with ISA smelting furnace, because smelting of copper scraps depends mainly on external heating, and combustion reaction in bath between coal proportioned to feed and oxygen makes heat conductivity high. For the smelter, one furnace with two stages is originally designed for intermittent operation. The first stage is reduction smelting where feed, flux and coal is added into ISA furnace for feed melting and slag producing from reduction reaction. Then pig copper with 80% copper content and furnace slag is produced. Zinc in feed is volatilized and recovered as dust. After the first stage, furnace slag is tapped and pig copper remains in furnace for the second stage, i.e. oxidation converting. Pb and Sn in pig copper are oxidized to produce blister and Pb&Sn oxide slag.

Some improvements are also made for this smelter. ISA furnace operation is improved to be continuous with only reduction smelting stage. One oxygen side-blowing rotary furnace is added for oxidation converting operation.

ENFI is now researching replacing ISA furnace with SKS furnace with the above process as reference. Bottom-blowing lances successfully invented by ENFI are used to inject natural gas and oxygen for feed melting, oxidation and reduction. The process is characterized with high heat efficiency, high operational rate and low OPEX.

Hoboken Cu & Pb Smelter in Belgium has gone through a transformation recently. An ISA furnace is adopted processing Cu & Pb complex concentrate and copper scraps (one third concentrate and two-thirds copper scraps) with two stages. In the first stage, Cu-Pb copper matte is produced after oxidation reaction. Then furnace slag is tapped and matte remains in furnace for the second stage operation, i.e. oxidation to produce blister copper. Pb-bearing slag is added into existing blast furnace to recover Pb after reduction.

Onahama Copper smelter in Japan, built in the 1960s, is originally equipped with a reverberatory smelting furnace and a converter. After transformation in 2009, a Mitsubishi smelting furnace is added to process Cu concentrate. The produced copper matte and furnace slag is settled and separated by former reverberatory furnace, which is also used for processing copper scraps and household waste. Former converter is used to produce blister. After modification, the Cu concentrate throughput of Mitsubishi smelting furnace is 664,000t/a, and copper scraps throughput of reverberatory furnace is 180,000t/a, with Cu production is expanded to 300,000t/a. Little Cu concentrate (3600t/a, to protect reverberatory furnace wall and prevent generation of dioxin) is still added into reverberatory furnace even when it is used for processing copper scraps and

Mitsubishi smelting furnace is put into operation.

Horne Smelter of Noranda Company in Canada once had its own mine, whose resources had been exhausted by now. Unlike nearby Kidd Creek Smelter, Horne Smelter is still in operation by purchasing some concentrate and importing waste copper materials (mainly electronic components containing Pt&Pd) from USA. After crushing and screening, the fine material is sent to Noranda furnace, and the coarse material is added into converter to maintain profitable production of the smelter.

Smelters processing waste copper materials must take in account the prevention and elimination of dioxin. Dioxin, as a kind of organic compounds containing chlorine with great toxicity, may be generated during smelting process if there is chlorine-containing plastic, rubber and paint in waste copper material. Dioxin exists in waste offgas and ash vented from smelting process.

According to experience of Copper smelters in each country processing waste copper materials, different methods of preventing and eliminating dioxin are as follows for choosing in different cases:

(1) Disposing after classification: waste copper material, like wires, cables and enamelled wires which containing oil, plastic, rubber and wrappers, should be disposed separately with that doesn't containing such organic compound like components of machines, vehicles and ships.

(2) Large amount of waste wires and cables must be disposed with shearing, crushing, screening and wrappers removing, before sent to the smelters.

(3) If offgas from smelting furnace processing waste copper materials would produce dioxin, WHB can be used to cool down the offgas temperature to 600°C before it is cooled down to 200°C with quench tower to prevent the generation of dioxin (because dioxin is generated at 300～325°C).

(4) If there is dioxin in dust, high-efficiency dust collector can be used to prevent dioxin from going into vented gas by improve dust collection efficiency; or activated carbon can be injected before dust collection because activated carbon can absorb dioxin,

which will benefit dust collection.

(5) Some sulfide can be added into smelting furnace processing waste copper materials to prevent the generation of dioxin. In Onahama Copper smelter in Japan, after former reverberatory furnace is modified to process copper scraps, little copper sulfide concentrate is still added to prevent the generation of dioxin.

(6) Dust containing dioxin can also be treated separately in a small closed furnace to decompose dioxin in dust by heating.

In China, Technologies recovering copper scraps are varied. In processing high grade copper scraps (Cu content >95%), rotary anode furnace can be used with oxygen combustion and purging brick technologies; tilting furnace or LNG furnace can also be used separately to process copper scraps. There are still many smelters with reverberatory furnaces processing copper scraps even if being ordered to shut down by Chinese government. In 2013 when copper price fell, these smelters had seen huge losses. It is predicted that copper scraps would be gathered and processed in large-scale Copper smelters in the future. ENFI is now developing a new kind of ENFI furnace for processing copper scraps. The furnace is modified based on rotary anode furnace by adding oxygen bottom-blowing lances, to improve treatment efficiency of high grade copper scraps, increase heat efficiency and reduce energy consumption.

ENFI furnace can be used to process low grade waste copper materials (Cu content 30%-60%) and e-waste. With newly-added oxygen lances, feed melting, oxidization and reduction processes can be realized to treat secondary copper materials in a large scale in a more energy-saving and environmental way. ENFI furnace can replace those small-scale and backward furnaces which cause serious pollution in processing copper scraps and e-waste. The technology will promote the further development of Chinese Cu smelting industries. China also embraces advanced technologies in treating low grade waste copper materials and e-waste. Different technologies will compete to promote technological progress, help realize energy saving and emissions reduction, and push forward sustainable development of copper scraps recovery and utilization.

CONCLUSION

In recent years, copper production from new and expanded smelters has increased by 80~90kt/a, which is nearly meet the requirements of Chinese government in weeding out backward Copper smelters. In future five years, the copper production in China will remain the status quo. Capacity of newly built Copper smelters maintains balance with that of eliminated backward smelters. Now sulfuric acid from Copper smelters in China is unmarketable, which will restraint copper production in future years. The buyers' market pattern will be remained in Cu concentrate supply.

REFERENCES

[1] Engin Ozberk, Eric Partelpeog, "NICKOLAS THEMELIS SYMPOSIUM ON PYROMETALLURGY AND PROCESS ENGINEERING" [M]. COM, SEPTEMBER 28 TO OCTOBER 1, 2014, VANCOUVER.

[2] Nozaki Tsutomu, translated by Zhang Baiting and Zhang Jinsong, "Introduction of Bottom-blowing Converter Process • Bubbling Effects • Combination of TSL and Bottom-blowing" [M]. Beijing: Metallurgical Industry Press, 2008.

[3] He Shanchi, Li Dongyuan, "Discussion on Theoretical Issues of SKS Copper Smelting Process" [J], China Nonferrous Metallurgy, 3:9-12, 2005.

[4] R.Bassa, R.Parra, A.Luraschi, S.Demetrio, "NICKOLAS THEMELIS SYMPOSIUM ON PYROMETALLURGY AND PROCESS ENGINEERING" Chile Copper International conference, December1-4, 2013.

Engineering Solutions for Sustainability: Materials and Resources II
Edited by: Jeffrey W. Fergus, Brajendra Mishra, Dayan Anderson, Emily Allyn Sarver, and Neale R. Neelameggham
TMS (The Minerals, Metals & Materials Society), 2015

ENERGY-EFFICIENT SUSTAINABLE PROCESSES BY "THIOMETALLURGY"

Neale Neelameggham[1]; Robert Brown[2]; Brian Davis[3];
[1]Ind LLC; [2]Magnesium Assistance Group; [3]Brian Davis Consultant Associates

Keywords: Thiometallurgy, Sulfur as fuel, reductant, Low effluent – low cost, sustainable process

Extended Abstract

Free or combined sulfur has been used as both the raw material and the energy material in carrying out economical manufacture of several metals in millions of tons per year quantities over the past century. This has controlled carbon emissions in an unintentional fashion and out of necessity as the ores in many cases have been sulfides to start with. The inherent benefits of "thiometallurgy," which uses sulfur in the extraction of metals in alleviating CO2 and water vapor–greenhouse gas (GHG) emissions, and its ability to provide a cost-effective energy material solution equates to sustainability of the process. Such solutions are not only applicable to existing base metal production but, as the authors will show, also are applicable to newer processes in the production of other metals and chemicals, The concept of generating "thiopower" as an alternative energy approach isalso introduced.

A full length paper on this subject can be seen in September 2014 *JOM*.[1] A set of presentation slides provided courtesy of IND LLC © are included for the proceedings.

Acknowledgment

We thank the review efforts by Indra Neelameggham, and IND LLC facilitating this research. Thanks also to Peter Pinfold, an international smelter process consultant for a critical review of the paper.

References

1 Neelameggham, N.R; Brown, R.E.; and Davis, JOM, **66**, Issue 9, 2014, 1622-1629

Introduction

The word sulfur has the Sanskrit origin 'sulvere ' adapted into Latin as 'sulfur'. Greeks called it 'thio' and many of the radicals and compounds of sulfur are referred as thio-compounds.

Sulfur, even though it is in group VI of the periodic table below oxygen [the universal oxidant in air], is also a reducing agent like carbon instead of being an oxidant alone at its elemental state. This happens mainly from its multi -valence and the affinity for elemental sulfur to oxygen.

Sulfur metabolizing bacteria are referred to as thiobacillus, etc.

SULPHUR
Atomic Number: 16
Atomic Symbol: S
Atomic Weight: 32.06
Electron Configuration: [Ne]3s23p4
Total Isotopes 25
Total Isomers 1
Radioactive Isotopes 20
Stable Isotopes 4

X-Ray Energies
Kα 2.31
Kβ 2.46
LL 0.15
Lα –
Mz –
Mα –

SULVERE IN SANSKRIT
http://www.radiochemistry.org/periodictable/elements/16.html

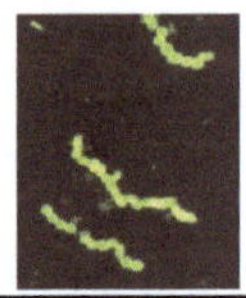

Sulfur aerosol

Ramanathan , Scripps Institute of Oceanography, modeling the benefits of global cooling by radiative forcing by aerosols from natural events such as volcanic eruptions as well as anthropogenic aerosols.

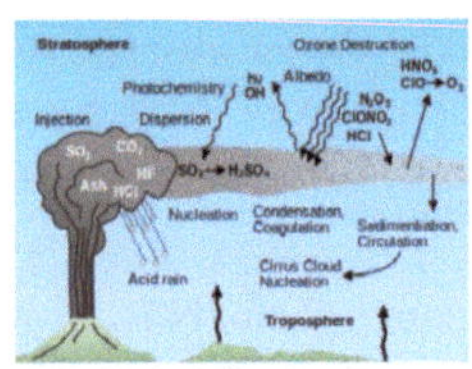

http://en.wikipedia.org/wiki/Stratospheric_sulfur_aerosols

The results of the Indian Ocean [INDOEX] experiments in the late 1990's were summarized by Vogelman , Ramanathan and S.K.Satheesh '… A persistent haze layer that spread over most of the northern Indian Ocean during wintertime was discovered.

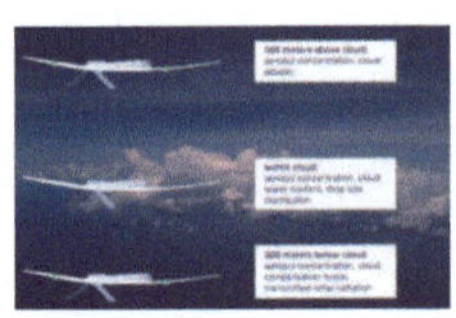

Sulfur aerosol Smog

The optical thickness of this layer ranges from as high as 0.5 to 0.6 (Arabian Sea and Bay of Bengal) to about 0.1 to 0.2 (equatorial Indian Ocean).

The layer, a complex mix of organics, black carbon, sulfates, nitrates, and other species, subjects the lower atmosphere to a strong radiative cooling and a larger reduction in the solar heating of the ocean.'

This is the same observation earlier of smog from smoke stacks and dilute sulfur oxides during the 60's which lead to the clean-up of environmental pollution.

Abate by Dilution

'Albedo Enhancement By Stratospheric Sulfur Injections: A Contribution To Resolve A Policy Dilemma' suggested that injecting 1 million tonnes of sulfur a year would cool down the climate so the greenhouse effect is wiped out. [P.Crutzen2006]

Hypothesis is that an added layer of sulfates in the stratosphere, approximately 16 kilometres above the earth, would reflect sunlight back into space and reduce the amount of solar radiation reaching the Earth's surface.

Most of us must have heard news items that Bill Gates, Microsoft founder has been funding some of this research since 2010.

Solution is not Dilution but Concentration

Kennecott -twin stacks -1970s

These suggested solutions continue in the similar line as 'solution to pollution is dilution' – as in the 1970s by dispersing the pollution by increasing the height of stack [here they want to go to stratosphere – some 50000 feet above earth's surface], in this case to cause unclean air which will allow energy inefficiencies to continue at the surface.

The authors want to note that it is more prudent to handle the 'solution by concentration at the source' as extractive metallurgists have done it by improving process efficiency, and energy efficiency by capturing at the source in a concentrated form and minimizing waste.

Courtesy of Ind LLC, ©

Sustainability = Minimal Cost

Global warming abatement discussions give little credit to the best practices of metal extraction, especially the base metals. These metals are produced in millions of tons per year and have been done with low amounts of carbon dioxide emissions and staying in business.

How?

Copper is extracted from sulfides, lead and zinc are extracted from sulfides, nickel, cobalt, and molybdenum are extracted from sulfides. The fuel value of the sulfides is effectively utilized in this metal extraction.

The scientists working on this for several decades developed efficient techniques – to see the benefits of flash smelting technologies in **cutting down the costs** of operation as well as cutting down the pollution resulting from release of dilute gaseous emissions – which could have aided in the radiative forcing technique mentioned earlier.

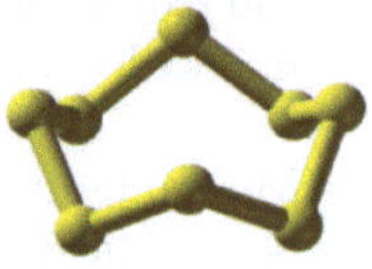

Courtesy of Ind LLC, ©

Valence States of Sulfur-1

Sulfur exhibits several valence states from -2, -1, 0, +4, and +6. Negative divalency is seen in sulfides in compounds such as H2S, Na2S, FeS, PbS, ZnS, etc.

Negative mono -valency is seen in compounds such as sodium disulfide [Na2S2], and FeS2.

Zero valence state is the naturally occurring elemental sulfur. In this state it occurs in orthorhombic as well as in monoclinic forms. It is denoted as S8 to show the polymeric nature of the element.

The +4 valence state is seen in compounds such as sulfur dioxide, and sulfites; while the +6 valence state is seen in sulfur trioxide, sulfuric acid and sulfates.

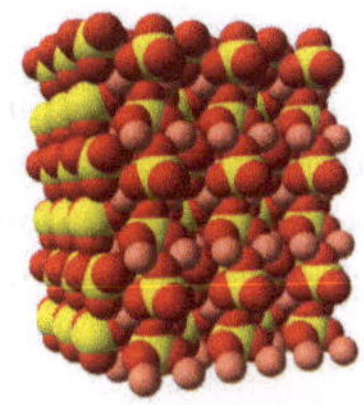

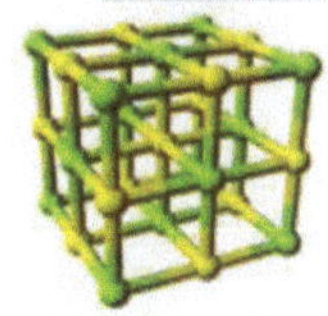

Courtesy of Ind LLC, ©

Valence States of Sulfur-2

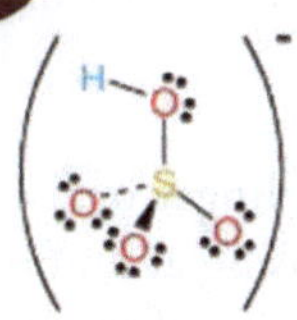

Thiosulfate ion, which is formed by addition of sulfur to sulfite ion, $S_2O_3^{-2}$ is not a +2 valence sulfur, but it has two sulfur atoms which are not equivalent. This is a sulfate ion SO_4^{-2} in which one of the oxygen atoms is replaced by a sulfur atom– the central sulfur atom may be assigned oxidation number +6, and the attached sulfur atom oxidation number -2 [similar to the other three oxygen atoms]. [Linus Pauling] Different interpretations are given by others now.

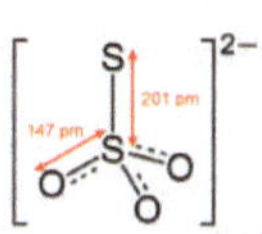

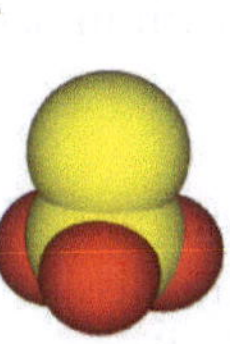

Thiosulfate ion is easily oxidized to tetrathionate ion, $S_4O_6^{-2}$ which is a complex ion. Yet another complex oxo-ion of sulfur is hexathionate, $S_6O_6^{-2}$.

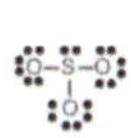

A Lewis structure for the sulfite ion that obeys the octet rule. This is not the best structure, due to formal charges on all the atoms.

The other Group VI elements selenium and tellurium behave similar to sulfur in their multivalence properties.

Courtesy of Ind LLC, ©

Sulfur as Fuel

The negative di to zero -valent sulfur can interact with oxygen atoms in further getting oxidized. Such a property makes the sulfides of metals to have the fuel like property of carbon. The sulfides get oxidized to sulfur dioxide and further to sulfur trioxide this property is also seen with elemental sulfur.

Heats of combustion of sulfur are comparable to carbon.

$S + O_2 \rightarrow SO_2$ DH=-296.8 kJ/mole ---[A]

$S + 1.5\ O_2 \rightarrow SO_3[v]$ DH=-396.1 kJ/mole [@ 318 K] --[B]

$C + 0.5O_2 \rightarrow CO$ DH=-110.5 kJ/mole ---[C]

$C + O_2 \rightarrow CO_2$ DH=-393.5 kJ/mole ---[D]

Carbon vs Sulfur

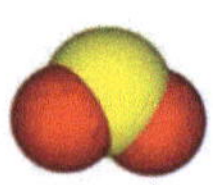

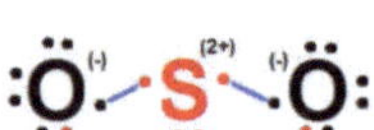

Elemental sulfur can become sulfur oxides – similar to carbon which can become carbon oxides.

Unlike carbon oxides which linger in the atmosphere without condensing at ambient and stratospheric conditions, sulfur oxides get dissipated easily by reaction with moisture in the atmosphere.

Sulfur oxides become sulfurous and sulfuric acid and forming the acid mist dissipation – alluded to by Paul Crutzen earlier.

Water Vapor GHG

Water vapor is a greenhouse. Increased water vapor in the atmosphere thus means further global warming. Willett, et al showed that the rise in humidity caused by people, in the air near Earth's surface, rose 2.2 percent in less than three decades since 1973 . The combined effort of CO_2 and water vapor can efficiently be handled – see –soda- fuel metallurgy discussions.

In this sense, hydrogen as a reductant falls in the category of global warming reductant like carbon, unlike sulfur…

We should, however, not discount the inefficient energy usage in processes – which may happen with sulfur as well by inadequate design.

Copper Flash Smelting-1

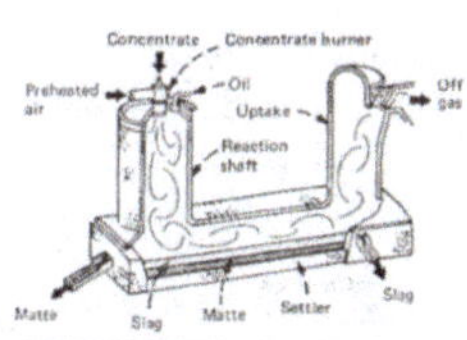

The technology of metal sulfide reduction has developed to be a clean air technology during the past six decades. These technologies were developed not to abate CO_2 – but to minimize the cost of production and be SUSTAINABLE.

Cost minimization came from minimizing waste, operating with little dilution and reduced pollution by capture at the source.

The economic benefits of operations with minimal dilution, forming concentrated sulfur oxides which are then captured and marketed as sulfuric acid have long been a norm in base metal industry.

Copper Flash Smelting-2

Kennecott's flash smelting, converting and refining of copper

$2\ CuFeS_2 + O2 \rightarrow Cu_2S$[matte] + 2 FeS [matte] + SO2---
[E]

$Fe_xS + O_{[2+y]} \rightarrow Fe_xO_y + \boldsymbol{SO_2}$ ---[F]

$Fe_xO_y + Z\ SiO_2 \rightarrow Fe_xO_y..\ Z\ SiO_2$ [slag] ---[G]

Cu_2S[matte] + O_2 = 2 Cu[blister] + $\boldsymbol{SO_2}$ ---[H]

Cu [blister] (elect.rorefining) → Cu [cathodes] ---[I]

Flash smelting [E]-slag [formation of iron with silica] is tapped ~ 1330 °C. Flash **smelting** uses **35 to 45%** $\boldsymbol{O_2}$. The matte ~62 to 68 % Cu, 23-25% S, 11-14% Fe, is granulated for processing in the flash **converter** with **75 to 85%** $\boldsymbol{O_2}$.
small amount of fuel [such as fuel oil] – about 1% the mass of contained sulfide in the charge is used to initiate reaction .

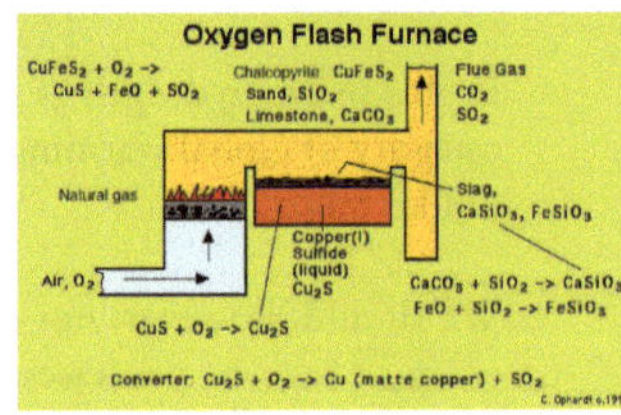

Copper Flash Smelting - 3

'The size of the gas handling equipment is also reduced, and the continuous flow of high strength SO2 gas allows optimization of acid plant design. .. Preliminary estimates indicate potential savings of up to 20% in both capital and operating costs for a "greenfield" smelting complex'. *[Note by Asteljoki in the 1980s - making process sustainable - instead of 'give up - shut down'].*

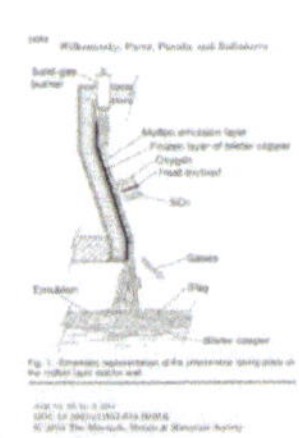

Energy consumption is further reduced by a one step continuous flash process taking place at a higher temperature to blister copper.., such as the Noranda process or other similar processes –(Andrzej Warczok and Gabriel Riveros).

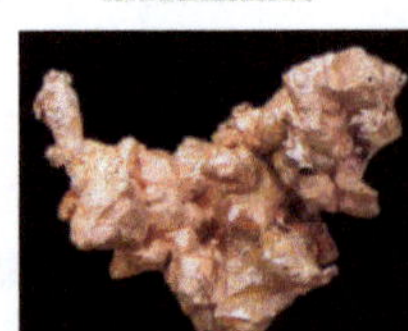

1970 processes of copper recovery from low grade ores by leaching, Solvent Extraction and Electrowinning have been already a low carbon process when it utilizes renewable electrical energy.

Gold – ThioHydro metallurgical Process

Thiosulfate ions are known to form complex ions, with noble metal gold such as $Au(S_2O_3)_2{}^{3-}$.

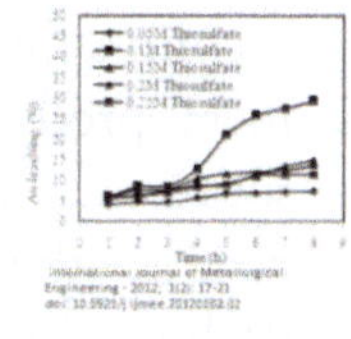

This property is used for cyanide free extraction of gold from refractory ores, followed by anion exchange removal of gold thiosulfate complexes from leach liquor.

'Cyanide free process' also means low carbon process. R.Y.Wan, J.D.Miller, and J.Li called this a 'Thio hydrometallurgical' process.

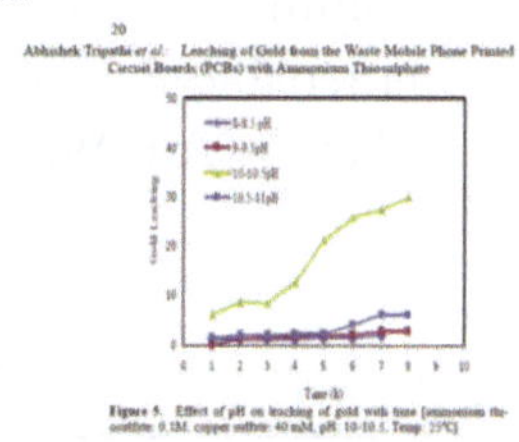

This 2005 paper discusses use of other sulfur complexes such as thio-urea $Au(SC(NH_2)_2)^{2+}$, and Thiocyanate $Au(SCN)^{2-}$ and compares with conventional cyanide complexation reactions

Courtesy of Ind LLC, ©

Zinc Sulfide to Zinc

Roast-Leach-Electrowin process, more widely used than the carbothermal reduction of ZnO. The carbon dioxide penalty may occur if the electricity is generated using carbonaceous fuels, instead of using hydro-power or other non-carbon alternate energy.

2 ZnS + 3 O2 → 2 ZnO + 2 SO2 ---[J]

2 SO2 + O2 + 2 H2O → 2 H2SO4 ---[K]

ZnO + H2SO4 + H2O→ ZnSO4[aq] + 2 H2O -[L]

Red sphalerite, Redlands Quarry, Niagara Falls, Niagara County,

Courtesy of Ind LLC, ©

Nickel Sulfide Smelting

'….. the output of flash smelters accounts for nearly 70% of the primary metal produced from nickel sulfide sources … Electric - furnace smelters produce the balance. The key merits of flash smelting are very low electrical and fossil fuel energy consumption and generation of a continuous, low-volume, SO_2-rich process gas stream amenable to processing in an acid plant.' [A.E.Warner, et al]

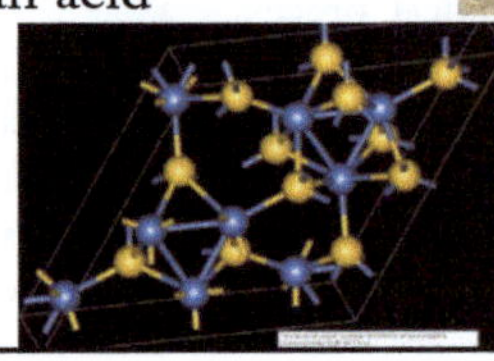

Lead Sulfide Hydrometallurgical Process

Hydrometallurgical conversion of lead sulfide to lead and elemental sulfur. [Agnes Y. Lee, et al – USBM]

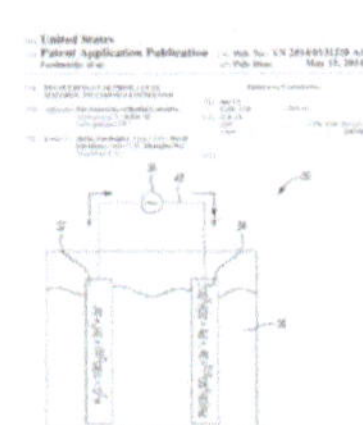

The steps were (a) leaching with H_2O_2 , PbO_2 , and recycled fluosilicic acid at 95 ° C to produce a solution of $PbSiF_6$ and a residue containing elemental S, (b) electrowinning of the $PbSiF_6$ solution at 35 ° C to produce 99.99 pct Pb metal and H_2SiF_6 , and (c) Sulfur recovery by solvent extraction, leaving a residue containing Cu, Ag, and other metal values.

Dissolve PbS in methane sulfonic acid & Electrolyse -2014 patent – Fassbender , et al

Thio-metallurgy for Green Alkaline Earth Metals-1

Green Alkaline Earth Metals and chemicals can be made from naturally occurring sulfates of magnesium and calcium in a *sustainable* fashion [Neelameggham and Brown, 2013]

Naturally occurring sulfates of magnesium include epsomite [$MgSO_4.7H_2O$], kieserite [$MgSO_4.H_2O$], as well as mixed sulfates with potassium – schoenite, langbeinite, etc. We all know gypsum, the naturally occurring sulfate of calcium $CaSO_4.2H_2O$ and anhydrite $CaSO_4$.

Use of sulfates as a raw material overcomes the CO_2 penalties arising from the use of alkaline earth carbonates such as calcite ($CaCO_3$), magnesite ($MgCO_3$) or dolomite ($CaCO_3.MgCO_3$).

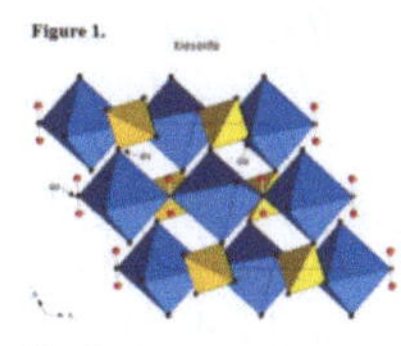

Thio-metallurgy for Green Alkaline Earth Metals-2

Magnesium chloride brines , calcium carbonate is used in making the calcium chloride used in removing the sulfates in the brine – result in CO2 emissions. Further, purified magnesium chloride brine is evaporated using fossil fuels to make anhydrous magnesium chloride –which creates additional carbon dioxide emissions.

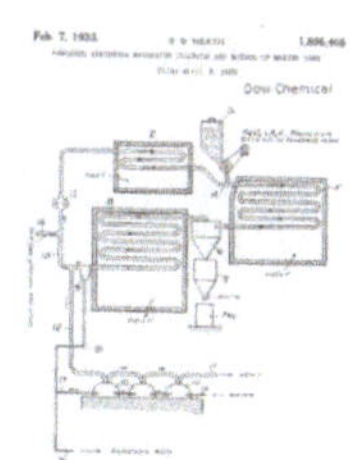

MgCl2[aq] + H2O [aq]→ MgCl2 + zH2O[crystal] + y MgO + y HCl +H2O[1-z] [v] .. [M]

[M] takes place around 300°C . The hydrolysis of magnesium chloride creates residual H2O [crystal] and MgO [crystal] – this can be removed by dehydration with additional HCl vapors [~400°C] or by molten salt chlorination [800°C].

Thio-metallurgy for Green Alkaline Earth Metals-3

In 1808, Davy and in 1884– Gerhard and Smith, have been noted to have used magnesium sulfate aqueous solution as the raw material in trying to isolate elemental magnesium, …

There have been a few patents in the 1930-40 period of using carbonaceous fuel to convert magnesium sulfate into magnesium oxides, sulfur oxides and carbon oxides.
In ***many cases millions of tons of magnesium sulfate are wasted in evaporite basins without realizing their value.***

Such evaporite basins include ponds at the banks of Great Salt Lake, other bittern effluent ponds worldwide, Aral sea dried basin part of Kazakhstan – Uzbekistan area.

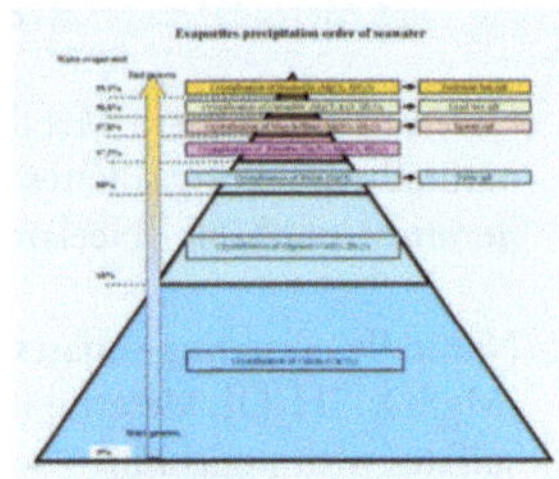

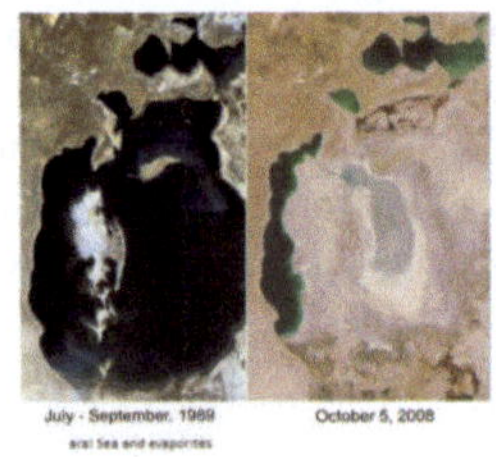

Thio-metallurgy for Green Alkaline Earth Metals-4

The following reactions show the decomposition temperatures of these anhydrous sulfates of alkaline earth metals by direct heat which can be reduced considerably using sulfur as the reductant.

MgSO4.7 H2O[epsomite]	→	MgSO4.H2O	(~ 80o C)	---[N]
MgSO4.H2O [kieserite]	→	MgSO4	(~ 250 o C)	---[O]
MgSO4	→	MgO + SO2 + 0.5 O2	(~ 1000 oC)	---[P]
MgSO4 + 0.5 S	→	MgO + 1.5 SO2	(~ 625 oC)	---[Q]
CaSO4	→	CaO + SO2 + 0.5 O2	(~1600 oC)	---[R]
CaSO4 + 0.5 S	→	CaO + 1.5 SO2	(~ 1200 oC)	---[S]
MgSO3	→	MgO + SO2		---[T]
CaSO3	→	CaO + SO2		---[U]

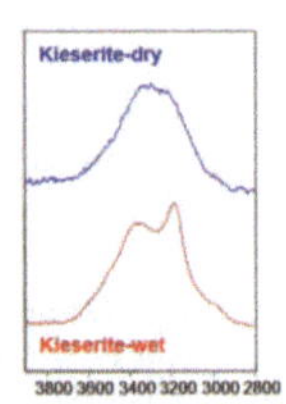

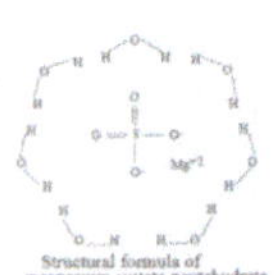

Thio-metallurgy for Green Alkaline Earth Metals-4

The following reactions show the decomposition temperatures of these anhydrous sulfates of alkaline earth metals by direct heat which can be reduced considerably using sulfur as the reductant.

$MgSO_4.7\ H_2O$ [epsomite]	→	$MgSO_4.H_2O$	(~ 80 °C)	---[N]
$MgSO_4.H_2O$ [kieserite]	→	$MgSO_4$	(~ 250 °C)	---[O]
$MgSO_4$	→	$MgO + SO_2 + 0.5\ O_2$	(~ 1000 °C)	---[P]
$MgSO_4 + 0.5\ S$	→	$MgO + 1.5\ SO_2$	(~ 625 °C)	---[Q]
$CaSO_4$	→	$CaO + SO_2 + 0.5\ O_2$	(~1600 °C)	---[R]
$CaSO_4 + 0.5\ S$	→	$CaO + 1.5\ SO_2$	(~ 1200 °C)	---[S]
$MgSO_3$	→	$MgO + SO_2$		---[T]
$CaSO_3$	→	$CaO + SO_2$		---[U]

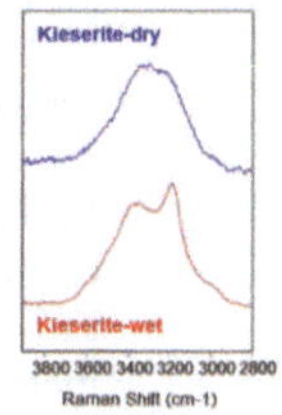

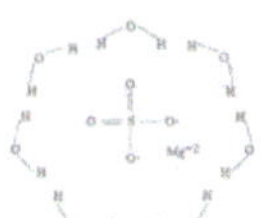

Structural formula of magnesium sulfate heptahydrate

Courtesy of Ind LLC, ©

Thio-metallurgy for Green Alkaline Earth Metals-5

Anhydrous $MgSO_4$ prepared at 250 °C vs anhydrous $MgCl_2$ ~ 450 to 800 °C. $MgSO_4.H_2O$ prepared at 80 °C unlike $MgCl_2.H_2O$ at 250 °C or > & with hydrolysis.

It is further noted that the 'green' MgO or anhydrous $MgSO_4$ can be converted to anhydrous magnesium chloride by Ind LLC's process for conventional electrolysis, with a lower energy cost.

Amenable for 'brownfield' conversions in existing magnesium facilities.

The 'green' MgO can be co -produced with green 'CaO' using gypsum or sulfite for use in processes such as silico-thermal process which claims hydro-electric powered Ferro silicon reduction production of 2.2 kg CO_2/kg Mg. Such a 'carbon credit' bearing process can be applied by present producers in their 'brown' field expansions. The 'green' MgO or 'CaO' process is suitable for regenerating the oxides from sulfites resulting in flue gas desulfurizing scrubbers.

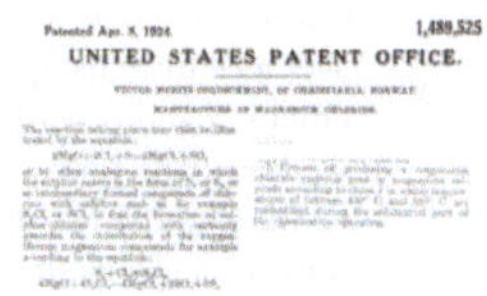
Patented Apr. 8, 1924. 1,489,525
UNITED STATES PATENT OFFICE.

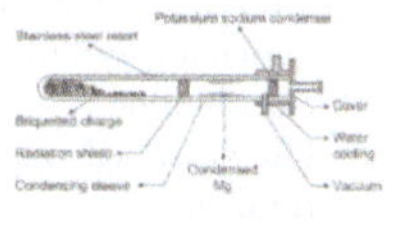

Courtesy of Ind LLC, ©

Thiometallurgical Cement & Calcium

Production of 'fully or partially green' Portland cement using calcium oxide generated from gypsum with clay to form the cement clinker – the endothermic heat being supplied by sulfur for fully green cement and carbonaceous fuel supplying the heat in the partly green cement [Neelameggham & Brown 2013].

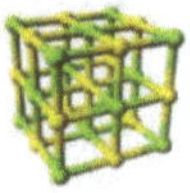

In the 1960's U.K. atomic energy used calcium sulfide as the feed to calcium metal eectrolyzer – this is another early application of thiometallurgy for alkaline earth metals. Sulfur removal from steel has been done mainly using either calcium carbide or magnesium during the last 30 years – showing the interaction between alkaline earth metals and sulfur in metallurgy.

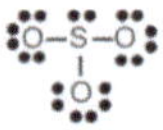

A Lewis structure for the sulfite ion that obeys the octet rule. This is not the best structure, due to formal charges on all the atoms.

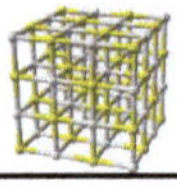

Courtesy of Ind LLC, ©

Known 'Thiometallurgical' conversion of titanium oxide minerals.

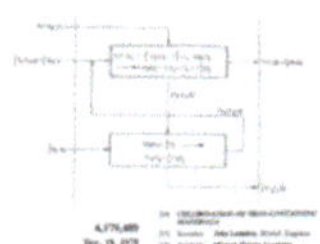

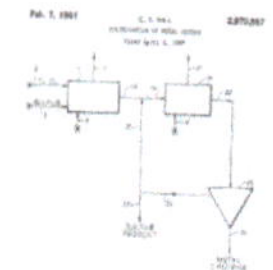

Carbon free titanium dioxide has been made for over a century using sulfuric acid digestion of ilmenite, and other iron bearing titanium oxides, by the sulfate process.
J.Miller and B.R.Davis disclosed an improved sulfuric acid digestion practice using a liquid phase approach.
Lumsden discussed reactions of sulfuryl chloride in making ferrous chloride along with titanium tetrachloride .
Hill described suspending ores such as TiO_2 – rutile, in molten sulfur [excess sulfur] and chlorinating the ore to make volatile chlorides and sulfur dioxide. The reactions are carried out using ferric chloride, sulfuryl chloride or chlorine and the temperatures being in the range 250 to 350 °C while keeping sulfur in the molten liquid state.

Courtesy of Ind LLC, ©

Thiometallurgy of Titanium Compounds-1

A proprietary green titania process - avoid the formation of waste copperas and produce marketable sulfuric acid from the sulfur oxides byproducts. [Neelameggham and Davis 2013] .

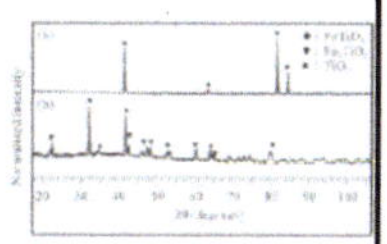

Sequential partial oxidation of iron as a water soluble ferrous chloride or sulfate if a market for that exists for water treatment. These processes use sulfur instead of sulfuric acid in controlling the products which are formed. [Neelameggham and Davis 2013]

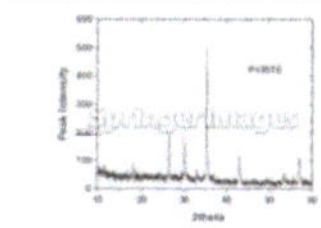

$FeTiO_3 + Cl_2 + 0.5\ S \rightarrow FeCl_2 + 0.5\ SO_2 + TiO_2$ ---[V]
$FeTiO_3 + 2.5\ O_2 + 2\ S \rightarrow FeSO4 + SO_2 + TiO_2$ ---[W]
$FeTiO_3 + S + O_2 \rightarrow FeO + SO_2 + TiO_2$ ---[X]

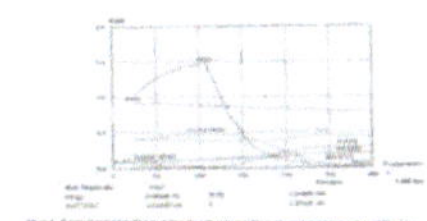

Thiometallurgy of Titanium Compounds-2

These reactions are easily carried out in by the sulfate process equipment for TiO2 – such as in rod mills or by other mechano chemical approaches using high energy milling, followed by leaching and separation steps.

The need for partial chlorination may become essential in removing other impurities in the ore such as silica, vanadium, etc.

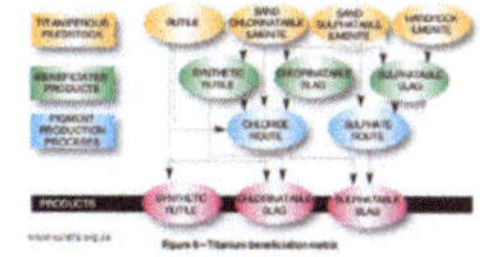

There have been patents of using sulfur chlorination of ilmenites in the 1930 – 1980 periods; those processes also fall under 'thiometallurgical ' conversion of iron bearing titanites .

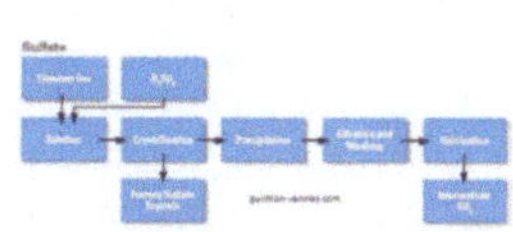

Mechanochemical upgrading of ilmenite with sulfur to form impure rutile is known; but this has not been studied in detail to facilitate better sustaining technologies as yet.

Thiometallurgy in Aluminum Production

Aluminum sulfide as feed instead of aluminum oxide -----
Nguyen Q Minh, et al showed in 1982, a *process to make metallic aluminum by the electrolysis of Al_2S_3 at 700°-800°* C in a chloride melt. THIS IS IMPROVED BY VAN DER PLAS & YANPING XIAO – ALERIS-NL- IN 2011 SHOWING HOW TO MAKE CELL FEED Al2Susp3 [7867373]

(a) lower cell voltage compared to typical Hall Heroult Process, thus lowering energy costs by the combination of operations at a lower temperature and lower cell voltage.

(a) (b) Anode produced sulfur vapors without attacking the graphite electrode unlike in the HallHeroult cell the oxygen release at carbon anodes resulting in carbon oxides.

Courtesy of Ind LLC, ©

Thiometallurgy of Iron

Electrolytic decomposition of ferrous sulphate in pickle waste liquor CA 1063548 A1

Inventors Jozsef Kerti, Andor Mandoki, Miklos Szeky
Applicant "Licencia" Talalmanyokat Ertekesito Vallalat, Jozsef Kerti, Andor Mandoki, Miklos Szeky

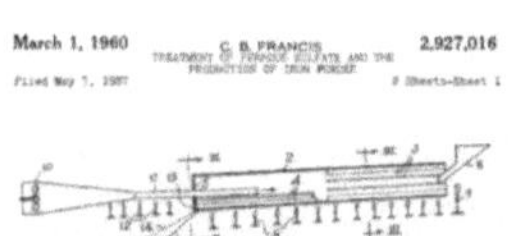

ULCOS in European Union Study has evaluated several techniques of making steel with low carbon dioxide emission. One of which is to electrolyze iron oxide converted to iron chlorides by electrolysis.

Neelameggham and Davis have analyzed techniques of making iron sulfates from thiometallurgical conversion of iron ores, which can be electrolyzed to iron.

Applicable to making ~ 5 million tpy iron Powder from sulfate TiO2 process –while recycling sulfuric acid

Courtesy of Ind LLC, ©

By-product Credit Improves Metal Production Sustainability

The cost of sulfur is competitive with that of carbon
Petroleum refineries have become suppliers of sulfur, besides the natural gas wells supplying sulfur.

When sulfur oxides released in concentrated make marketable sulfuric acid as done in smelters.
This by-product credit will pay for the sulfur used as a reducing agent or as an energy matter.
Sulfuric acid consumption will increase with population.

Such a by-product credit is not available when using carbon as a fuel or a reducing agent. Thus are less sustainable unless the carbon fuel costs are kept very low.

Power plants are still slow in reducing their costs while abating pollution. Thus Thio-metallurgy provides unique cost effective and sustainable solutions.

Courtesy of Ind LLC, ©

Engineering Solutions for Sustainability: Materials and Resources II
Edited by: Jeffrey W. Fergus, Brajendra Mishra, Dayan Anderson, Emily Allyn Sarver, and Neale R. Neelameggham
TMS (The Minerals, Metals & Materials Society), 2015

PRODUCTION OF (Mn,Fe)-CARBIDE CONTAINING LOW PHOSPHORUS BY CARBOTHERMIC REDUCTION OF Mn OXIDE AND Fe OXIDE

Dong-Yuk KIM[1], Hyun-soo KIM[2] and Sung-Mo JUNG[1]

[1]Graduate Institute of Ferrous Technology (GIFT), POSTECH;
77 Cheongam-Ro, Nam-gu, Pohang, Gyeongbuk, 790-784, Korea
[2]Iron-making Research Group, Technical Research Laboratories, POSCO;
Donghaean-Ro, Nam-gu, Pohang, Gyeongbuk, 790-300, Korea

Keywords: dephosphorization, carbothermic reduction, carburization, manganese iron carbide, metallic phosphide, solubility, distribution.

Abstract

The dephosphorization process for high P iron ores have recently been studied employing several methods. However, the researches of phosphorus removal about the high P manganese ores have seldom done as many as which of the high P iron ores. As a potential approach for the removal of phosphorus contained in the composite, the formation of (Mn,Fe)-carbide by the carbothermic reduction of the oxides of manganese and iron was investigated at 1200°C in H_2 employing thermogravimetric analysis (TGA). It was confirmed that the phosphorus did not dissolve into the manganese iron carbide, even though it has the high reactivity with Fe and Mn. According to EPMA line mapping, (Mn,Fe)-carbide and (Mn,Fe)-phosphide coexist in the reduced particle. The current results could be applied for a promising treatment of dephosphorization of high phosphorus ores due to the low solubility of phosphorus in carbide.

Introduction

Due to the increased consumption of manganese in steel industry, high grade manganese ores have gradually depleted with the great increase of its price. Thus, manufacturers of manganese-based alloys will have no choice but to use the low grade ores with impurities. One of the principal impurities in manganese ores is phosphorus which is difficult to be removed in smelting reduction process.[1] Most of the phosphorus is removed in oxidizing conditions by the conventional method such as BOF (Basic Oxygen Furnace), however, whose process can bring out a significant loss of manganese because of the preferential oxidation of manganese.[2]

Carbothermic reduction of manganese ore in solid state, if its rate is acceptable, can improve not only the process efficiency in the extraction of Mn but also a decrease of gaseous phosphorus dissolution into liquid Mn.[3] In the process, the manganese iron carbide can be formed, which could be the one of the various dephosphorization methods because the solubility of phosphorus in carbide might be expected to be low. The presence of carbon can enhance the dephosphorization of ferromanganese due to the increase in the activity coefficient of phosphorus with increasing carbon content. In addition, the carbon activity can be increased by phosphorus due to the repulsive force exerted between them (interaction parameter in iron alloys, $e_P^C = 0.126$), which results in the low solubility of phosphorus in carbide with high content of carbon.[4] Above all, it is not possible to carburize phosphorus because it requires huge amount of energy for the formation of phosphorus carbide. That is, in contrast with the computational prediction[5] estimating the possible crystal structure of phosphorus carbide, in practice, it is extremely difficult to synthesize phosphorus carbide in general experimental conditions of reduction system.

However, the hypothesis should need the precise experimental proofs for the commercialization of dephosphorization process. Therefore, based on the above evidences, the current research aims to investigate the behavior of phosphorus during the formation of (Mn,Fe)-carbide with the intention of applying it as dephosphorization treatment.

Experimental

The reagent-grade chemicals of MnO_2, Fe_2O_3 and β-$Ca_3(PO_4)_2$ were prepared typically to represent most of the high phosphorus manganese ores. MnO2 (Aldrich chemistry, > 99% purity) was in size of greater than 63μm and less than 250μm, Fe_2O_3 was in size of 1μm with 99.9% purity. Because phosphorus exists mainly as apatite in manganese ores, β-$Ca_3(PO_4)_2$ of 10μm diameter was used as phosphorus-containing compound in the current experiment. These reagents were mixed with synthetic graphite (99.7% purity, mean particle size: 5 μm) as carbothermic reducing agent for 30min employing the mixing apparatus (SPEX SamplePrep). The mass ratio of MnO_2, Fe_2O_3, $Ca_3(PO_4)_2$ and graphite in the mixture is 6:1:0.5:2, which was normalized from the composition of high phosphorus manganese ore.

Reduction experiments were conducted in the high temperature TGA (RUBOTHERM, Germany) with the 0.5g of pellets weight in alumina crucible. The gas atmosphere in TGA was high purity H_2 with flow rate 1000mL/min regulated by mass flow controller. The use of H_2 was for the formation of (Mn,Fe)-carbide as sooner as possible due to the increased diffusivity of CO in atmosphere gas.[6] The reduced samples were analyzed by XRD, FE-SEM with EDS and FE-EPMA for the identification of phases and morphology observation.

Results and Discussion

A. Formation mechanism of manganese iron carbide from MnO_2 and Fe_2O_3.

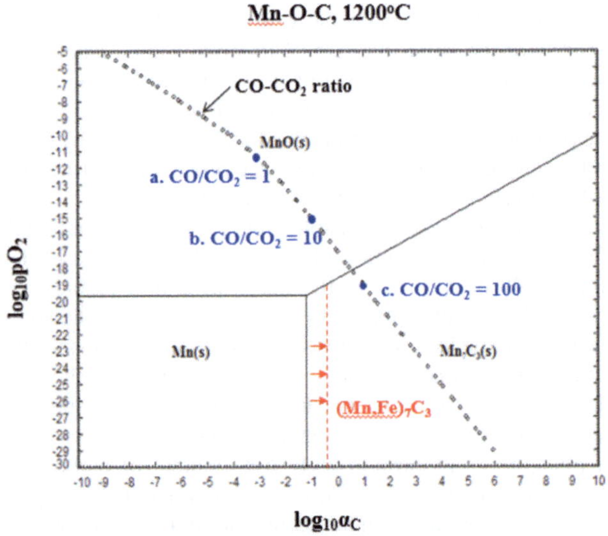

Figure 1. Mn-O-C phase stability diagram at 1200°C as a function of oxygen partial pressure and the activity of carbon.

The manganese iron carbide has thermodynamic characters differ from iron carbide, cementite. The cementite is very metastable thermodynamically, and will decompose to metallic iron and graphite due to the carbon deposition under strongly carburizing atmosphere.[7]

However, the cementite can be stable state as the manganese with similar element to iron is substituted in carbide. The carbon activity decreases by Mn substitution due to the formation of Mn-C dipoles, resulting from a strong attractive interaction between Mn and C.[8] In the case of manganese ore containing much manganese oxide relatively more than iron oxide, the Mn-oriented carbide with a small amount of iron must be formed.

It is the stable carbide structure from the carburization of MnO by intermediate CO-CO_2 gas at 1200°C as shown **Fig. 1**. In the figure, the carbon activity (a_c) and oxygen partial pressure (pO_2) can be changed dependently along the line with the different ratio of CO to CO_2 once temperature of the system are fixed, for example, the points a, b and c are set by some typical values of CO-CO_2 ratio. From this figure with the line of CO-CO_2 ratio, Mn_7C_3 could be formed more easily in condition of high CO-CO_2 ratio under thermodynamic aspect. The carbon activity for formation of manganese iron carbide, $(Mn,Fe)_7C_3$, can be increased depending on the amount of iron substitution. Since the bonding of substituted Fe and C is weaker than Mn-C bonding in the carbide lattice,[9] the formed $(Mn,Fe)_7C_3$ is more unstable than Mn_7C_3 with the higher carbon activity. It can affects also the activity of Mn to increase relatively.

B. Carbothermic Reduction of MnO_2, Fe_2O_3 and $Ca_3(PO_4)_2$.

The carbothermic reduction of MnO_2, Fe_2O_3 and $Ca_3(PO_4)_2$ was carried out at several temperatures in H_2 atmosphere. **Fig. 2** shows the change in the fractional reduction of the sample with time, which was divided into two stages across the fractional reduction of 0.55.

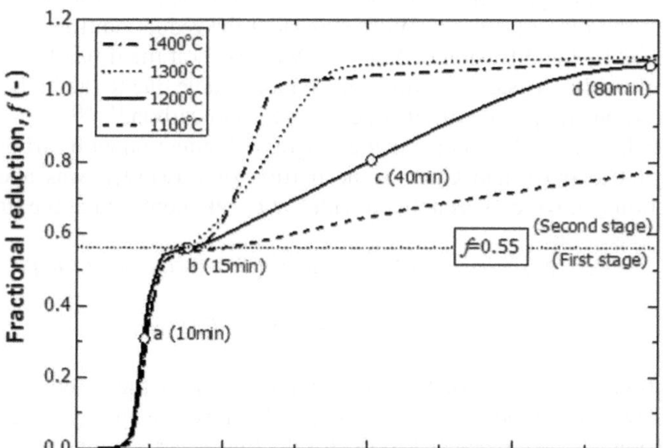

Figure 2. Carbothermic reduction of MnO_2, Fe_2O_3 and $Ca_3(PO_4)_2$ in H_2 atmosphere at the temperature range of 1100 to 1400°C.

The critical fractional reduction was calculated for the reduction stage where the oxygen contained in both MnO_2 and Fe_2O_3 was only removed by H_2.

That is, the first stage represents the reduction of MnO_2 to MnO with that of Fe_2O_3 to metallic Fe by H_2 according to the following reactions:

$$MnO_2 + H_2 \rightarrow MnO + H_2O \quad (1)$$

$$Fe_2O_3 + 3H_2 \rightarrow 2Fe + 3H_2O \quad (2)$$

As shown in **Fig. 2**, these reactions started while raising temperatures and were completed as soon as reaching the target temperatures, and the reduction rates in the first stage were almost similar in the temperature range investigated. On the other hand, the second stage corresponds to the carburization of MnO and Fe by carbon, resulting in the formation of manganese iron carbide by Reaction (3):

$$x\text{MnO} + y\text{Fe} + \left\{\frac{3(x+y)}{7} + x\right\}\text{C} = \frac{(x+y)}{7}(\text{Mn,Fe})_7\text{C}_3 + x\text{CO} \quad (3)$$

The rate of carburization increased with increasing temperature, which is believed to be limited by the interfacial Boudouard reaction. Therefore, the formation of carbide was completed in the temperatures above 1100°C where CO_2 becomes significantly unstable in the presence of carbon, resulting in the regeneration of CO.[6,10]

C. Progress in the carbothermic reduction of MnO_2, Fe_2O_3 and $Ca_3(PO_4)_2$ at 1200°C.

The isothermal reduction experiments at 1200°C in H_2 atmosphere were carried out to investigate the reduction mechanism for the points (a, b, c and d) marked in **Fig. 2**. First of all, XRD pattern of each point was analyzed as shown in **Fig. 3**. The manganese iron carbide started to be formed with the nucleation of Fe at the reduction time of 15 min, which was followed by the carburization of MnO to carbide gradually. With Fe as the nuclei, $(Mn,Fe)_5C_2$ was then formed in the stable lattice of carbide at the time of 40 min. With increasing the reduction time, the peak intensity of $(Mn,Fe)_7C_3$ increased by the diffusion of Mn into $(Mn,Fe)_5C_2$ at the time of 80 min. It is believed that the carbide preferably formed with Mn is much more stable due to the higher interactive force between Mn and carbon compared with that between Fe and carbon.[11,12]

In addition, the behavior of phosphorus in the progress of reduction with carbon can be estimated by XRD pattern of $Ca_3(PO_4)_2$ and CaO. Some portion of $Ca_3(PO_4)_2$ was about to decompose according to Reaction (4), which was noticed with CaO peak identified at the time of 40 min:[13]

$$Ca_3(PO_4)_2(s) + 5C(s) = 3CaO(s) + P_2(g) + 5CO(g) \quad (4)$$

$$\Delta G^\circ = 1{,}750{,}831 - 1{,}035.1T \text{ (J/mol)}^{[14]} \quad (5)$$

That is, the reaction temperature of $Ca_3(PO_4)_2$ with carbon at the standard conditions can be calculated to be 1418°C from the value of the standard Gibbs free energy of reactions by Eq. (5). However, the reaction can take place at 1200°C for the lower partial pressures of product gases in the current experimental conditions. XRD peak of CaO at 40 min indicates that small amount of P_2 gas was generated in the course of carburization although the experimental temperature was lower than 1418°C. As shown in **Table 1**, the amount of remained phosphorus in the sample with time shows that a lot of P_2 gas was vaporized between Step b (15min) and Step c (40min) which is the starting point of carburization.

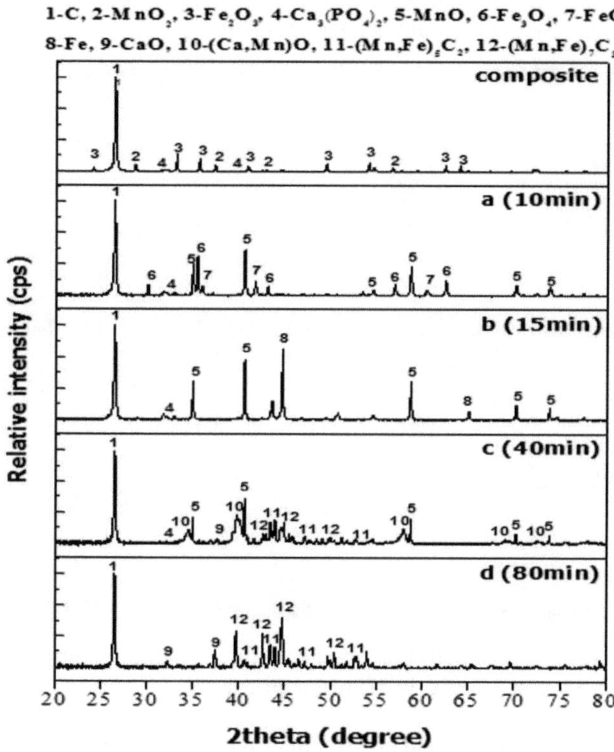

Figure 3. XRD patterns of the carbothermically reduced sample comprising MnO_2, Fe_2O_3 and $Ca_3(PO_4)_2$ at 1200°C in H_2 atmosphere.

Table 1. Phosphorus concentration in reduced samples comprising MnO_2, Fe_2O_3 and $Ca_3(PO_4)_2$ at 1200°C in H_2.

Stages	a (10 min)	b (15 min)	c (40 min)	d (80 min)
P (mass%)	1.33	1.23	0.93	0.91

It could also be supported by the fractional reduction value higher than unity, which was resulted from the additional weight loss due to P_2 volatilization from the sample.[15,16] In considering the chemical reaction between $Ca_3(PO_4)_2$ with H_2 as the atmosphere gas , on the other hand, there is no possibility at the experimental temperature according to Eq. (6):[17]

$$2CaO \cdot P_2O_5(s) + H_2(g) = 2CaO(s) + P_2(g) + O_2(g) + H_2O(g) \quad (6)$$

$$\Delta G^\circ = 1{,}942{,}600 - 530.2T \text{ (J/mol)}^{[14]} \quad (7)$$

The morphology of MnO_2, Fe_2O_3 and $Ca_3(PO_4)_2$ in the sample reduced and carburized at 1200°C in H_2 was observed by FE-SEM. As shown in **Fig. 4**, the change of morphology was investigated for the points a, b, c and d shown in **Fig. 2** and the elemental compositions of selected points in first stage were analyzed by EDS as summarized in **Table 2**.

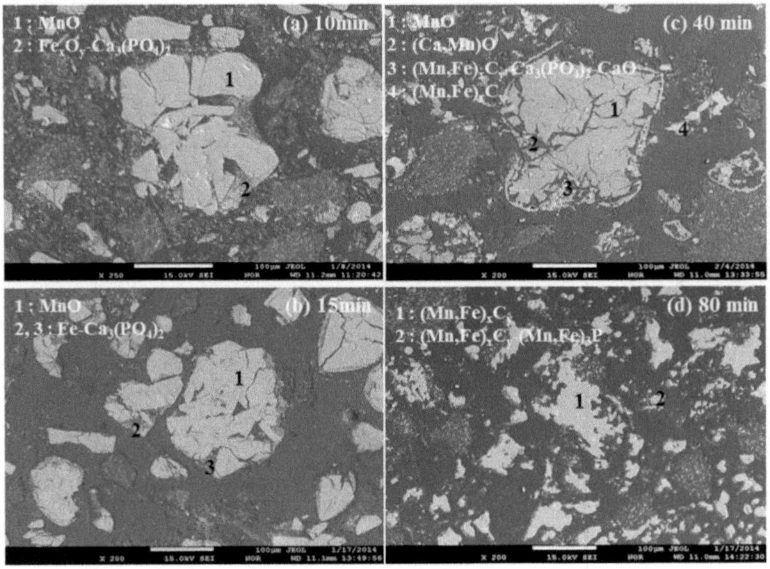

Figure 4. SEM images of the carbothermically reduced sample comprising MnO_2, Fe_2O_3 and $Ca_3(PO_4)_2$ at 1200°C in H_2 atmosphere.

Table 2. Elemental analyses of the carbothermically reduced sample comprising MnO_2, Fe_2O_3 and $Ca_3(PO_4)_2$ in the first stage (a (10 min), b (15 min)) by EDS.

Time(min)	Position	at%				
		Mn	Fe	P	Ca	O
a (10 min)	1	54.83	-	-	-	45.17
	2	-	36.97	3.40	6.32	53.31
b (15 min)	3	54.67	-	-	-	45.33
	4	-	16.42	10.81	18.13	54.65
	5	-	28.36	7.40	13.93	50.31

The elemental analyses of the points on FE-SEM images by EDS were assumed to correspond to some compounds with the help of XRD patterns. In the beginning of reduction, the reduced MnO with dense and large blocks exists as a main dominant phase and the particle mixture of Fe oxide, $Ca_3(PO_4)_2$ and carbon surrounds the reduced MnO. Further reaction reduced Fe oxide to metallic Fe by H_2 from Step (a) to Step (b) where the reduction in the first stage was finished.

Elemental analysis results of the selected points in the second stage at the reduction time of 40 and 80 min were presented in **Table 3**. In this stage, the carbide was about to form with the nucleation of Fe surrounding the reduced MnO.

Table 3. Elemental analyses of the carbothermically reduced sample comprising MnO_2, Fe_2O_3 and $Ca_3(PO_4)_2$ in the second stage (c (40 min), d (80 min)) by EDS.

Time (min)	Position	at%				
		Mn	Fe	P	Ca	O
c (40 min)	1	57.55	-	-	-	42.46
	2	47.03	-	-	7.53	43.56
	3	47.53	9.08	3.21	7.28	32.90
	4	81.25	18.75	-	-	-
d (80 min)	5	86.85	13.15	-	-	-
	6	78.57	15.21	6.22	-	-

In the **Fig. 4**, it was observed that MnO particles broke into smaller oxide blocks and generated numerous cracks, which indicates that MnO might be reduced. The carbide and MnO phases were simultaneously identified in XRD pattern for Step (c), which implies that the reduction of MnO and carbide formation could have simultaneously proceed. At the completion of reduction for Step (d), it appeared that all the MnO was reduced and that the formed manganese iron carbides were sintered. It is believed that most of the particles comprise the mixture of (Mn,Fe)-carbide and (Fe,Mn)-phosphide surrounded by CaO and $Ca_3(PO_4)_2$ according to the EDS results and XRD patterns. In addition, $(Fe,Mn)_3P$ could be resulted from the reaction between metallic (Fe,Mn) and P_2 gas generated from $Ca_3(PO_4)_2$.[18]

D. Estimation of Carbothermic Reduction Mechanism of MnO_2, Fe_2O_3 and $Ca_3(PO_4)_2$ at 1200°C by EPMA Mapping Analysis.

The mixture of Fe and carbon around particles cannot form the carbide without Mn due to the carbon deposition and unstable state of iron carbide at high temperatures.[7] However, it can be stabilized by the substitution of Mn into the lattice.[8] It indicates that the carbide should be formed along with the diffusion of Mn, which can proceed by overcoming the space existing between MnO and Fe-C mixture. It is believed that Fe-C mixture could easily contact the surface of MnO since it was in liquid state at 1200°C. This mechanism can be supported by the overall formation of Mn carbide in terms of the following reactions:[19]

$$7MnO(s) + 10C(s) = Mn_7C_3(s) + 7CO(g) \tag{8}$$

$$7MnO(s) + 13CO(g) = Mn_7C_3(s) + 10CO_2(g) \tag{9}$$

$$10CO_2(g) + 10C(s) = 20CO(g) \tag{10}$$

In other words, CO supplied by the Boudouard reaction reduced MnO to produce Mn which diffused to the melt of Fe-C mixture. Then Mn could react with Fe and carbon to form manganese iron carbide. The formation for final stable phase of manganese iron carbide at the equilibrium state was shown in the following reaction;

$$7MnO(s) + (Fe - C)melt\,(l) = (Mn,Fe)_7C_3 + 7CO(g) \tag{11}$$

It can be considered that the Fe-C melt can react with P_2 gas more easily than the solid state, and the formation of (Fe,Mn)-phosphide might be accompanied by (Mn,Fe)-carbide formation simultaneously. However, this possibility was considered based on the previously reported results.[20,21] More in-depth understanding of the direct formation mechanism requires further investigation.

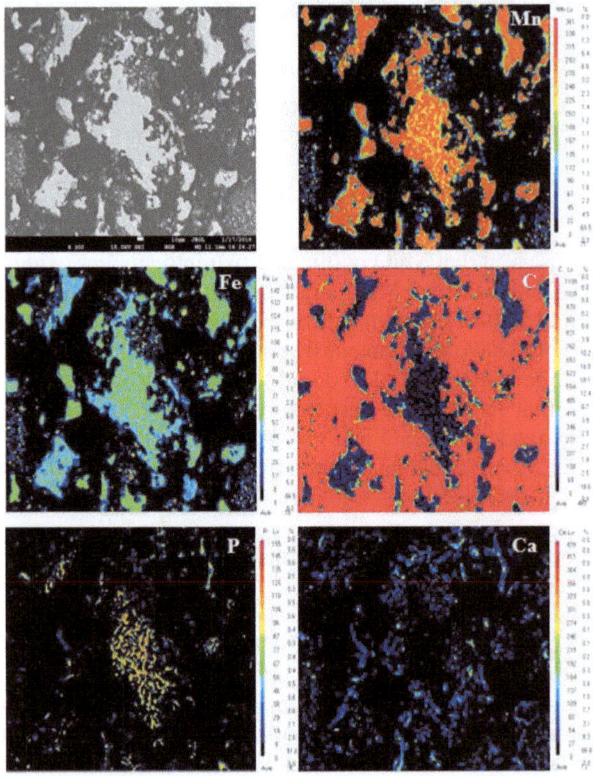

Figure 5. EPMA area mapping of the carbothermically reduced sample at 1200°C in H_2 atmosphere (d (80 min)).

Around the completion of reduction at the reduction time of 80 min, the elemental distribution through the EPMA mapping was obtained for the products of carburized MnO_2, Fe_2O_3 and $Ca_3(PO_4)_2$ as shown in **Fig. 5**. Mn and Fe were found at the similar location, which indicates that most of the particles consisted of (Mn,Fe)-carbide in the presence of carbon and (Mn,Fe)-phosphide in the presence of phosphorus. In the FE-SEM images, the sintering of (Mn,Fe)-carbide and (Mn,Fe)-phosphide indiscriminately in one particle can be assumed. The diffusion of Mn to carbide phase and the reaction of vaporized P_2 gas with Fe-C melt was carried out simultaneously. The area of carbon existing with Mn and Fe was believed to represent carbide even though the distribution of carbon in the particle was not distinctive due to the relative high amount of carbon in resin. On the other hand, the phosphorus exists in the form of two compounds, either $Ca_3(PO_4)_2$ with Ca or $(Fe,Mn)_3P$ with Fe and Mn. It is believed that $Ca_3(PO_4)_2$ usually distributes around particles, which was identified from the overlapped area of P and Ca distribution. On the other hand, Ca identified in the mapping seemed to originate from the CaO decomposed from $Ca_3(PO_4)_2$.

There is no clear distinction between the distribution of (Mn,Fe)-carbide and that of (Fe,Mn)-phosphide in the EPMA area mapping. Hence, the EPMA line mapping was carried out for the clearer distribution of the elements in the particle containing carbide and $(Fe,Mn)_3P$ as shown in **Fig. 6**. From the opposite distribution of carbon and phosphorus in the line profile, it is believed that the carbide does not overlap the phosphide although two compounds coexist within one particle. That is, phosphorus does not dissolve into carbide, which strongly indicates that phosphorus is not involved in the formation of carbide. The magnetic properties of manganese iron carbide is changed by the ratio of Mn to Fe, and the ferromagnetic (Mn,Fe)-carbide can be separated from the antiferromagnetic (Mn,Fe)-phosphide with the help of the magnetic separator with mechanical crushing. In conclusion, the formation of carbide with high carbon content might be helpful not only for the removal of phosphorus, but also for the incorporation of Mn and Fe into the lattice of carbide.

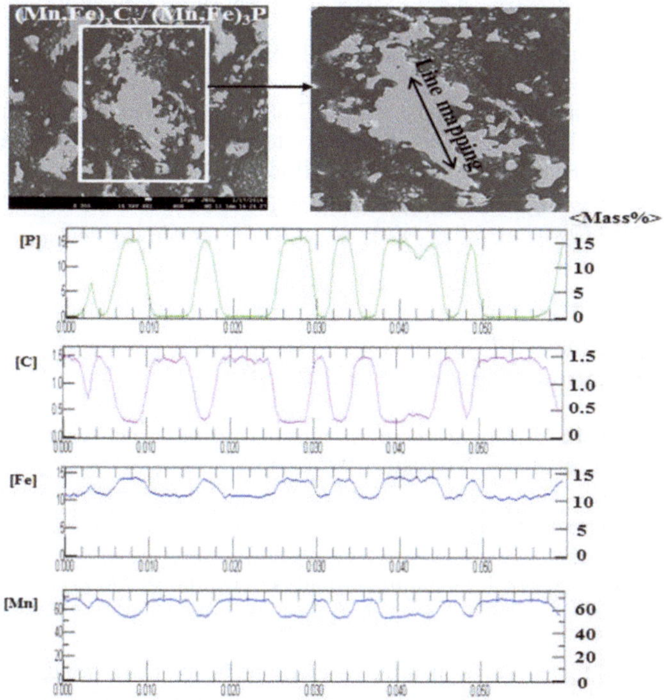

Figure 6. EPMA line mapping of the carbothermically reduced sample at 1200°C in H_2 atmosphere (d (80 min)).(mass%)

Conclusions

The carbothermic reduction of Mn oxide, Fe oxide and phosphorus compounds was investigated to utilize the manganese iron carbide for the dephosphorization of high P manganese ores. From the findings, the following conclusions were obtained:

1. The manganese iron carbide, $(Mn,Fe)_7C_3$, formed by MnO_2 and Fe_2O_3 was very stable thermodynamically differ from the iron carbide, Fe_3C. It is due to the strong attraction force between Mn and C, resulting in the decrease of carbon activity. The manganese iron carbide can be get easily as product from high phosphorus manganese ores containing most of manganese oxides.

2. The carbothermic reduction of MnO_2, Fe_2O_3 and $Ca_3(PO_4)_2$ was carried out in the temperature range 1100 to 1400°C in H_2 atmosphere. The rate of carburization increased with increasing temperature, which is believed to be limited by the interfacial Boudouard reaction. The overall carbide mechanism was divided by the two stage, the carbide formation was started in center of metallic Fe as nuclei with MnO diffusion in the second stage.

3. As the behavior of phosphorus in the progress of reduction with carbon can be estimated by XRD pattern of $Ca_3(PO_4)_2$ and CaO, the most stable gas P_2 generation can be possible at the experimental temperature during the formation of manganese iron carbide.

4. P_2 gas generated from the decomposition of $Ca_3(PO_4)_2$ reacted with diffused Mn to form $(Fe,Mn)_3P$ using Fe in Fe-C melt as a nucleation site. From the opposite distribution of carbon and phosphorus in the EPMA line profile, the carbide does not overlap the phosphide although two compounds coexist within one particle. This might suggest that phosphorus is not involved in the formation of carbide.

References

1. S. E. Olsen, M. Tangstad and T. Lindstad, *Production of manganese ferroalloys* (Trondheim, Norway: SINTEF and Tapir Academic Press, 2007), 46.

2. S. C. Shim and N. Sano, "*Thermodynamics of phosphorus in carbon-saturated manganese-based alloys*", FFF, Trondheim, Norway, (1995), 611.

3. R. Kononov, O. Ostrovski and S. Ganguly, "*ISIJ Int.*", 49 (2009), 1107-1114.

4. B. C. DeCooman, J. G. Speer, I. Pyshmintsev and N. Yoshinaga, *Material Design* (GRIPS, media GmbH, 2, 2007).

5. F. Claeyssens, G. M. Fuge, N. L. Allan, P. W. May and M. N. R. Ashfold, "*Phosphorus carbides: theory and experiment*", Dalton Transactions, (2004), 3085.

6. M. Yastreboff, O. Ostrovski and S. Ganguly, *"ISIJ Int.",* 43 (2003), 161.

7. J. Zhang and O. Ostrovski, "Cementite formation and its stability in Iron carbide Process", 59th ironmaking conference proceedings, ISS, (2000), 339.

8. J. Briki and S. Ben slima, *Journal of Metallurgy*, Hindawi Publishing Corporation, (2012), 1.

9. B. C. DeCooman, J. G. Speer, I. Pyshmintsev and N. Yoshinaga, *Material Design*, GRIPS media GmbH, 2 (2007).

10. A. K. Biswas, *Principles of Blast Furnace Ironmaking*, SBA Pub., Calcutta, India, (1981), 70.

11. A. Brahmi and R. Borrelly, "Scripta Metallurgica et materialia", 32 (1995), 365.

12. Yunchang Zhang and Donal M. Schleich, "Journal of solid state chemistry", (1994), 270.

13. X. H. Huang, *Iron and Steel Metallurgy Principles*, Metallurgy Industry Press, Beijing, (1981), 437.

14. R. J. Fruehan, The Making, Shaping and Treating of Steel, Steelmaking and Refining Volume, 11th ed., AISE Steel Foundation, Pittsburgh, (1998), 20.

15. E. Matinde and M. Hino, "*ISIJ Int.*", 51 (2011), 220.

16. J. Yin, X. Lv, C. Bai, G. Qiu, S. Ma and B. Xie, "*ISIJ Int.*", 52 (2012), 1579.

17. H. Lee, Chemical thermodynamics for metal and materials, Imperial college press, 284.

18. Y. E. Lee, "M*etall. Trans. B"*, 17B (1986), 777.

19. R. Kononov, O. Ostrovski and S. Ganguly, "*ISIJ Int.*", 49 (2009), 1099.

20. R. Kononov, O. Ostrovski and S. Ganguly, "*ISIJ Int.*", 49 (2009), 1115.

21. J. R. Wynnyckyj and W. J. Rsukin, "*Metall. Trans. B"*, 19B (1988), 73.

Engineering Solutions for Sustainability: Materials and Resources II
Edited by: Jeffrey W. Fergus, Brajendra Mishra, Dayan Anderson, Emily Allyn Sarver, and Neale R. Neelameggham
TMS (The Minerals, Metals & Materials Society), 2015

PILOT-SCALE DECHLORINATION OF CuCl RESIDUE FROM ZINC HYDROMETALLURGY BY MICROWAVE ROASTING

Guo Zhanyong[1,2,3], Ju Shaohua[1,2,3]*, Lei Ting[4], Peng Jinhui[1,2,3], Zhang Libo[1,2,3], Jiang Feng[1,2,3]

1. National Local Joint Laboratory of Engineering Application of Microwave Energy and Equipment Technology, Kunming, Yunnan 650093, China;
2. Key Laboratory of Unconventional Metallurgy, Ministry of Education, Kunming University of Science and Technology, Kunming, Yunnan 650093, China;
3. Faculty of Metallurgical and Energy Engineering, Kunming University of Science and Technology, Kunming, Yunnan 650093, China;
4. Kunming Metallurgy College, Kunming, Yunnan 650093, China.

Keywords: Microwave roasting; Zn hydrometallurgy; CuCl residue; Removal of Cl; Pilot-scale experiment

Most Zn hydrometallurgy factories adopt Cu_2SO_4 as a dechlorination reagent from zinc solution recently, thus much CuCl residue was produced. The present process of treating this residue is washing CuCl residue with water or sodium carbonate solution, which would bring a lot of trouble for water treatment and waste discharge. In this study, a method of roasting the residue with microwave energy is adopted for not only dechlorination of the CuCl residue and also recycle Cl therein as HCl. The pilot-scale equipment system includes a 20 KW microwave roasting equipment, a dust collection and a tail gas adsorption system. The experiments result shown that under following parameters of microwave roasting: 108 Kg CuCl residue with a grain size of -60 mesh, stirring speed at 10 r/min, roasting at 350 ^{0}C for 4 h, the dechlorination rate is as high as 78.54 %, and a product of CuO with a Cl concentration of less than 3.7% can be obtained. What's more, the tail gas was analyzed as air and HCl, which can be absorbed as byproduct of hydrochloride. The power consumption for treating such CuCl residue with microwave energy is only about 488.88kW • h/t.

Introduction

Cl element can be accumulated gradually in the solution of Zn hydrometallurgical system. When Cl content in electrolysis solution comes up to 100 mg/L, the stability of electrodeposits process would be severely affected[1]. It would not only accelerate the consumption of cathode, anode plate and elevate the power consumption, but also cause serious corrosion of device, increasing production costs and lowering the quality of electrolytic zinc, as demonstrated [2,3]. Thus, Cl- in the solution should be removed before electrodeposits.

At present, there are two methods of dechlorination. The first one is roasting the raw material with high Cl content, as demonstrated, the second one is removing Cl from zinc sulfate solution by adding Cu^+ to form CuCl. Now most of zinc hydrometallurgical smelters select the second method for its high removal rate and low cost.

However, the CuCl residue produced during this process is still hard to treat. Most smelters wash the residue with alkaline solution such as Na_2CO_3. CuCl containing in the residue will transform to NaCl solution subsequently, which will result in a lot of troubles to water treatment and waste discharge.

Microwave heating process is a green energy supply method which can selectively transfer the required energy to the reaction molecules or atoms by the means of dialectic loss of the material themselves. Jones, Some researchers[4-6] have concluded that microwave heating offers a number of advantages over conventional heating such as: non-contact heating; energy transfer, not heat transfer; rapid heating; material selective heating; volumetric heating; quick start-up and stopping; heating starts from interior of the material body; higher level of safety and automation; no waste gas giving out itself. Researchers have carried out many investigations in fields of drying[7], roasting[8], reduction[9,10], oxidation[11], leaching strengthening of the processes[12] by microwave energy.

In this research, the pilot-scale experiments of microwave roasting of CuCl residue were conducted for removing Cl, the experimental system was introduced, and the roasting condition, Cl removal effect, and the smoke composition of the removed Cl were investigated.

1 Experimental

1.1 Material

The CuCl residue cake was supplied by a zinc hydrometallurgy smelter in Yunnan Province, China. The chemical composition of the dried material is shown in Table 1.

Table 1. Chemical Composition of CuCl residue after drying at 100 ^{0}C for 12 h

Composition	Zn	Cu	Cl	S
Content (%)	9.55	54.68	14.27	4.74

The Cl content of the dried CuCl residue is 14.27%, the main phases of the dried residue are CuCl and $ZnSO_4$.

For pilot-scale study, a 2000 Kg CuCl residue cake was firstly dried with a 54 KW microwave dryer, and the final moisture content is about 4.5%. Then the dried CuCl residue was crushed and grinded to -60 mesh for the following roasting experiment.

1.2 Experimental Equipment

In this research, a rotary microwave roasting system, which was researched and developed by the Key Laboratory of Unconventional Metallurgy, Ministry of Education, Kunming University of Science and Technology, was used as experimental device. The schematic diagram and photo of the device is shown in Figure 1.

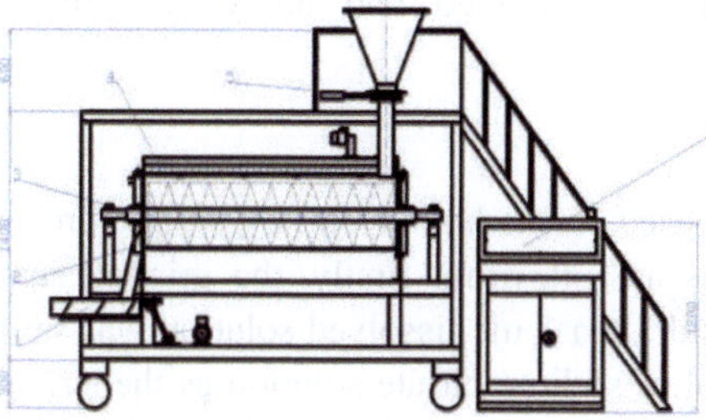

(a) Schematic diagram

(b) Photo of the system

1. Material Conveyor; 2. Discharge Port; 3. Spiral Sirring Shaft; 4. Stainless Steel Chamber; 5. Feed Port; 6. Controlling Operation Station

Figure 1. The rotary microwave roasting system used in the pilot experiment

The performance data of the equipment contains: total power of 20 KW, chamber made of stainless steel, a volume of 300 L, material stirring equipment with stirring speed at 0-30 r/min, et al. Its features include rapid-treatment, uniform roasting, easy to control, easy to operate and so on.

1.3 Experimental Procedure

The process chart of microwave roasting for dechlorination is shown in Figure 2.

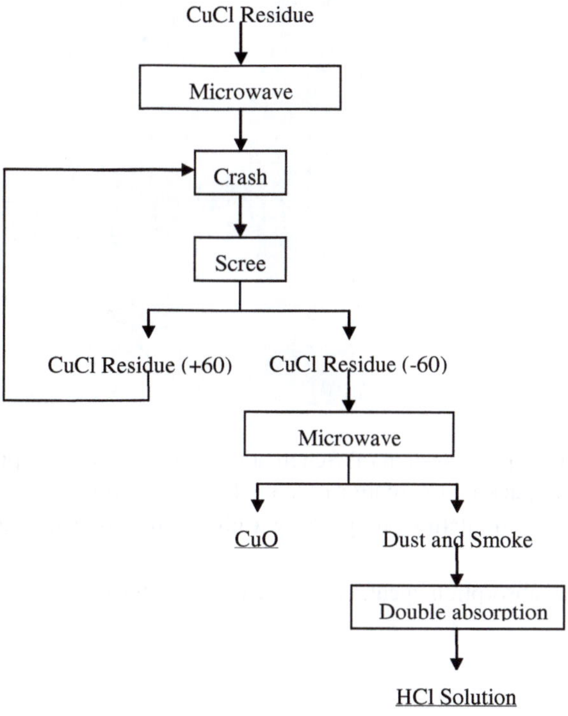

Figure 2. The process chart of pilot-scale dechlorination experimental of CuCl residue

A double vortex-dust-collecting system, an exhaust fan and a double-absorption systems were set for treating the waste gas, which was sampled and analyzed with a gas chromatography.

1.4 Analysis Method

The method of Cl^- selective electrode was used for its high accuracy, simple procedure and anti-interference. The specific procedure is as following: firstly, the grinded samples were weighted and dissolved in diluted nitric acid; then 1 ml dissolved solution was sampled and controlling the pH at about 5.5; adding 5 ml of sodium citrate solution as the strong ion agent, and finally measure the potential value by using Cl- selective electrode and mercury-mercurous sulfate as reference electrode. Then the content of Cl in solid material can be calculated.

Atomic absorption spectrometry (PE AA700, PerkinElmer) was used for detecting the content of Cu and Zn in samples. XRD (XRD-7000, Shimadzu Corporation) was selected to study the phase of samples. The dust and smoke generated from roasting process was analyzed by gas chromatography (GC-2010, Shimadzu Corporation) so that the gas products can be determined.

1.5 Waste Gas Absorption System

For gas absorption, a double-class spray washing system was utilized shown in Figure 3.

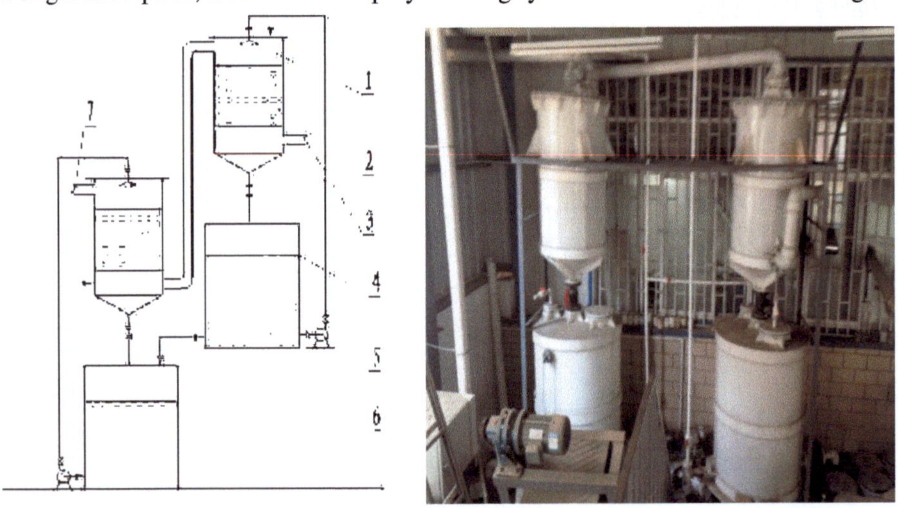

(a) (b)

Figure 3. The gas absorption system (a) Schematic diagram; (b) Equipment photo
1 – Water nozzle, 2 - packing (Raschig ring), 3 - flue gas import, 4 - washing liquid circulation tank, 5 -, circulating pumps, 6 - washing liquid, 7- Water gas export

Pure water was used as absorption agent, each tank with a water volume of 500 L.

Results and Discussion

2.1 Dechlorination effect of microwave roasting

At first, 108 Kg CuCl residue was put into the pilot microwave equipment. Then open the microwave source system, when the temperature of the material increased to 100 ^{0}C, started the exhaust fan and the double-absorption system, the roasting temperature were kept at 350 ^{0}C for 4 h.

The temperature curves were recorded and shown in Figure 4.

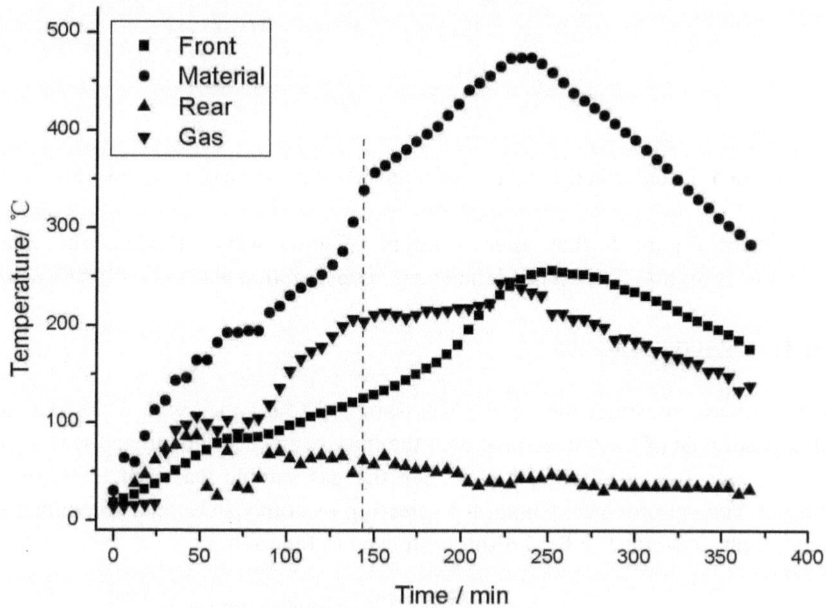

Figure 4. Temperature curve of the pilot experiment

It is shown in Figure 4 that after about 120 min of microwave heating, the temperature of the material went up to 350 ^{0}C, then the microwave source was shut down, but the stirring system was still working. Because the oxidation reaction of CuCl into CuO is an exothermic reaction, the temperature raised continuously until to about 475 ^{0}C, then the temperature went down slowly because of the heat balance of the reaction and heat loss changed. Finally, after reaction for 4 h, the temperature of the material was still at 350 ^{0}C. Thus, the energy consumption of dechlorination by microwave roasting is very low.

During the reaction, the material was sampled 4 times for analysis of the dichloride effect. The results are shown in Figure 5.

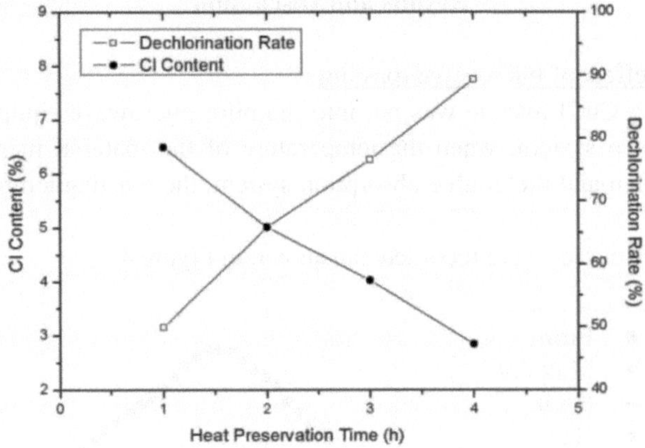

Figure 5. Dechlorination results of microwave roasting of CuCl residue

It is shown in Figure 5 that after 4 hours of microwave roasting, the rate of dechlorination was as high as 78.54%, and the energy consumption was only 488.88kW·h/t.

<u>2.2 Smoke and gas absorption results</u>

During microwave roasting, the smoke was sampled four times with a 500 mL glass syringe from a small hole of the smoke pipe near the microwave equipment, and was injected together into two tinsel gas collecting bags. Then the gas sample was sent to testing and analysis center of Yunnan province, China, for detecting its composition and containing with a gas chromatography system. The final result is showed in Figure 6.

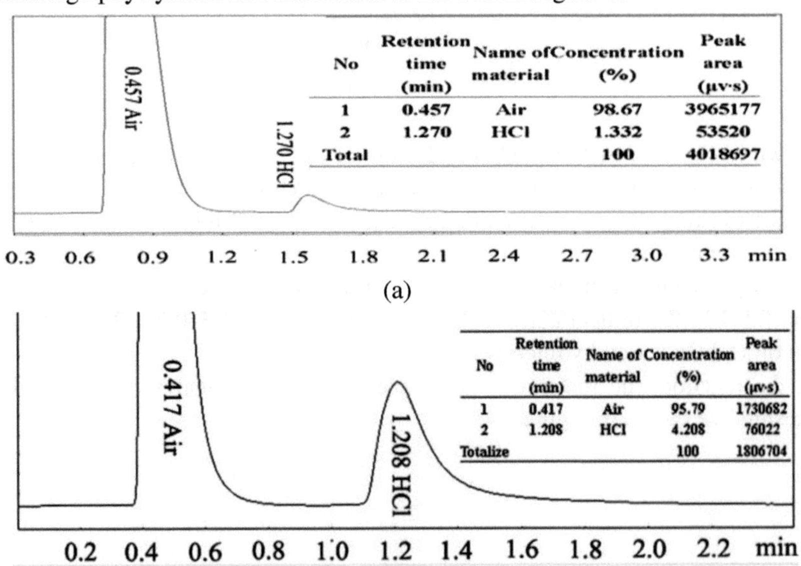

(a)

(b)

Figure 6. Gas chromatography pattern of gas sample obtained under atmosphere of water vapor with air

From results of gas chromatography showed in Figure 6 (a) and (b), HCl gas was generated during the microwave roasting. So, it's easy to be absorbed into water to generate hydrochloride. Thus, the process not only achieved the goal of dechlorination, but also can recover the removed chlorine into hydrochloride.

2.3 Phase transformation of material in dechlorination process

The residue samples before and after microwave roasting were analyzed with a x-ray diffractometer, and the result was showed in Figure 7.

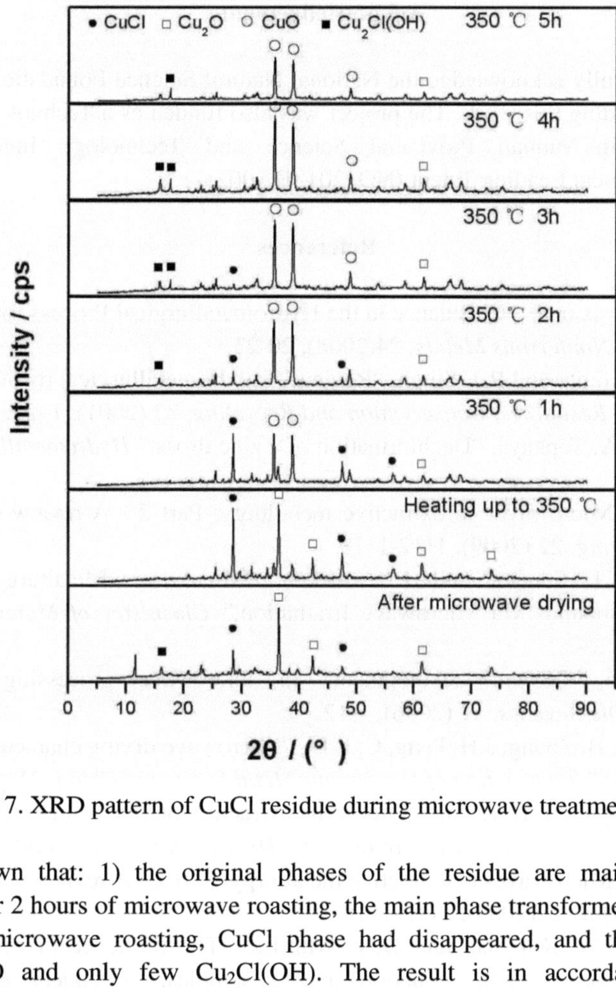

Figure 7. XRD pattern of CuCl residue during microwave treatment

Figure 7 shown that: 1) the original phases of the residue are mainly CuCl and $Cu_2Cl(OH)$; 2) after 2 hours of microwave roasting, the main phase transformed into CuO; 3) after 4 hours of microwave roasting, CuCl phase had disappeared, and the main phase changed into CuO and only few $Cu_2Cl(OH)$. The result is in accordance with the dechlorination rate in Figure 5.

Conclusions

Through pilot-scale experimental study of dechlorination of CuCl residue, it can be concluded as following:

1. A 20KW microwave roasting system with stirrer was used for dechlorination of a CuCl residue from zinc hydrometallurgy, the result showed that after 4h of roasting at 350 ^{0}C, the dechlorination rate can be over 75%, and the gas can be absorbed as hydrochloride.
2. The electrical consumption of dechlorination is very low, as treating per-kilo CuCl residue by microwave roasting is just 0.5KW·h.
3. Microwave roasting not only can achieve the goal of dechlorination, but also can recover the removed chlorine as hydrochloride. Thus, it has wide industrial prospect.

Acknowledgements

The authors gratefully acknowledge the National Natural Science Foundation of China (NO. 51104073) for funding this work. The project was also funded as a Technology of People' s Republic of China.Yunnan Provincial Science and Technology Innovation Talents scheme-Technological Leading Talent (NO. 2013HA002)

References

[1]J. Chen, "Analysis on F - Cl Balance in the Hydrometallurgical Process for Zinc Smelting," *Hunan Nonferrous Metals*, 24(2008), 20-23.

[2]M.K. Jha, V. Kumar and R.J. Singh, "Review of hydrometallurgical recovery of zinc from industrial wastes," *Resources, Conservation and Recycling,* 33 (2001), 1–22.

[3]N. Güresin, Y. A. Topkaya, "Dechlorination of a zinc dross." *Hydrometallurgy*," 49(1998), 179–187

[4]C.A. Pickles, "Microwaves in extractive metallurgy: Part 2 - A review of applications," *Minerals Engineering,* 22 (2009), 1112-1118.

[5]J.W. Lekse, T.J. Stagger and J.A. Aitken, "Microwave Metallurgy: Synthesis of Intermetallic Compounds via Microwave Irradiation," *Chemistry of Materials,* 19 (2007), 3601-3603.

[6]S.W. Kingman, "Recent developments in microwave processing of minerals," *International Metals Reviews,* 51 (2006), 1-12.

[7]Y. Li, Y. Lei, L.B. Zhang, J.H. Peng, C. L Li, "Microwave drying characteristics and kinetics of ilmenite," *Transaction of Nonferrous Metal Society of China,* 21 (2011), 202-207.

[8]W. Li, J.H. Peng and L.B. Zhang, "Pilot-scale extraction of zinc from the spent catalyst of vinyl acetate synthesis by microwave irradiation," *Hydrometallurgy,* 92 (2008), 79–85.

[9]C.A. Pickles,. "Microwaves in extractive metallurgy: Part 2 - A review of applications," *Minerals Engineering,* 22(2009), 1112-1118.

[10]D.A. Jones, S.W. Kingman and D.N. Whittles, "The influence of microwave energy delivery method on strength reduction in ore samples," *Chemical Engineering and Processing*, 46(2007), 291-299.

[11]S.W. Kingman and N.A. Rowson, "Microwave Treatment of Minerals - A Review," *Minerals Engineering*, 11(1998), 1081-1087.

[12]M.A Harahshen and S.W. Kingman, "Microwave-assisted leaching: a review," *Hydrometallurgy,*. 73(2004), 189-203.

Engineering Solutions for Sustainability: Materials and Resources II
Edited by: Jeffrey W. Fergus, Brajendra Mishra, Dayan Anderson, Emily Allyn Sarver, and Neale R. Neelameggham
TMS (The Minerals, Metals & Materials Society), 2015

OPTIMIZATION OF PROCESSING CONDITIONS LEADING TO DANGEROUS EMISSIONS IN STEELMAKING PLANTS

P. Cavaliere, A. Perrone

Department of Innovation Enginnering, Univerisity of Salento,
Via per Arnesano, 73100 Lecce, Italy

Keywords: Sintering, Blast furnace, Greenhouse emissions, Optimization

Abstract

Sintering is a process for ironmaking operations; it represents one of the main sources of production emissions of polychlorinated dibenzo-p-dioxins, polychlorinated dibenzo-furans, NOx and SOx. In the present study, the operating conditions through which a reduction of dangerous emissions can be achieved are defined through experimental and numerical analysis. Through such analysis, a broad range of processing parameters affecting the development of PCDD/Fs in the sintering process has been evaluated. The first aim was the possible reduction of dangerous emissions through numerical and experimental analyses allowing the definition of the optimal conditions for the minimization of pollutants. Blast furnace represents the dominant hot metal making production process worldwide and one of the main Modern research in the field focuses on the increase in plant productivity through energy saving and on the greenhouse emissions reduction compatible with legal limits. Productivity is mainly governed by relevant input parameters such materials rates, materials properties and operating conditions. All the dominant input parameters and their variation have been analyzed in the present study and they have been optimized in order to increase the plant productivity and reduce the greenhouse emissions.

Introduction

Ironmaking and steelmaking are a highly material and energy intensive industrial operations. More than half of the mass input becomes outputs in the form of off-gases and solid products. The most dangerous emissions are those to air [1, 2]. Iron ore fines, other iron-bearing wastes and coke dust are blended and combusted [3-5]. The process of sintering to improve the physical and chemical properties of iron ore for use in blast furnaces is well documented [6, 7]. The agglomeration process gives rise to many different physical and chemical phenomena. At the end of the grate a sinter breaker is placed. It reduces the sintered material to the desired size [8]. Here PCDD/Fs form in the presence of carbon-containing materials [9, 10], the process is favored by the presence of specific organic compounds or a carbonaceous matrix-sand sources of chlorine and oxygen. The temperature increase favors the PCDD/Fs formation in the range 200-800 °C, at higher temperatures they rapidly decompose. It was observed that the presence of catalytic metals (Cu) can be essential at modest temperatures [11, 12]. The gas temperature inside the wind boxes and wind legs is lower (100-500 °C) with respect to the sintering grate; such conditions lead to the optimal physical and chemical conditions for the formation of pollutants such as PCDD/Fs, NOx and SOx [13-18]. The PCDD are generally measured in terms of toxicity equivalent (TEQ) relative to TCDD as a reference, being the most polluting and dangerous. The poly dibenzo-dioxins have different toxicities in relation to their structure. TEQ expresses the quantity of a "toxic" substance as the concentration of the reference substance. It is also possible to obtain the concentration of a PCDD with its toxic equivalency through the use of toxic equivalency factor (TEF). The TEF for TCDD is assigned equal to one, while the other dioxins have a factor less than one. This dimensionless parameter, multiplied by the actual concentration, results in the TEQ [19]. In the present study, a broad range of processing parameters affecting the development of PCDD/Fs in the sintering process has been evaluated. The main aim was the possible reduction of dangerous emissions through numerical and experimental analysis allowing the definition of the optimal conditions for the minimization of the pollutants. Starting from a database, built by employing experimental and literature data, a computational model (n-dimensional virtual surfaces), capable of reproducing at best the actual process, was developed. The analysis performed led to the minimization of the output variables (PCDD/F, NOx and SOx). For PCDD/F, it was necessary to apply a filtering system in order to obtain quantities of emissions below the legal limit of 0.4 ng I-TEQ/Nm3 as actually required by legislation [20-23]. The main objectives of sintering ore plant operations include maximizing grate productivity (expressed in tons per square meter of grate area of sintering machine per day), minimizing fuel consumption, maintaining fired pellet quality within limits specified by blast furnace or reduction furnace requirements and to minimize greenhouse gas and sulfur emissions to the environment. Some interesting results are available in literature belonging to studies performed to reach an optimization of sintering operations coupled with a reduction in dangerous emissions [24-30].The blast furnace productivity depends strongly on the rate and degree of reduction; they are directly related to the reactivity of coke and to the composition and reducibility of the iron-bearing charge. Actually, one of the fundamental physical parameter for the reduction optimization is the mean dimension of the ores, pellets and sinters, in general the rate of reduction decreases with increasing size of the raw materials. Complex gas-solid, solid-solid and liquid-liquid interactions take place in different zones of the blast furnace. Productivity increase and CO2 emission decrease depend strongly on the high temperature properties of sinter and coke [31], particle size distributions of sinter and coke and coke and ore mixed charging.Ironmaking through blast furnaces is strongly influence by many parameters including, operating conditions, process and plant design. All the variables related to the raw materials largely affect the BF performances. Consequently the different quantities of raw materials and additives influence the operating conditions and the performance indices of the overall production. By deeply analyzing the materials properties, the input parameters and performance indices it is possible to obtain those relationships useful for the industrial practice. Control and modification of BF operations is fundamental in optimizing efficiency and reduce dangerous emissions and energy consumption. The main goal of the present work is the minimization of greenhouse emissions from the blast furnace through holding acceptable levels of productivity compatible with the industrial needing. Such goals have been reached through a deep analysis and control of the main parameters affecting the process.

Experimental procedure

The analysis of the sintering process were performed on a sintering plant belonging to an italian steel company (Dwight-Lloyd sinter). For the development of such analysis, the emissions levels of the last 6 years and the corresponding inputs have been recorded to define the starting point of the problem. The corresponding productivity of the plant and the quality of the sintered material have been taken into account for comparison. The main goal is the reduction of dangerous emissions coupled with acceptable levels of productivity and quality of the material.

Each windbox was equipped with thermocouples (k-type) in order to monitor the off-gas temperature during sintering. The flue gases composition was monitored, according to EN1948 parts 2 and 3, EN-1948SS (sampling standards, Wellington Laboratories EN-1948ES (extraction standards, Wellington Laboratories) and EN-1948IS (injection standards, Wellington Laboratories), by employing an high resolution gas chromatograph and an high resolution selective mass detector. The output variables (PCDD/F, NOX and SOX) define a multi goal analysis and have been minimized taking into account some constraints or limitations typical of the actual process of sintering.Many of the blast furnace quality and performance indices are strictly dependent on sinter plant ones, the quality and properties of raw materials was analyzed before each blast furnace production cycle. The working and useful volumes of the blast furnace are 1209 m3 and 937m3 respectively. Pellets, limestone, coke compositional and quality ranges were analyzed at each cycle and the data were reported in the present paper. The main independent variables have been underlined, the weight of such parameters on productivity and CO2 emissions has been analyzed.

Results and discussion

From deep numerical analysis belonging to experimental set-up of the plant many fundamental results have been obtained in terms of input parameters (in a broad range of existence) influencing the dioxin, NOx and SOx emissions.Dioxin emission shows a maximum in the windbox 19, in the temperature range 350-480 °C, it increases also with increasing O_2 content up to very high levels, from the graph it is clear how the emissions are very sensitive to such parameters.To obtain a sintered material of uniform quality it is necessary to schedule heating and cooling on the grate in order to ensure that the thermal history of all the components is the same. This leads to the perfect control of temperature on the grate taking into account that at temperatures below the red heat the main mechanism of heating is through convection and conduction while at higher temperatures the main mechanism is radiation, such behavior is strongly related to the gas flow rate in the windboxes and consequently the gas flow rate deeply influences the finale quality of sintered grains. Dioxin reaches highest values for high temperature and high moisture presence; In such conditions also the CO_2 formation reaches its maximum values. Dioxin emissions reach high values for high levels of flow rate in the windboxes, high values of dioxin correspond to low values of carbon monoxide.Many elements and compounds are very active in dioxins formation and, in particular, they lead to an increase in the levels of dangerous emissions, Cu and Cl lead to a strong increase in the levels of dioxins emissions.On the contrary, the addition of urea and CaCO3 leads to a strong efficiency in dioxins emission reduction (Figure 1).

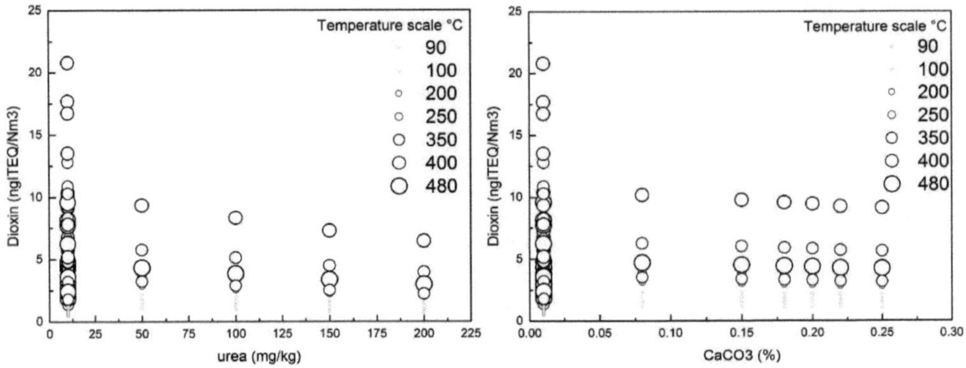

Figure 1. Dioxin emission in the sintering plant monitored in the present study as a function of urea and temperature a), and as a function of $CaCO_3$ and temperature b).

The addition of urea leads to a decrease in dioxin level but, at the same time, such reduction of dioxin coincides with an increase in NOx levels. The influence of urea is very important in the reduction of polluting emission. In particular, the emissions levels are reduced as the urea levels increase. For the PCDD/F it occurs by means of a physical deposition or by poisoning the catalytic sites (Figure 2a).Another strong example is represented by the addition of Sufphur, it is very efficient in reducing the dioxin emissions but at the same time produces a strong increase in SOx emissions (Figure 2b). The role of sulfur in the reduction of emissions in the sintering can be noted. In all three ways previously observed, there has been a reduction of PCDD/F, especially in the second case, and it is probably due to the presence of SOx in the flue gas. It is believed that these sulfides can be converted to SO_2, reducing the chlorine in HCl.

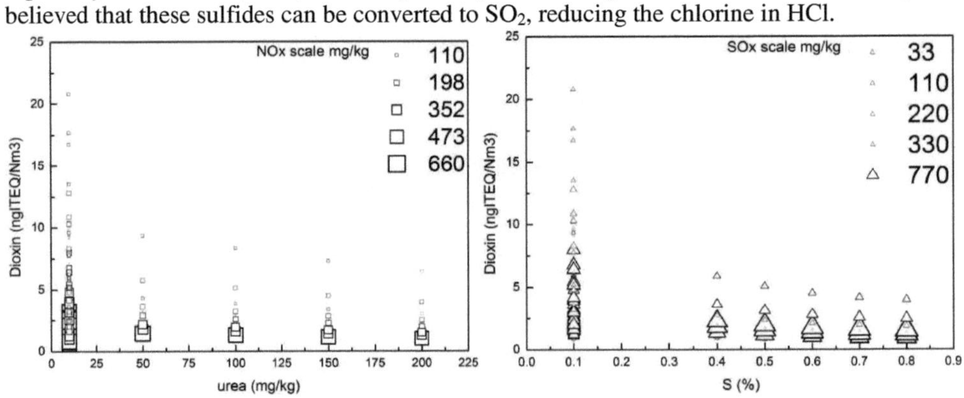

Figure 2. Dioxin emission in the sintering plant monitored in the present study as a function of urea and NOx emissions a).Dioxin emission as a function of S and SOx emissions b).

The importance of lime in the reduction of PCDD/F have been largely outlined in the present paper. The emissions levels are reduced as the lime quantity in the raw materials increases. The lime also brings to a strong reduction of SOx and a moderate reduction of NOx. Although the resultant optimal combination of input parameters able to reduce the dangerous emissions from the plant, it is very important to examine the impact of the chosen input parameters on the sinter productivity. As a matter of fact, many different studies were performed on the productivity measurements for selected designs with controlled levels of dioxin, NOx and SOx emissions.high levels of dioxin emissions are found in correspondence of very low levels of productivity, then productivity increases with increasing dioxins and then increases with decreasing dioxin emissions. By analyzing also the oxygen addition it can be concluded that a decrease in O_2 shifts the same productivity to lower dioxin emissions.Actually, temperature and oxygen flow influences moisture behavior in the sinter bed and, as explained later in the paper, it is strongly related to the quality of the sinter, all the input parameters ranges were set in order to fix the quality of the pallets at a well known level.High levels of productivity and low levels of dioxins can be reached with high level of moisture, high levels of gas flow rate leads to an increase of productivity but unfortunately to an increase in dioxin emissions. Such parameter results very important because it is strongly linked to the sinter permeability which can be directly related to the ore quality in terms of granule size and distribution, such parameter was taken strongly into account during the definition of the parameters ranges in order to set an acceptable pellets quality for all the analyzed input conditions. Also moisture strongly influences the sinter quality because it is strongly related to voidance, also the range of such parameter was set to fix a given range of the product quality Generally, low levels of moisture lead to a decrease in the spread of granule size distribution and to an increase in the spherical aspect of the product, such behavior leads to a material with improved voidance and consequently to good levels of productivity. In the conditions of higher levels of moisture, normally the sinter thickness decreases and this leads to a tendency for voidance to decrease, in this way the productivity tends to decrease.Very low levels of dioxin and productivity are in correspondence oh high emissions of SOx and NOx (Figure 3).

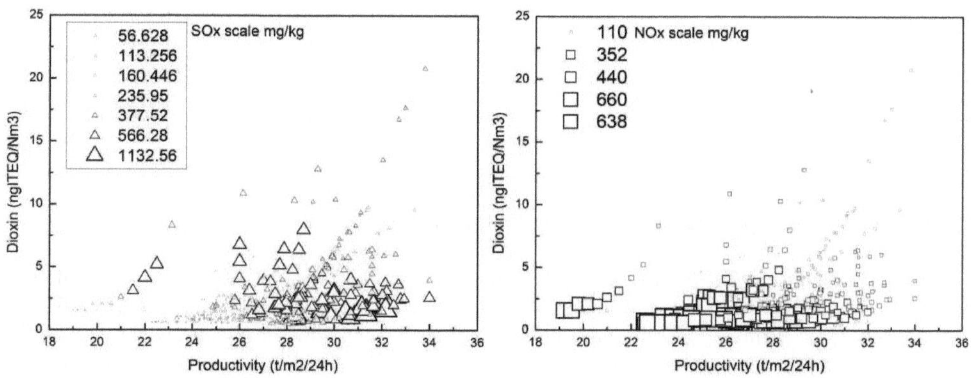

Figure 3. Dioxin emission in the sintering plant monitored in the present study as a function of productivity and SOx a), and as a function of productivity and NOx b).

By increasing the lime content it is possible to reach good values of productivity and, at the same time, to reduce the dioxin emissions (Figure 4a). Lime is very effective in improving productivity with contemporary reduction of SOx emissions, the addition of lime leads to a moderate reduction of NOx for a given value of productivity (Figure 4b).

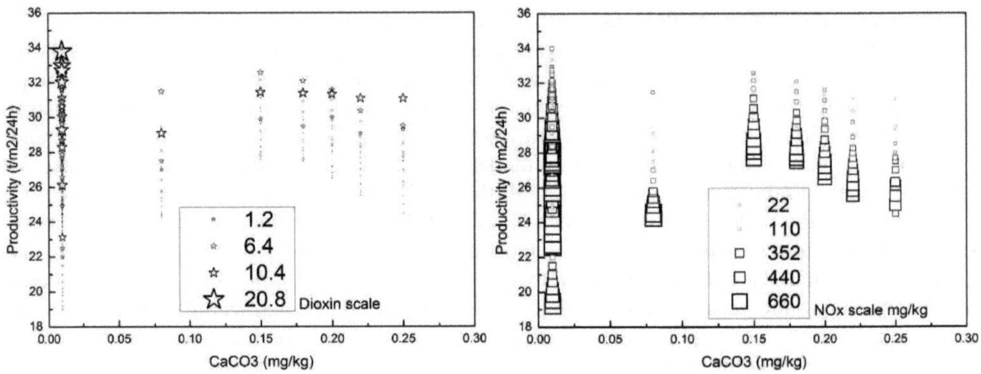

Figure 4. Productivity as a function of Lime and dioxin a), Productivity as a function of Lime and NOx c).

In table 1 the results of the different designs are compared.

	SOx *mg/kg*	**NOx** *mg/kg*	**PCDD/F** *ng I-TEQ/Nm^3*	**Productivity** *t/m2/24h*
New Design	297	201	0,45	29.8
Design 1	291	177	0,43	29.0
Legal limit	400	400	0,4	

Table 1. Emissions and productivity belonging to the plant optimization.

the level of PCDD/F was found to vary from 0,43 to 0,45 ng I-TEQ/Nm3, with also a contemporary increase of NOx and SOx. With the exception of the latter, the PCDD/F still exceeded the legal limits. Despite an increase of emissions, the application of the set of parameters of the New Design was chosen because technologically simpler to realize. In addition it was possible to obtain a value of productivity close to 30 in agreement with the industrial desires. This study made it possible to implement a setting of the system through which it is possible to obtain a clean reduction of the emissions of polluting substances.

The further analyses started from the results previously obtained. It was performed an analysis on the sinter plant operations in terms of productivity vs. reduction rate for different values of productivity of the blast furnace (Figure 5).

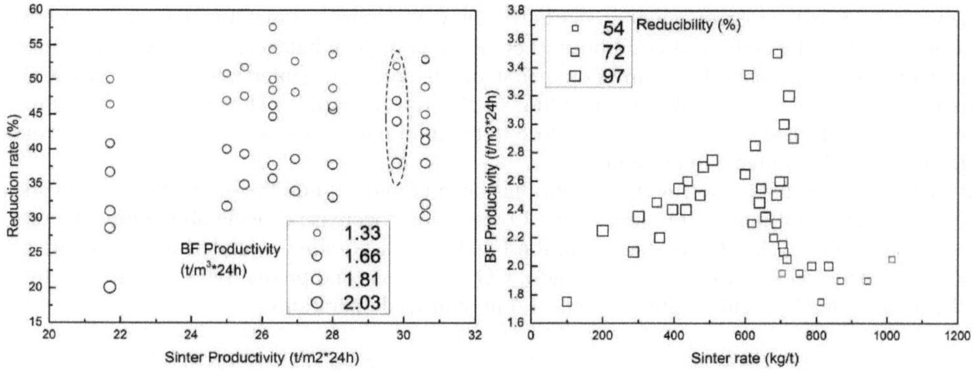

Figure 5. Reduction rate as a function of sinter productivity and BF productivity a). BF productivity as a function of sinter rate and reducibility b).

Coke rate and sinter rate have a strong effect on BF productivity (Figure 5b). The furnace performance indices depend on the behavior of the coke throughout the entire volume of the furnace. It is necessary to analyze how an optimization of raw materials utilization is compatible with the productivity of the blast furnace; one of the more direct way is to couple the analysis of materials flow and consequent productivity with the specific reducibility as a function of the variation of coke and sinter rate.The present analysis shows that productivity increases with decreasing of coke rate and increasing sinter rate up to a limit in which an inversion of the trend can be underlined. In the case of sinter rate, productivity increases with sinter reducibility, such behavior is due to a decrease in reduction time in the BF and an increase in the indirect reduction operations. In this way it is possible to identify operating conditions coupling both a reduction in coke consumption and a good level of productivity in the blast furnace; with a good quality of the sinter materials and an optimal level of sinter rate it is possible to achieve such goals.
Temperature and oxygen addition influences BF productivity are shown in Figure 6.

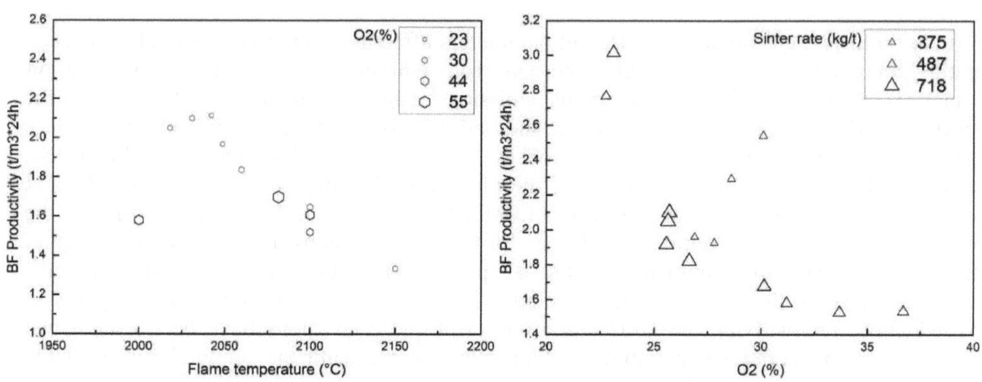

Figure 6. BF productivity as a function of flame temperature and oxygen a); as a function of sinter rate b).

The maximum increase in oxygen content is also limited by the fact that the pressure of the blast air is theoretically and practically restricted. Basically CO_2 emissions depend on different parameters such as the improvement of shaft efficiency, the charge of metallic materials, the heat control in the BF, the control of equilibrium FeO/Fe then oxygen balance. Actually all the main input parameters influence CO_2 emissions, first of all coke and sinter rate (Figure 7a); CO_2 emission factor was calculated corresponding to the raw materials. The minimum measured values are around 1100 kg/t which is very close to the minimum theoretical values resulting from large valid models available in literature.

For further analysis coke rate/sinter rate ratio has been employed into the graphs, such ratio and limestone addition influence CO_2 emissions,CO_2 emissions decrease considerably with coke rate decrease and low levels of limestone addition.Finally, it is possible to underline an important relationship between sinter rate/coke rate ratio, CO_2 emissions and productivity (Figure 7b), that can be interpolated by a cubic spline very useful in industrial operations.

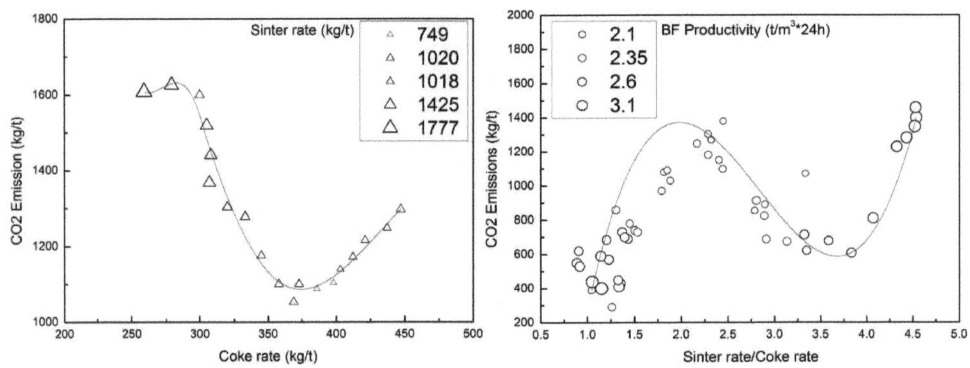

Figure 7. Carbon dioxide emissions as a function of coke and sinter rate a); as a function of BF productivity and Sinter/Coke rate.

Conclusions

The study outlines the influence of different parameters affecting it in order to establish a set of operating conditions capable of reducing the dangerous emissions.A large attention was put on the analysis of the reduction of dangerous emissions coupled with an acceptable level for productivity. All the input parameters, before the optimization, were chosen in order to fall in a range guaranteeing an acceptable quality of the sintered material in terms of dimension, sphericity and voidance. The productivity of the plant, resulting from the chosen optimal input parameters, showed no significant differences from the required ones. The results confirmed the applicability of the obtained optimal conditions for the ordinary industrial production.The effect of the changing operating parameters on the BF productivity and greenhouse emissions have been investigated. In the study, several efficiency improvements were identified as regards to coke reduction, productivity and CO2 emissions. Multiple regressions have been established between the main operational parameters of the blast furnace and the productivity and CO2 emissions.As a general point of view, the precise control of all the main processing parameters from the sintering operations to the BF ones lead to a strong reduction in dangerous emissions and to good levels of plants productivity.

References

[1] P. Cavaliere, A. Perrone, P. Tafuro, V. Primavera, "Reducing emissions of PCDD/F in sintering plant: numerical and experimental analysis" *Ironmaking and Steelmaking*, 38 (2011), 422–431.
[2] P. Cavaliere, A. Perrone, "Analysis of dangerous emissions and plant productivity during sintering ore operations", *Ironmaking and Steelmaking*, 40 (2013), 9-24.
[3] M. Nakano, K. Morii, T. Sato, "Factors accelerating dioxin emission from iron ore sintering machines", *ISIJ International*, 49 (2009), 729–734.
[4] D. Senk, H. W.Gudenau, S. Geimer, E. Gorbunova, "Dust Injection in Iron and Steel Metallurgy", *ISIJ International*, 46 (2006), 1745–1751.
[5] E. Aries, D. R. Anderson, R. Fisher, T. A. T. Fray, D. Hemfrey, "PCDD/F and dioxin-like PCB emissions from ironore sintering plants in the UK", *Chemosphere*, 65 (2006), 1470–1480.
[6] C. Xhrouet, E. De Pauw, "Formation of PCDD/Fs in the sintering process: Influence of the raw materials", *Environmental Science and Technology*, 38 (2004), 4222-4226.
[7] L. Hsieh, "Effect of Raw Material Composition on the Sintering Properties", *ISIJ International*, 45 (2005) 551-559.
[8] D. R. Anderson, R. Fisher, "Sources of dioxins in the United Kingdom: the steel industry and other sources", *Chemosphere*, 46 (2002), 371-381.
[9] S. P. Ryan, E. R. Altwicker, "Understanding the role of iron chlorides in the de novo synthesis of polychlorinated dibenzo-p-dioxins/dibenzofurans", *Environmental Science and Technology*, 38 (2004), 1708-1717.
[10] N. Tsubouchi, S. Kuzuhara, E. Kasai, H. Hashimoto, Y. Ohtsuka, "Properties of Dust Particles Sampled from Windboxes of an Iron Ore Sintering Plant: SurfaceStructures of Unburned Carbon", *ISIJ International*, 46 (2006), 1020-1026.
[11] T. Kawaguchi, M. Matsumura, E. Kasai, Y. Ohtsuka, H. Noda, *Tetsu-to-Hagane*, 88 (2002), 12-19.
[12] M. K . Cieplik, J . P. Carbonell, C. Munoz, S. Baker, S. Kruger, P. Liljelind, S. Marklund, R. Louw, "On Dioxin Formation in Iron Ore Sintering", *Environmental Science and Technology*, 37 (2003), 3323-3331.
[13] A. Iosif, F. Hanrot, D. Ablitzer, "Process integrated modelling for steelmaking Life Cycle Inventoryanalysis", *Environmental Impact Assessment Review*, 28 (2008), 429–438.
[14] N. Menad, H. Tayibi, F. Garcia Carcedo, A. Hernandez, "Minimization methods for emissions generated from sinter strands: a review", *Journal of Cleaner Production*, 14 (2006), 740-747.
[15] P. Tan, I. Hurtado, D. Neushutz, G. Eriksonn, "Thermodynamic Modelling of PCDD/Fs Formation in Thermal Processes", *Environmental Science and Technology*, 35 (2001), 1867-1874.
[16] P. S. Kulkarni, J. G. Crespo, C. A. M. Afonso, "Dioxins sources and current remediation technologies - A review", *Environment International*, 34 (2008), 139–153.
[17] K . Raghunatan, B. K . Gullet, "Role of sulfur in reducing PCDD and PCDF formation" *Environmental Science and Technology*, 30 (1996), 1827-1834.
[18] K. Suzuki, E. Kasai, T. Aono, H. Yamazaki, K. Kawamoto, "De novo formation characteristics ofdioxins in the dry zone of an iron ore sintering bed", *Chemosphere*, 54 (2004), 97-104.
[19] United States Enviroenmental Protection Agency, EPA/100/R-10/005, December 2010. www. epa.gov/osa.

[20] M. Altarawneh, B. Z. Dlugogorski, E. M. Kennedy, J. C. Mackie, "Mechanisms for formation, chlorination, dechlorination and destruction of polychlorinated dibenzo-p-dioxins and dibenzofurans (PCDD/Fs)", *Progress in Energy and Combustion Science*, 35 (2009), 245-274.
[21] P. Tan, D. Neuschutz, "Study on polychlorinated dibenzo-p-dioxin/furan formation in iron ore sintering process", *Metallurgical Transactions*, 35B (2004), 983-990.
[22] S. Kasama, Y. Yamamura, K. Watanabe, "Investigation on the dioxin emission from a commercial sintering plant", *ISIJ International*, 46 (2006), 1014-1019.
[23] GazzettaUfficialeItaliana, 2005, 163.
[24] Yu-Cheng Chen, Yi-Min Sun, Jin-LuhMou,Perng-Jy Tsai, "Application of Orthogonal Array Tests Method to Optimize Operating Conditions for Iron Ore Sintering", *ISIJ International*, 49 (2009), 743–748.
[25] S. Machida, T. Higuchi, N. Oyama, H. Sato, K. Takeda, K. Yamashita, K. Tamura, "Optimization of coke breeze segregation in sintering bed under high pisolite ore ratio", *ISIJ International*,49 (2009), 667-675.
[26] T. Kang, S. Gupta, V. Sahajwalla, "Characterizing Swelling Behaviour of Iron Oxides during Solid State Reduction for COREX Application and their Implications on Fines Generation", *ISIJ International*, 47 (2007), 1590-1598.
[27] R. P. Bhagat, U. S. Chattoraj, S. K. Sil, "Porosity of Sinter and its relation with the sintering indices", *ISIJ International*, 46 (2006), 1728-1730.
[28] N. Sakamoto, "Ironmaking - Iron ore granulation model supposing the granulation probability estimated from both properties of the ores and their size distributions", *ISIJ International*, 42 (2002), 834-843.
[29] Ken-ichi Higuchi, T. Kawaguchi, M. Kobayashi, Y. Hosotani, K. Nakamura, K. Iwamoto, M. Fujimoto, "Improvement of productivity by stand-support sintering in commercial sintering machines", *ISIJ International*, 40 (2000), 1188-1194.
[30] R.P. Bhagat, "Factors affecting return sinter fines regimed and strand productivity in iron ore sintering", *ISIJ International*, 39 (1999), 889-895.
[31] P. Cavaliere, A. Perrone, "Optimization of Blast Furnace Productivity Coupled with CO_2 Emissions Reduction", *Steel ResearchInternational*, 85 (2014), 89-98.

Engineering Solutions for Sustainability: Materials and Resources II
Edited by: Jeffrey W. Fergus, Brajendra Mishra, Dayan Anderson, Emily Allyn Sarver, and Neale R. Neelameggham
TMS (The Minerals, Metals & Materials Society), 2015

TWINS EVOLUTION DURING THE RECRYSTALLIZATION INDUCED BY ELECTRIC CURRENT PULSES IN A Cu-Zn ALLOY

X. Zhao[1], X. L. Wang[2], W. B. Dai[3], M. S. Liu[1], N. Wu[1]

[1]Key Laboratory for Anisotropy and Texture of Materials (Ministry of Education), Northeastern University, Shenyang 110819, China
[2]Research Institute, Northeastern University, Shenyang 110819, China
[3]School of Materials and Metallurgy, Northeastern University, Shenyang 110819, China

*To whom correspondence should be addressed.
Email address: zhaox@mail.neu.edu.cn

Keywords: Twin, Recrystallization, Electric current pulse, Cu-Zn alloy

Abstract

A high current density (up to 10^4 $A \cdot mm^{-2}$) electric current pulse (ECP) is used to cause recrystallization of a cold-rolled single phase Cu-Zn alloy directly. Since the whole treating time is the discharge time of ECP, and the duration time is very short, it is very easy to remain the non-equilibrium phase at high temperature to the room temperature to investigate the microstructural evolution during the recrystallization process. By the determination of TEM, it is found that due to the application of ECP, the original deformed twins are greatly decreased while a lot of incomplete annealing twins are emerged. Interestingly, the various morphologies of annealing twins are revealed, and the different growth directions are also observed. This phenomenon would contribute to study the variant selection of the twin in the early stage of recrystallization process.

Introduction

The research about the application of electric current pulses (ECP) on materials is mainly about the electroplasticity[1], electromigration[2], crack healing[3], nanostructured transformation[4], and so on. With an intensive study of the effect of ECP on materials, Dai et al. proposed that the ECP treatment should be a special and effective method to control the recrystallized nucleation orientation by changing the exerted current direction[5]. Their results prove the importance of the exerted current direction. In addition, since ECP treatment is an instantaneous nonequilibrium processing technique, it is possible that some unstable high temperature phase may be remained to room temperature. Zhang et al. reported the martensitic transformation from α-Ti to β-Ti on rapid heating by using ECP[6]. Wang et al. presented the up-transformation variant selection obtained by ECP treatment[7]. Therefore, ECP can be as a kind of method to investigate the unstable microstructure during phase transition, including recrystallization and phase transformation.

As known, annealing twins observed in a variety of recrystallized f.c.c. metals and the formation of annealing twins can be broadly classified into two different groups, growth accidents and nucleation of twins by stacking faults or fault packets. Gleiter developed a comprehensive model for the formation of annealing twins[8]. He assumed that grain growth occurred by the transfer of atoms from a shrinking grain to a growing grain. Dash and Brown proposed that for the first time that twin nuclei consisted of stacking fault packets having complicated morphology[9]. Meyers and Murr proposed the "pop-out" model to explain the formation of stacking fault packets[10]. Pande et al. carried out the formation of twins was caused by grain boundary force for migration[11]. Though lots of clarifications have been presented, the instantaneous twin evolution by experiment is still an open question. The aim of this work is to investigate the twin evolution during recrystallization process induced by ECP.

Experimental process

A 40% reduction in thickness direction of a cold-rolled commercial Cu-Zn alloy sheet with a composition of Cu 63.8 mass% and Zn 36.2 mass% was used as the research object in this study. By using the electrospark discharge technique, dog-bone-shaped samples with effect parts of 10 mm length, 5 mm width, and 1 mm thick were prepared, and the sizes of the two ends were much larger than the size of the effective part. A single ECP was produced through a discharge of capacitor banks under ambient conditions. During the ECP process, the two ends of each sample were put into copper electrodes. Then the temperature rise of the two ends was very small and could be regarded as room temperature. Thus, due to the cooling effect of the two cool ends, a higher cooling rate could be obtained in the effective part of sample during cooling. The waveform of the ECP was detected to be a damped oscillation wave using a Rogowski coil and a TDS3012 digital storage oscilloscope, the waveform of the current density (j) versus the applying time (t) was approximately described as damping sinusoids with the maximum current density j_{max}=9.86 kA mm^{-2}. And the period of ECP (t_p) was 220 μs and the pulse duration (t_d)

was about 1.5ms. The microstructures of samples after ECP treatment were characterized by a JEOL-2100F high-resolution transmission electron microscopy (HRTEM) operated at 200kV.

Results and discussion

Planar-view TEM observations of the rolled sample shows a heavily deformed microstructure. A high density of dislocations is noticed with a non-uniform distribution, as shown in Fig.1. Dislocation cells with sizes ranging from submicron to micron are visible. Also a layered microstructure with an extremely fine layer thickness can be observed through the dislocations. The selected area electron diffraction (SAED) pattern shows a typical twin relationship among the lamellae corresponding to face-centered cubic (fcc) metals. There is an obvious elongation detected from the diffraction spots shown in Fig.1b. Therefore, it is reasonable to say that these thin lamellae are ascribed to the deformation twins.

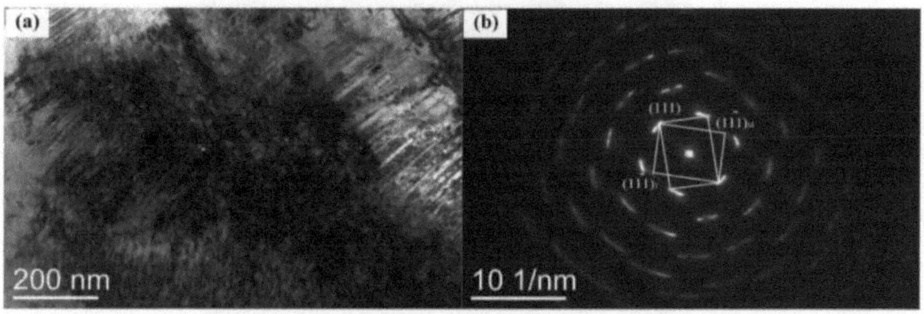

Figure 1. (a) Bright-field TEM image of the rolled sample, (b) the corresponding SAED pattern

Figure 2 presents the TEM microstructures of samples treated after ECP. Obviously, the dislocation density is greatly decreased after ECP treatment. Detailed TEM observations shown in Fig.2a to c reveal that the different lamellae microstructures all are typical twins with twin boundary steps. The twins distribute randomly in a grain, and the growth direction is parallel to the <111> direction, as shown in Fig.2d. Due to the rapid heating and high-rate cooling, the twin boundary steps (in Fig.2c) are remained from high temperatures. Moreover, it can be observed that the growth of the twin is limited and the different twin variants are obtained. From Fig.2e, it can be found that the boundary steps are incoherent, and the further investigation reveals that the boundary steps are arranged by lots of stack faults. In addition, it should be mentioned that no residual deformed twins can be detected.

During current passing, the temperature increment of the samples can be induced by Joule heating, and it is dominated by the energy input by ECP, the temperature increment can be evaluated according to the waveforms of ECP based on the following equation[12],

$$\Delta T = \frac{\int_0^{t_d} j^2(t)\rho dt}{c_p d} \tag{1}$$

where $j(t)$ is the electric current density at time t, ρ is the resistivity ($0.062\times10^{-6}\,\Omega\cdot m$), t_d is the pulse duration (220 μs), c_p is the specific heat (378 $J\cdot kg^{-1}\cdot K^{-1}$), and d is the density (8.40 $g\cdot cm^{-3}$) of the specimen. The temperature increment can be calculated as 417 K at room temperature 293 K, and the ECPed samples are lower than the phase transformation temperature (953 K). Thus, the microstructural evolution induced by ECP can be ascribed to the recrystallization process.

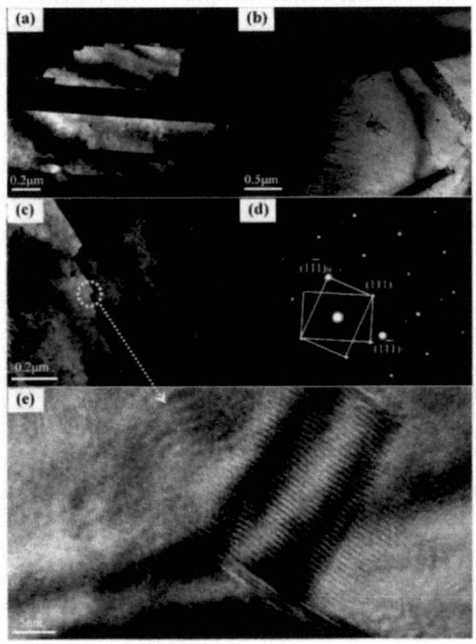

Figure 2. (a) and (b) The random distribution of the twins in a grain, (c) The twin boundary step, (d) The corresponding SAED from (c) area, (e) The HRTEM (high resolution TEM) of the white circle area from (c).

From Fig.2a, the dislocation densities are greatly decreased after treated by ECP. Therefore, the glide and climb of dislocations play very important roles in the formation of twins. During current passing, the mobility of the dislocations can be enhanced due to the effects of the electron wind force and the transient thermal compressive stress. Conrad emphasized that electron wind force exerted by drift electrons on dislocations can accelerate the dislocations motion[13]. Wang et al. also emphasized that the electron wind force was the principal driving force to accelerate the formation of recrystallization[14]. When a current passing a current-carrying metallic material, the force resulting from the momentum transfer as the electrons collide with atoms is called electron-wind force[13]

$$F_{ew} = (\rho_D / N_D) e n_e j \qquad (2)$$

where F_{ew} is the electron-wind force per unit dislocation length, ρ_D/N_D is the specific resistivity per unit dislocation length, e is the electron charge, N_D is the dislocation density, n_e is the electron density and j is the current density. By using the data given by Conrad[13], in the case of Cu, $\rho_D/N_D = 2.05\times10^{-19}\Omega cm^3$, $n_e = 8.76\times10^{22} Acm^{-3}$, so $F_{ew} \approx 1.6\times10^{-7} Ncm^{-1}$ at $j = 1.658\times10^6 Acm^{-2}$. Obviously, F_{ew} is proportional to current density and the mobility of the dislocations is advanced by the ECP treatment. Due to the enhancement of the dislocation mobility, the recrystallization nucleation rate is greatly increased by using ECP.

Besides the effect of the electron wind force, the transient thermal compressive stress is also regarded as an important factor to increase the mobility of dislocations. In our experiment, heating rate can get to 10^6 °C·s^{-1} during ECP. Such a high-rate heating can generate non-synchronous change of temperature rise and dynamic thermal expansion[15], then a transient thermal stress comes into being, and a possible average maximum thermal compressive stress can be estimated by[15]

$$\delta_{max} = E\alpha\Delta T \tag{3}$$

where E is Young's modulus, and α is the thermal expansion coefficient. For Cu-Zn alloy, E=191 GPa, and $\alpha = 20.5\times10^{-6}$ °C^{-1}[16], the matrix will receive a maximum compressive stress of about $\delta = 1.63$ GPa, which will strongly enhance the dislocation mobility, and alter the original stress state of the alloy. Consequently, the nucleation rate of recrystallization would be enhanced with the increased mobility of dislocation produced by ECP. In addition, a large overheating can be also obtained for a deformed material during recrystallization under such a high-rate heating, and the recrystallization nucleation rate will be enhanced with the increased overheating. With the nucleation rate increasing, the recrystallized grain size is decreased. Since a major driving force for the growth of newly recrystallized grains is the stored energy of the residual dislocation density, the lower dislocation density in the electric pulsed samples provides a smaller driving force and thus a smaller growth rate. Therefore, when current passing, the newly recrystallized grains have not enough time to grow, and then the unstable microstructures are obtained in the room temperature. Simultaneously, due to the strong thermal stress, the strain energy may be formed between the crystal lattice, which also provides a free energy for the formation of twins. Also it is combined with the results shown in Fig.2e, where stacking faults can be observed in the boundary steps.

Basically, three different selective principles have been suggested to account for the observed restrictions of twin formation: one is grain boundary energy to minimize the energy of the moving boundary, one is twinning mechanism to operate most easily for a particular arrangement of boundary orientation and dislocation structure, and another is a grain boundary mobility to maximize the growth rate of the migrating boundary. In our work, the unstable non-equilibrium state is remained to room temperature after ECP treatment. Therefore, it is reasonable to say that the grain boundary energy is an effective method to minimize the energy of the unstable system formed by current passing. However, it is suggested that twin boundaries form during grain growth when the free energy of the boundaries between a grain's neighbors and its twin would be less than that of the boundaries between the neighbors and the grain itself. As known, the formation of a twin ($\Sigma3$ boundary) is accompanied by the conditions required for a decrease in free energy. Considering that the potential reduction in free energy for a Σ3 is far in excess of that for any other type of boundary since very low energies, <0.1J/mm^2 are achieved for Σ3s on {111}, which is much less than for any other CSL[17]. Meanwhile, during the application of ECP, an instantaneous higher energy is input to the matrix with a change of microstructure; then, the formation of Σ3s rapidly reduced the system's energy, which is favor to the equilibrium of the system after ECP. Therefore, lots of Σ3s on {111} are observed after ECP treatment. In copper, the ratio of the interfacial free energy of coherent twin boundaries to that of a average grain boundaries is only 0.035, much less than the variability of grain boundary free energy, so that twin boundary formation are frequent and always occupy the greatest CSL population after ECP treatment. Thus, a lot of twins might then form in a corner of a growing grain and extend as the grain continues to grow as a twin of the original crystal, as that shown in Fig.2b. However, due to the high cooling-rate, there is enough time for the twins to grow up, then lots of twins with boundary steps are formed. In addition, from Fig.2c, lots of twins carry out inter the grains without an enough growth, which implies that a higher energy is introduced into the sample system during current passing, and to minimize the system energy, lots of twins are formed inter grains not only from the corner of a grain.

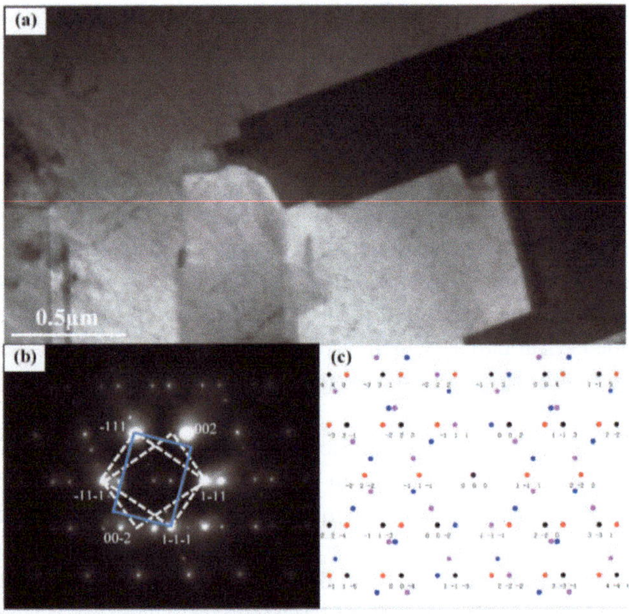

Figure 3. (a) The twins variant observed by TEM, (b) The corresponding SAED, (c) The schematic diffraction patterns of the twins with three direction, twining axis is parallel to [110], the three twinning planes are (1-11), (-111) and (1-1-1), respectively.

In order to reveal the twin variants remained from the high temperature unstable state[18], a twin family with three orientations is detected by TEM, as presented in Fig.3a, and the corresponding selected area diffraction is illustrated in Fig.3b. According to the schematic twinning diffraction with three twinning direction, it can be sure that the three twin variants are obtained due to the application of ECP. Commonly, with the grain growth, there will be a competition among the different orientations, and the preferred one with more low energy will grow up. In fact, the present results explain the possible reason that there is a random nucleation and an oriented growth during recrystallization induced by ECP, as mention in our previous results[14].

Conclusion

ECPs can increase the nucleation rate of the recrystallization due to the electron wind force and the thermal compressive stress. As soon as the current termination, the unstable high temperature system will be remained to the room temperature with a high cooling rate. In order to reduce the system energy, lots of twins are formed. The formation of the different twinning variant provides a strong evidence to support our viewpoint that there is a random nucleation and an oriented growth during recrystallization induced by ECP.

Acknowledgement

This work was supported by the National Nature Science Foundation of China, Grant No.51471047, the 111 Project of China (Grant No.B07015), and the Fundamental Research Funds for the Central Universities No.N130418001. The authors thank Professor J.D. Guo of Institute of Metal Research, Chinese Academy of Sciences, for many helpful discussions.

References

1. A.F. Sprecher, S.L. Mannan, and H.Conrad, "On the temperature rise associated with the electroplastic effect in titanium," *Acta Metallurgica*, 34 (1986), 1145-1162.

2. P.S. Ho and T. Kwok, "Electromigration in metals," *Reports on Progress in Physics,* **52** (1989), 301-348.

3. Y.Z. Zhou, Y. Zeng, G.H. He, and B.L. Zhou, "The healing of quenched crack in 1045 steel under electropulsing," *Journal of Materials Research*, **16** (2001), 17-20.

4. R.S. Mishra, S.H. Risbud, and A.K. Mukherjee, "Influence of initial crystal structure and electrical pulsing on densification of nanocrystalline alumina powder," *Journal of Materials Research*, **13** (1998), 86-89.

5. W.B. Dai, X.L. Wang, H.M. Zhao, and X. Zhao, “Effect of electric current on microstructural evolution in a cold-rolled 3%Si steel,” *Materials Transactions*, **53** (2012), 229-233.

6. W. Zhang, W.S. Zhao, D.X. Li, and M.L. Sui, “Martensitic transformation from α-Ti to β-Ti on rapid heating,” *Applied Physics Letters*, 84 (2004), 4872-4874.

7. X.L. Wang, W.B. Dai, R. Wang, X.Z. Tian, and X. Zhao, “Enhanced phase transformation and variant selection by electric current pulses in a Cu–Zn alloy,” *Journal of Materials Research*, 29 (2014), 975-980.

8. H. Gleiter, “Theory of grain boundary migration rate,” *Acta Metallurgica*, 17 (1969), 853-918.

9. S. Dash, N. Brown, “An investigation of the origin and growth of annealing twins,” *Acta Metallurgica*, 11 (1963), 1067-1075.

10. M. A. Meyers, L. E. Murr, “A model for the formation of annealing twins in F.C.C. metals and alloys,” *Acta Metallurgica*, 26 (1978), 951-962.

11. B. B. Rath, M. A. Imam, and C. S. Pande, “Nucleation and growth of twin interfaces in FCC metals and alloys,” *Materials Physics Mechanics*, 1 (2000), 61-66.

12. H.Conrad and A.F. Sprecher, in *Dislocations in Solids* (ed. F. R. N. Nabarro, Amsterdam, Elservier, 1989), 497-499.

13. A.F. Sprecher, S.L. Mannan, and H. Conrad, “Overview no. 49: On the mechanisms for the electroplastic effect in metals,” *Acta Metallurgica*, 34 (1986), 1145-1162.

14. X.L. Wang, W.B. Dai, C.W. Ma, and X. Zhao, “Effect of electric current direction on recrystallization rate and texture of a Cu-Zn alloy,” *Journal of Materials Research*, 28 (2013), 1378-1385.

15. D.W. Tang, B.L. Zhou, H. Cao, and G.H. He, “Thermal stress relaxation behavior in thin films under transient laser-pulse heating,” *Journal of Applied Physics*, 73 (1993), 3749-3752.

16. E.A. Brandes, *Smithells Metals Reference Book* (Butterworth, Washington, DC, 1983), 13–17.

17. C.B. Thomson, and V. Randle. "Fine tuning" as $\Sigma 3^n$ boundaries in nickel," *Acta Metallurgica*, 45 (1997), 4909-4916.

18. X.L. Wang, W.B. Dai, R. Wang, X.Z. Tian, X. Zhao, J.M. Li, "A method to distinguish twinning variants and second twins by EBSD measurements," *Acta Metallurgica Sinica (English Letters),* 27 (2014), 267-271.

Engineering Solutions for Sustainability: Materials and Resources II
Edited by: Jeffrey W. Fergus, Brajendra Mishra, Dayan Anderson, Emily Allyn Sarver, and Neale R. Neelameggham
TMS (The Minerals, Metals & Materials Society), 2015

REACTION BETWEEN CARBONACEOUS MATERIALS CONTAINING HDPE AND STEEL-MAKING SLAG

Lan Hong, Huihua Wang* Binna Song and Dong Chen
School of Iron and Steel, Soochow University, Suzhou, China, 215021
*corresponding author

Abstract

Since waste plastics have been applied extensively in many high temperature processes such as iron- and steel-making, coke-making and carbon nano-structure manufacturing etc, the investigation about the reactivity becomes increasingly necessary. In this study, a mixture of coke and a typical component of waste plastics, high density polyethylene (HDPE), at a ratio of 1:1 was oxidized in an atmosphere of low oxygen partial pressure of 10^{-1} Pa at 1823 K to evaluate the reactivity of HDPE. The oxidation mechanism of HDPE was analyzed using the monitored CO and CO_2 contents in the off-gas by an infrared (IR) gas analyzer in combination with the oxidation results of graphite and coke. It is clear from the results that in the presence of HDPE, the Boudouard reaction can reach equilibrium more easily compared to coke and graphite.

Introduction

Value-added reutilization of waste plastics has attracted tremendous attention world-widely. Waste plastics have been studied for applications in iron- and steel-making processes[1)] such as injection into blast furnaces[2)] and electric arc furnaces[3)] as a replacement of coke partially, in coke-making industry[4-5)], and in manufacturing advanced carbon nano-structures[6)] cited as carbon nano-tubes or graphenes. In many countries, waste plastics have been considered as one of national strategic resources rather than only a waste polluting environment. Searching ways to utilize waste plastics more efficiently or make essential value-added products from waste plastics becomes increasingly focused.

Most of the applications of waste plastics mentioned above are accomplished by heating the materials to high temperatures, which implies that a comprehensive understanding about the characteristics of waste plastics at high temperatures is one of the key issues for the most appreciate applications. Of the characteristics, the reactivity of waste plastics is a necessary chemical property in assistance in applying waste plastics more efficiently and economically.

The reactivity of carbonaceous materials is usually assessed by measuring its reaction with oxygen or carbon dioxide. In this study a similar test process is employed to evaluate the reactivity of HDPE, a typical component of post-consumer plastics. A mixture of HDPE and coke is designed to be oxidized by oxygen in atmosphere at high temperature. A low oxygen partial pressure is selected to give satisfied results since the reaction proceeds very quickly in high oxygen partial pressure environment when the temperature is high as 1823 K in the current case. For comparison, pure graphite and coke are tested reactivity under the same experimental conditions as well.

Experimental

Experimental apparatus

A horizontal resistance furnace was employed in the current study to provide the high temperature for the reaction, of which schematic diagram is shown in Fig.1. Both ends of the furnace tube were sealed to air-tighten the furnace and make the atmosphere controllable. An alumina sample holder, connected to a stainless steel rod, was used to transport samples to and from the isothermal zone of the furnace. The outlet gas was led to a CO/CO_2 IR gas analyzer to monitor the off-gas composition for further analyzing the reaction mechanism.

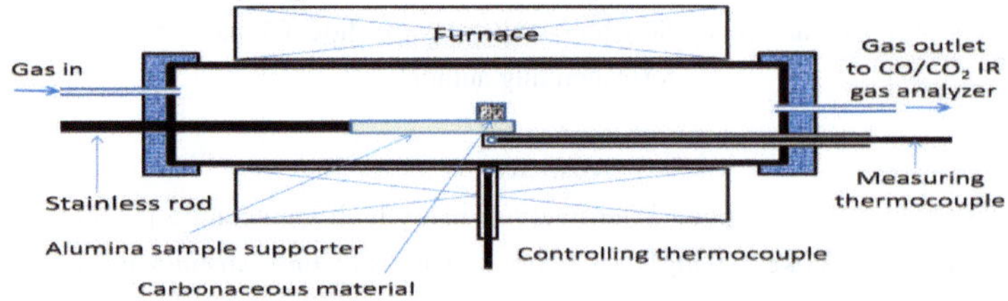

Fig.1 Schematic diagram of experimental set-up

Experimental materials

Synthesized graphite powder, industry coke fines and HDPE particles were chosen as the starting raw materials for the study. The purity of graphite (mean size 44 μm) and HDPE (mean size 3 mm) is more than 99.9%. The composition of the coke is listed in Table 1. The coke was grinded to get an average size of about 100 μm before use. Three types of carbonaceous blocks around 10 g each were prepared for the experiment by hydraulically pressing under a high pressure of 2.2×10^8 Pa: graphite, coke and the mixture of HDPE and coke with a mixing ratio of 1:1. The mixture of HDPE and coke was made by mixing the pre-weighed grinded coke powder and HDPE at 473 K homogeneously before hot-pressed at 423 K. An Ar gas having an oxygen partial pressure of 10^{-1} Pa was used as the atmosphere and the gas flow rate was set at 1.67×10^{-5} m^3/s.

Table 1 Compositions of industry coke (%mass)

Fixed carbon	83	Ash composition: $50.1SiO_2$, $38.2Al_2O_3$, $5.3Fe_2O_3$, $2.0CaO$, $1.7TiO_2$, $1.5P_2O_5$, $0.4MgO$, $0.3Na_2O$, $0.4K_2O$, $0.1SO_3$
Ash	15	
Volatile	1	
moisture	1	

Experimental procedure

After a pre-weighed sample block was placed on the alumina sample holder installed through the front flange, the furnace tube was sealed. The sample was pulled back to the cold zone close to the water-cooled flange to keep it at low temperature before the start of the experiment. Then Ar gas was purged to remove the air inside the furnace tube for 30 min before the heating program was started. When the temperature of the isothermal zone was pre-set one of 1823 K, the sample was pushed to the middle of the furnace tube to commence the experiment, and at this moment the CO/CO_2 IR gas analyzer was started as well to monitor the CO and CO_2 contents in the off-gas. After a certain time of period, the reaction was stopped by pulling the sample holder, on which the sample was located, back to the cold zone again.

Experimental results and discussion

The measured CO and CO_2 contents in the off-gas are plotted against reaction time for the three carbonaceous substrates in Fig.2. As using similar initial weight for all the carbonaceous substrates, it is reasonable to compare the reaction rate qualitatively by the CO and CO_2 amounts in the off-gas shown in the graphs for different carbonaceous materials with an assumption that the measured gas composition was the local one. There are two distinguished reaction patterns classified according to the dominant reaction products, which are called graphite type and coke type in the present study. The output of the oxidation reaction of graphite was mainly CO_2 in a relatively long period of the reaction time, whereas the dominant reaction product for the oxidation reaction of coke or the mixture of HDPE and coke was CO almost during the whole reaction time except in the very beginning. The amounts of CO and CO_2 originated from the oxidation of coke and the mixture of HDPE and coke were much greater than those from the oxidation reaction of graphite. Both CO and CO_2 contents rose to their respective peaks quickly after the start of the reaction and fell down steeply to a very gradually changing stage in the case of graphite substrate. In contrast to this, the CO content in the off-gas of coke or the mixture of HDPE and coke oxidation reaction rose to a near plateau maximum value zone lasting for around 600 s, then from the maximum value zone dropped to a gradually changing zone in a very short time; on the other hand, CO_2 content increased to a peak value shortly after starting the reaction, and then decreased slowly along the reaction time.

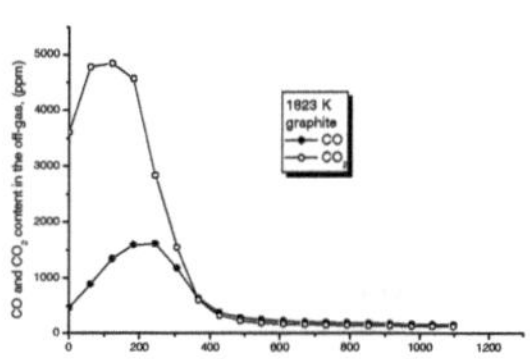

Fig.2a Change of CO and CO_2 contents in the off-gas for graphite

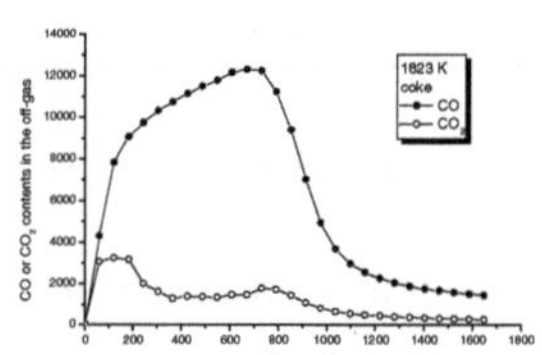

Fig.2b Change of CO and CO_2 contents with time for coke

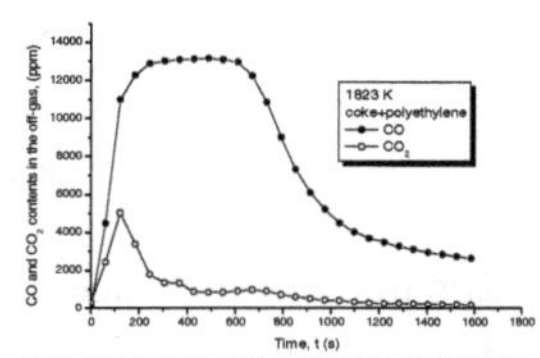

Fig.2c Change of CO and CO_2 contents in the off-gas with time for the mixture of coke and polyethylene

The oxidation process of carbonaceous materials by oxygen is generally accepted following the elemental steps as indicated below.

(i) Surface chemisorption of oxygen on the surface of carbon

$$2C + O_2 = 2C\text{-}O \tag{1}$$

(ii) Desorption of C-O complexes from the surface of carbon

$$C\text{-}O = CO \tag{2}$$

$$O\text{-}C\text{-}O = CO_2 \tag{3}$$

Surface reactions (2) and (3) have been proved to be primary ones by isotopic studies[7]. However, the following two reactions may occur but not necessarily depending on the conditions.

$$C + CO_2 = 2CO \tag{4}$$

$$CO + \frac{1}{2}O_2 = CO_2 \tag{5}$$

Both coke and the mixture of HDPE and coke are oxidized more quickly than graphite because graphite is chemically more stable than coke. The differences in the crystalline structure of graphite and coke are considered to be one of the causes. There are two types of planes in crystallized or partially crystallized carbonaceous materials: basal planes and prismatic or edged planes. The surface energies of these two types of crystallographic direction differ largely: about $0.11 J/m^2$ in the basal planes and $5 J/m^2$ in the prismatic planes[8]. Therefore basal planes are very stable and prismatic or edged planes get involved in reactions much more actively. The crystalline size of graphite is larger than that of coke and usually crystalline direction of coke is highly scattered as shown in Fig.3. The very sharp peak of graphite suggests the crystalline size in graphite is very large and the crystals are well-oriented. On the other hand, the peak of coke is very broad, corresponding to the fact that in coke only a small portion is graphitized while its dominant part is amorphous and the crystals are small-sized. As a result, more edged planes are available for reaction in coke than in graphite and hence more oxidation products CO and CO_2 are generated in the case of coke compared to graphite. The oxidation of prismatic or edged planes may correspond to the quick reaction period, while the slow reaction stage may be the result of the oxidation of basal planes which occurs after most of the prismatic or edged planes have been consumed. Another reason for the faster reaction rate of coke than graphite may exist in the fact that in coke there is 15% ash composed of oxides acting as catalysts during the oxidation process.

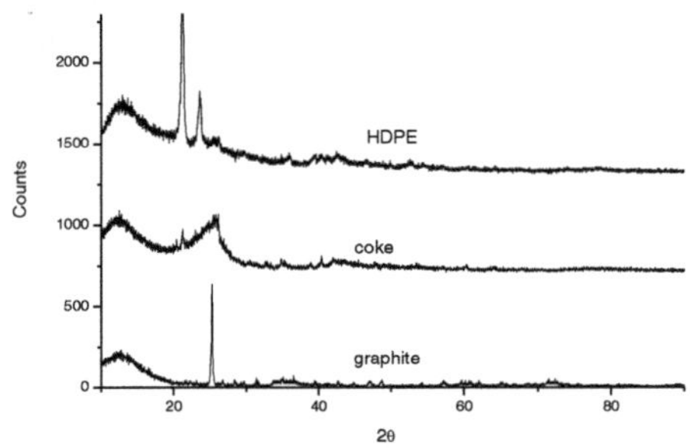

Fig.3 XRD results of various carbonaceous materials

The standard free energy changes of the two primary reactions in the reaction system are expressed by Eqs. (6) and (7) and plotted in Fig.4.

$$C(s) + \frac{1}{2}O_2(g) = CO(g) \qquad \Delta G^0 = -114391-85.77T \text{ J/mol} \qquad (6)^{9)}$$

$$C(s) + O_2(g) = CO_2(g) \qquad \Delta G^0 = -395346-0.54T \text{ J/mol} \qquad (7)^{9)}$$

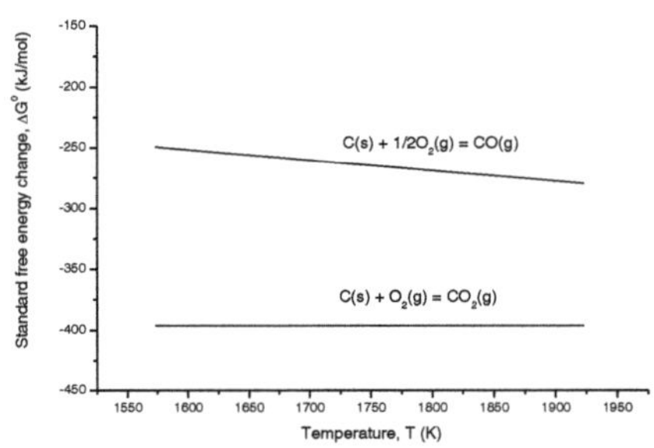

Fig.4 Standard free energy change as a function of temperature

It is clear from this figure that the primary reaction (7), the formation reaction of CO_2, proceeds preferably than the formation reaction of CO as expressed by Eq.(6). Thus, more CO_2 was generated than CO as the reaction product in the beginning of the oxidation for all the carbonaceous substrates studied in the current research. However, the duration of CO_2 overwhelming CO was long for graphite whereas in the cases of coke and the mixture of HDPE and coke this phenomenon only appeared for seconds in the very beginning of the reaction. Afterwards CO was evolved more than CO_2 especially for coke and the mixture of HDPE and coke, in which cases CO was predominantly generated compared to CO_2.

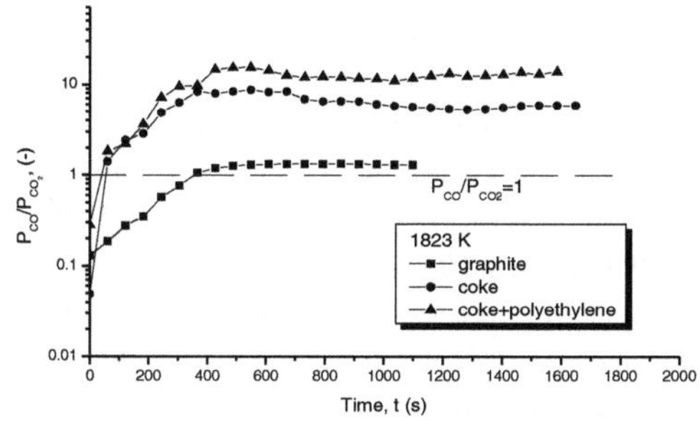

Fig.5 Change of P_{CO}/P_{CO2} value with time

The index of Pco/Pco_2 , an indicator for the ratio of CO to CO_2 partial pressure, is plotted as a function of time in Fig.5. The turning points at $Pco/Pco_2=1$ for all of these three carbonaceous substrates have no correlation with their respective turning points at which the oxidation reaction transits from the quick reaction rate zone to the slow reaction rate zone of the dominant oxidation product (CO for graphite and CO_2 for coke and the mixture of HDPE and coke).

Since carbon was excess near the reaction sites, the reaction (8) may not proceed necessarily while Boudouard Reaction or the solution-loss reaction (9) could occur unavoidably. Therefore the solution-loss reaction of carbon is considered to be the reason of CO overtaking CO_2 later on.

$$CO(g) + \frac{1}{2}O_2(g) = CO_2(g) \qquad \Delta G^0 = -280956+85.23T \text{ J/mol} \qquad (8)$$ [9]

$$C(s) + CO_2(g) = 2CO(g) \qquad \Delta G^0 = 166565-171.00T \text{ J/mol} \qquad (9)$$ [9]

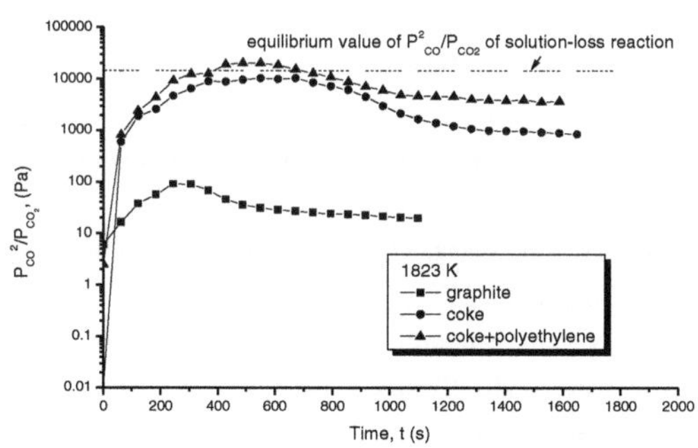

Fig.6 Comparison of P^2_{CO}/P_{CO_2} of carbonaceous substrates with the equilibrium value of solution-loss reaction

Solution-loss reaction (9) plays a key role in many carbothermic reduction processes of oxides and is judged at least one of the main rate controlling steps by many researchers[10]. The value of P_{CO}^2/P_{CO_2} for the oxidation results of graphite, coke and the mixture of HDPE and coke is calculated and plotted in Fig.6. In this graph, the dotted line is the value of P_{CO}^2/P_{CO_2} when the solution-loss reaction is at equilibrium. The oxidation of graphite is far below the equilibrium of the solution-loss reaction, while the oxidation of coke and the mixture of HDPE and coke is almost at the equilibrium of the solution-loss reaction. Specifically the mixture of HDPE and coke, its value of P_{CO}^2/P_{CO_2} is larger than the equilibrium value of the solution-loss reaction once in a short time period. This may be due to the fact that hydrogen, one of the pyrolysis products of HDPE, can react further with CO_2 according to Eq. (10) to give a higher CO content and a lower CO_2 content compared to the respective equilibrium values of the solution-loss reaction. The overall pyrolysis reaction of HDPE at high temperature is expressed by Eq. (11).

$$CO_2(g) + H_2(g) = CO(g) + H_2O(g) \qquad \Delta G^0 = 33472\text{-}29.37T \text{ J/mol} \qquad (10)^{9)}$$

$$\text{-}(CH_2)\text{-}_n = nC(s) + nH_2(g) \qquad (11)$$

The release of hydrogen gas is a very fast process starting from low temperatures as 650 K[11]. Therefore, it is highly possible that after 1000 s, there was no hydrogen in the reaction system any more. However, the P_{CO}^2/P_{CO_2} value of the mixture of HDPE and coke was still above that of coke, much closer to the equilibrium value of the solution-loss reaction than coke at the late stage of the reaction. This implies that in the mixture of HDPE and coke, after the pyrolysis of HDPE, porous solid carbon residue was left in the presence of ash in coke acting as catalysts[12] and the reactivity of the porous solid carbon residue was chemically more active than coke or graphite.

Conclusions

The reactivity of HDPE was compared to graphite and coke by oxidizing these carbonaceous materials in a low oxygen partial pressure atmosphere of 10^{-1} Pa at a high temperature of 1823 K. In the main reaction stage of graphite, CO_2 was released dominantly whereas CO was the main gas product in both cases of coke and the mixture of HDPE and coke. As a reducing agent, HDPE is more active than both graphite and coke probably due to the fact that in the early stage of the reaction, more reductive pyrolysis product hydrogen was present in the reaction system and at the late stage of the reaction another pyrolysis product of HDPE, porous solid carbon residue, accelerated the solution-loss reaction.

References

1. F. Hanrot, D.Sert, J. Delinchant, R. Pietruck; T. Bürgler, A. Babich, M. Fernández,R. Alvarez and M.A. Diez: Proc. of 1st Spanish National Conference on Advances in Materials Recycling and Eco – Energy, Madrid, 2009, 181.
2. For example, K. Wakimoto, H. Nakamura, M. Fujii, Y. Yamamoto, K. Nemoto and K. Tomioka: *NKK Tech. Rep.*, **78** (1998), 59.
3. V. Sahajwalla, L. Hong and N. Saha-Chaudhury: AISTech 2006 Proc., USA, 2006, 91.
4. S. Nomura, K. Kato, T. Nakagawa, and I. Komaki (2003): Fuel, 82(2003), 1775.
5. S. Nomura and K. Kato:Fuel, 85(2006), 47.
6. A. Bazargan and G. McKay: Chem. Eng. J., 195-196 (2012), 377.
7. P. Walker, Jr., F. Vastola and P. Hart: Fundamentals of gas-Surface Interactions (Editted by H. Saltsburg, J. Smith and M. Rogers), Academic Press, New York, (1967), 307.
8. J. Abrahamson: Carbon, 11(1973), 337.
9. E. T. Turkdogan: Physical Chemistry of High Temperature Technology,Academic Press, New York, (1980), 5.
10. For example, L. Hong and V. Sahajwalla: Metall. Mater. Trans. B, 2013, DOI 10.1007/s11663-013-9920-y.
11. Y.-S. Cho, M.-J. Shim and S.-W. Kim: Mater. Chem. Phys., 52(1998), 94.
12. Z. Jiang, R. Song, W. Bi, J. Lu and T. Tang: Carbon, 45(2007), 449.

DEVELOPING ALTERNATIVE INDUSTRIAL MATERIALS FROM MINING WASTE

Javier Flores B.[1], Juan Hernández A. [1], Miguel Pérez L. [1] Francisco Patiño C.[1], José Abacú Ostos S.[2], Norma Y. Trápala P. [3]

[1]Área Académica de Ciencias de la Tierra y Materiales, Universidad Autónoma del Estado de Hidalgo. Carretera Pachuca-Tulancingo Km. 4.5, C.P. 42184 Pachuca, Hidalgo, México., wik_1000@hotmail.com, herjuan@uaeh.edu.mx

[2]Área Académica de Ingeniería. Universidad Autónoma del Estado de Hidalgo. Carretera Pachuca-Tulancingo Km. 4.5, C.P. 42184 Pachuca, Hidalgo, México. Tel. y Fax: (01 771)7172-000 ext. 2501.

[3]Alfarería "Los Toños" S.A de C.V. Andador Miguel Hidalgo No.2 Mza. 2 Lote 5 Colonia Francisco Villa (Napateco), C.P. 43629 Tulancingo de Bravo, Hgo. México.
yacelit@hotmail.com

Keywords: Tailings, heavy clays, alternative industrial materials.

Abstract

The presence of mining waste (known as tailings or mining dumps) near residential zones has been a major health and environmental problem in the mining district of Hidalgo, Mexico, causing allergies and respiratory complications due to the strong winds that characterize the region. For this reason the development of industrial materials from mining waste is suggested in this paper. The stabilization of the tailings involved taking samples and determining their chemical composition and particle size. Afterwards, the alternative industrial materials were produced by using the tailings and heavy clay in order to give the composite a good green strength and plasticity during development, but above all to give it a compressive strength similar or higher than that of products derived from conventional processes.

Introduction

473 years of mining exploitation (mainly focused on gold and silver [1] in the state of Hidalgo have left a great amount of waste, such as mine drainage water, solid-transport water, mounds of abandoned, unprocessed minerals and mining waste, also known as tailings.

Regarding the latter, in the mining district of Pachuca-Real del Monte, there is a total of 108.1 million metric tons, covering approximately 1200 hectares [1,2,3] inside the urban sprawl of the municipalities of Pachuca de Soto and Omitlán de Juárez (Fig. 1). Due to the geography of this mining district, there are very strong wind currents, mainly between February and April, as well as in November (Fig. 2). This causes a considerable increase in respiratory diseases and eye complications of nearby inhabitants [1,2,3].

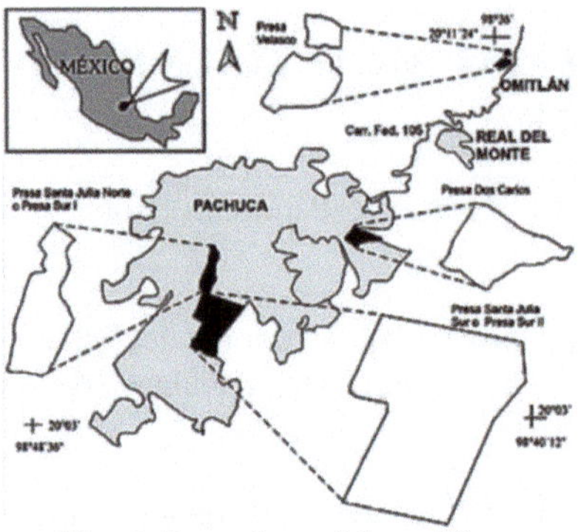

Fig. 1. Location of the tailings in the Pachuca-Real del Monte district.

Fig. 2. Tailings dust column caused by strong winds.

In this study we used the material of one of the oldest tailings in the mining district, which shows the use of different metallurgical processes, such as grinding-amalgamation (used for 353 years), grinding-cyanidation (used for 47 years), and grinding-flotation-cyanidation (used for 70 years). Undoubtedly, the oldest residues were found at the base; they bear witness from colonial times and the first years of an independent Mexico, to the arrival of the grinding-flotation-cyanidation technology, also known as "Tanques Pachuca". Inside the base of the tailings there is a higher concentration of gold and silver, as well as bigger particle size. This feature, when added to the use of heavy clays in order to conglomerate the tailings, helped to produce alternative industrial materials that contain similar or better properties than those of products derived from conventional processes.

Experimental procedure

Characterization of the Dos Carlos tailings was carried out by X-ray Diffraction and atomic absorption spectrophotometry (*AAS*) using a Philips X'PERT diffractometer and a *Perkin Elmer 2380* Atomic Absorption Spectrometer. Granulometric analysis was performed by using Tyler meshes with the following scales: 177, 149, 105, 74, 53 and 37 μm. 100g wet samples were sieved; the obtained fractions were dried at room temperature and then weighed.

For the production of the alternative materials, 13 ceramic composites were made by mixing the tailings material and heavy clays. In one case we used only the tailings material. The alternative materials were then produced by molding the ceramic composite in a roof tile mold. It was dried at room temperature and green sintering was finally applied.

Results and discussion

The results obtained by XRD, AAS and granulometric analysis from the Dos Carlos and Velasco tailings are presented as follows:

Regarding the Dos Carlos tailings, the diffractogram in figure 3 shows the presence of major mineral phases such as silica, orthoclase, albite, berlinite, potassium jarosite

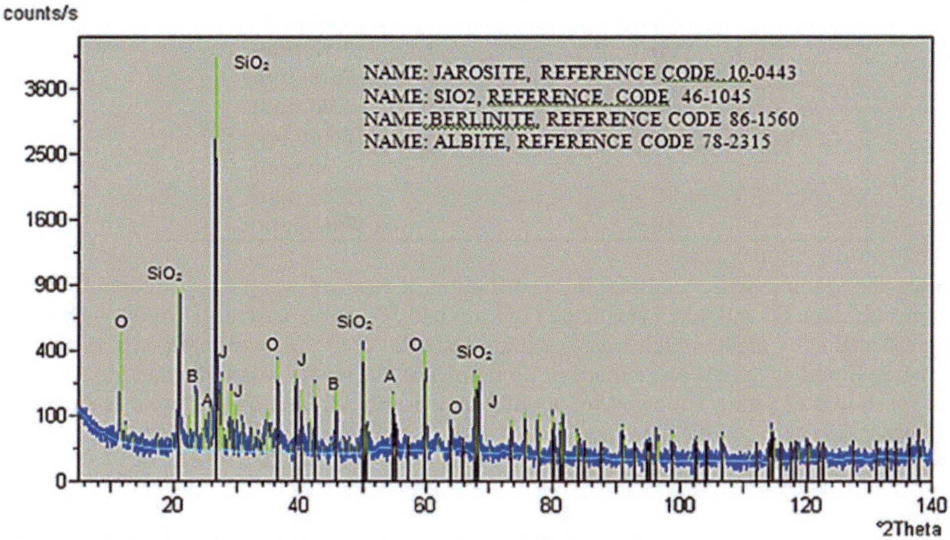

Figure 3. X Ray diffractogram of the Dos Carlos tailings, whose major mineral phases are silica (SiO_2), orthoclase (O), albite (A), berlinite (B) and jarosite (J).

The Dos Carlos tailings chemical characterization was carried out by taking a sample from each cardinal point. An average of the four points was obtained. The concentration amounts of each element in the tailings are shown in table I.

Table I. Characterization of the compounds contained in the Dos Carlos tailings.

Element/Compound	**Concentration**
SiO_2	**70.43 %**
Al_2O_3	**7.32 %**
K_2O	**0.08 %**
Na_2O	**2.32 %**
CaO	**0.69 %**
FeO	**2.41 %**
Fe_2O_3	**2.80 %**
MnO	**0.73 %**
MgO	**0.54%**
TiO_2	**0.53 %**
P_2O_5	**0.12 %**
Ag	**55 ppm**
Au	**0.58 ppm**
As	**21.60 ppm**
Ba	**658.00 ppm**
Be	**1.00 ppm**
Cd	**5.92 ppm**
Co	**8.80 ppm**
Cr	**68.40 ppm**
Cu	**88.10 ppm**
Mo	**9.50 ppm**
Ni	**64.20 ppm**

Pb	**420 ppm**
Sb	**0.55 ppm**
Sc	**7.00 ppm**
Sn	**4.86 ppm**
Sr	**106.40 ppm**
U	**0.76 ppm**
W	**5.85 ppm**
Zn	**900.00ppm**

Regarding the granulometric analysis, the distribution of retained weight was studied for the mesh sizes between 80 and 400 Tyler scale (177 μm and 37 μm respectively). Figure 4 shows the highest retained weight, which was found at mesh 100 (149 μm) and represents 19.18 % of all the analyzed material. Next there is a continuous decrease, which reaches its lowest peak at mesh 400 (37 μm), followed by another increase in retained weight at sizes smaller than 37 μm.

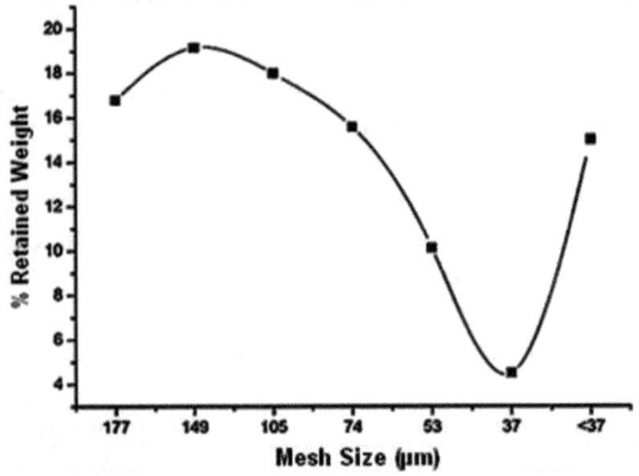

Fig. 4. Granulometric distribution. Dos Carlos tailings.

Later on, the ceramic composites were made from the previously stabilized tailings material (in the form of oxide or sulfide phases), and according to the granulometric analysis, the existing particle sizes and percentages were used. For the production of the alternative industrial materials, heavy clays were used as binding agents at a particle size of 200 mesh (74 μm) in order to give the materials similar properties to those of conventional processes.

Table II presents the composition of the different produced ceramic composites in weight percentage. In some cases two heavy clays were added, one of which has a higher plasticity. In the other cases only one heavy clay was added, except for the case where no extra materials were combined with the tailings.

Table II. Composition of the different ceramic composites used for the production of alternative industrial materials, weight percentage.

Ceramic composite	Tailings %	High plasticity clay %	Low plasticity clay%
1	66.67	26.67	6.67
2	63.00	31.00	6.00
3	63.00	31.00	6.00
4	60.00	34.00	6.00
5	60.00	40.00	---
6	60.00	40.00	---
7	60.00	40.00	---
8	55.00	45.00	---
9	50.15	49.85	---
10	45.30	54.70	---
11	40.45	59.55	---
12	35.60	64.40	---
13	100.00	---	---

Figure 5 shows the physical appearance of the mentioned composites; it can be affirmed that, at least regarding this aspect, different finishes and, above all, natural colors can be achieved. In the case of the last composite, a ceramic glaze finish was obtained without using enamels, ceramic fluxes or any other glaze-forming agent besides the own composition of the tailings.

Fig. 5. Ceramic composite samples for the production of alternative industrial materials.

Finally, the alternative industrial material was produced by using the different ceramic composites. It is worth mentioning that samples 8 to 12 are the most suitable ceramic composites for the manufacture of roof tiles, and, as observed in Fig. 6, the alternative industrial material was left with a natural rustic style.

Fig. 6. Tile made from ceramic composite number 9.

Conclusions

1. According to their chemical and mineralogical characterization, the Dos Carlos tailings are made up of 70.43% silica. The other compounds contained in the tailings are 7.32 % Al_2O_3, 2.32% Na_2O, 0.08% K_2O and 0.69% CaO. These compounds, together with a part of the percentage of the silica, form the major mineral phases orthoclase, albite and berlinite, as well as the aluminosilicate minor mineral phase anorthoclase. It is also worth mentioning that the tailings contain 2.14 % FeO and 2.8 % Fe_2O_3, which are compounds that form the phases hematite and magnetite.

2. According to the granulometric analysis of the Dos Carlos tailings, the highest retained weight corresponds to mesh 100 (149 μm); therefore the material contained in the tailings is suitable for the production of alternative industrial material with the use of heavy clays as binding agents.

3. 13 ceramic composites were made from homogenized tailings (obtained from the Dos Carlos mining dumps) and heavy clays. 5 ceramic compounds were sintered in the shape of roof tiles. It can be noticed that these ceramic composites have natural colors and a better particle binding when adding heavy clays than when using only the tailings material.

Acknowledgements

The authors would like to thank the FOMIX-Hidalgo for their support in this work as part of project 128491.

References

1. Carrera S., M. "Bartolomé de Medina y el beneficio de patio", Revista de la Cámara Minera de México, 1 (12) (1973), 12-31.
2. Hernández J., Patiño, F., Rivera, I., Salinas, E., Alamilla, M., Cerecedo, E., Reyes, M., Pérez, M., "Silver Dissolution From The Waste Mounds Of Pachuca, Hidalgo State, Mexico – Effect Of Cyanide Concentration On Dissolution Rate", European Metallurgical Conference EMC, (2009).

3. Patiño, F., Hernández , J., Rivera, I., Salinas, E., Longoria, L.C., Gonzalez, J.M. "Characterization and kinetics of the grinding of the Dos Carlos burrows in the state of Hidalgo", International Precious Metals Institute, (2007).

Engineering Solutions for Sustainability: Materials and Resources II
Edited by: Jeffrey W. Fergus, Brajendra Mishra, Dayan Anderson, Emily Allyn Sarver, and Neale R. Neelameggham
TMS (The Minerals, Metals & Materials Society), 2015

PREPARATION OF BLOCKS FROM TAILINGS

Javier Flores Badillo, Juan Hernández Ávila, Eleazar Salinas Rodríguez, Miguel Pérez Labra, Isauro Rivera Landero, Ister Mireles García and Eduardo Cerecedo Sáenz

Área Académica de Ciencias de la Tierra y Materiales. Universidad Autónoma del Estado de Hidalgo. Carretera Pachuca-Tulancingo Km. 4.5, C.P. 42184 Pachuca, Hidalgo, México. Tel. y Fax: (01 771)7172-000 ext. 2282; Fax: (01 71)7172-000 ext. 6730, Email: wik_1000@hotmail.com.

Keywords: Tailings, compression strength, block.

Abstract

This work is related with the preparation of block from mining wastes (tailings), since in the mining district of Pachuca – Real del Monte the presence of these residues is a factor considered of high pollution and this type of study allows mitigation of the problem and presents an economically feasible alternative for this purpose. The mineral phases existing in the material are quartz, anorthoclase, orthoclase, albite and berlinite; it was found that the tailings have a particle size of 60 % minor to mesh 270 (53 µm).

The blocks produced using these tailings as fine aggregated were reinforced using expanded perlite and pumucite as coarse aggregates and as cementing agents. The blocks so produced, presented at the end of set compression strengths equal or greater tan 8 MPa, and an absorption rate between 35 and 19 %; and the blocks can be used in internal or load walls.

Introduction

The Dos Carlos tailings, is one of the oldest tailings in the Mining District Pachuca-Real del Monte, establishing the tailings in 1912. Currently, Dos Carlos tailing has a corresponding amount of 14.3 million tons of material, [1]. This material has the following chemical composition; 70.43 % wt. of SiO_2, 7.32 % wt. of Al_2O_3, 2.32 % wt. of Na_2O, 0.08 % wt. of K_2O, 0.69 % wt. of CaO, 0.54 % wt. of MgO, 0.73 % wt. of MnO, 2.8 % wt. of Fe_2O_3, 2.41 % wt. of FeO, 0.53 % wt. of TiO_2, 0.12 % wt. of P_2O_5, 55 gram per ton of Ag and 0.58 grams per ton of Au, [2,3]. Showing a coarse granulometry of the order of 60 % accumulated to 200 mesh (53 µm), [2,3]. Within the mineral phases have been reported so far, oxides are present such as quartz, hematite, magnetite and pyrolusite; sulfates such as gypsum, jarosite, natrojarosite and hydronium jarosite; carbonates like calcite; phosphates like berlinite; silicates like ferrosilicite; aluminosilicates such as orthoclase, albite and anorthoclase, [3,4].

Moreover, studies have been performed using the material from tailings, to produce bricks or even like a pozzolanic material. Jaquays describes a procedure and a product by the flux of residues containing bauxite that are treated to neutralizing alkaline materials, and the resulting material is incorporated into a building material or other similar material to improve the properties, adding as binding material calcium silicates, [5]. Anicama Acosta *et al.* reported the formation of mixtures of cement with pozzolan added (2 – 25 % wt. of tailings), resulting that the best mixtures are in the range of 2 to 15 % of tailing and this does not significantly affect mechanical properties in cementing matrix, [6].

Note that to have a potentially puzzolanic material, according to Montaño Cisneros *et al.*, it is necessary a percentage in sum of SiO_2 and Al_2O_3 at least of 70%, [7]. Furthermore, with the sum of these two species, in the chemical composition of the material of Dos Carlos tailings, has a high pozzolanic material potential, which can be used to prepare block, with a fine aggregate, due to the grain brought.

Finally, define what is concrete and its components (cement, fine and coarse aggregates), because for purposes of this part of the investigation, will be a mix similar to concrete and it will be vibro compacted for forming building material shaped block, whereby the concrete is defined as a composite material consisting of cement, fine aggregate, coarse aggregate, water in proportion to corresponding mass 1:2:3:0.5.

According to ASTM C125 and the Portland Cement Association, cement is an inorganic material or a mixture of inorganic material which has mechanical strength through the chemical reaction with water by formation of hydrates, and which is capable of hardening by adding water, [8]. The most common hydraulic cement is Portland cement, which is mainly composed of calcium silicates ($3CaO\text{-}SiO_2$,$2CaO\text{-}SiO_2$), also having within their tricalcium aluminate composition ($3CaO\text{-}Al_2O_3$), aluminoferrite tetracalcium ($4CaO\text{-}Al_2O_3\text{-}Fe_2O_3$) and gypsum ($CaSO_4{\cdot}2H_2O$), [8,9,10].

Moreover aggregates are commonly used in a proportion of 60 to 80% of the total volume, [8,11], considering that the coarse aggregate is used to give strength to the matrix and fine aggregate used to fill the gaps created between coarse aggregate particles, having a good interfacial contact between aggregates [11]. The coarse aggregates are materials with particle size between 9.5 to 37.5 mm, and fine aggregates are materials with a particle size less than 9.5 mm, [8]. Among the materials used as aggregates are limestone, basalt, diabasite, granite, gravel, blast furnace slag, scrap iron, hematite, ilmenite, barite, pumicita, solidified lava and coke, [8].

Experimental Procedure

The general characterization of the material from Dos Carlos tailings was performed by X-ray diffraction (DRX); sample was sieved to a size between 74 and 53 microns (200 mesh series Tyler® and 270, respectively). Once sample was prepared to that size 1 gram of sample was weighted and packed in an Aluminium port samples for its analysis by XRD made with an Inel Equinox 2000 difractometer, with radiation of Cu-$K_{\alpha 1}$ (λ=1.540560 Å), to 30 kV, 20 mA and a scanning rate of 22 θ /min. The diffractogram obtained was subsequently indexed by the software MATCH! Version 1.10, for determining the mineral phases present.

Subsequently size analysis was conducted by wet sieving, using Tyler ® Series sieves with mesh sizes of 177, 149, 105, 74, 53, 37 μm, where 100 g of sample were sieved. Fractions obtained were dried at room temperature atmosphere and weighed.

For the production of block material from Dos Carlos tailings, a cementing agent (Portland Cement Type II) was used, coarse aggregates (pumicita and expanded perlite) and as fine aggregate material from Dos Carlos tailings just as coming from the tailings. With this, were subsequently developed three different mixtures for manufacturing block, to which before the vibro compaction were evaluated as follows: first the plasticity for each mixture, secondly the cost and finally the existence of reserves of material added to tailing as coarse aggregate (pumicita and expanded perlite), all above, in order to reduce costs, provide good plasticity, improve strength after curing mixture and use the maximum amount of material from the tailings.

Finally, the blocks produced were subjected to physical and mechanical tests such as sizing, absorption and resistance to simple compression, under the rule of standard SCT N-CMT-2-01-002/02, [12]. Likewise permeability and volumetric weight of each of the blocks produced was determined.

Results

Figure 1 shows the X-ray diffractogram of material from Dos Carlos tailings, where it is seen that there are majority phases such as quartz, albite, berlinite and potassium jarosite; Minor mineral phases, such as oxides (hematite, magnetite and pyrolusite), hydroxides (gibbsite) and sulfides (pyrite, chalcopyrite, galena, sphalerite and argentite) were found along with non metal mineral species in the form of carbonates (calcite), sulfides (anhydrite and gypsum) and aluminosilicates (anorthoclase).

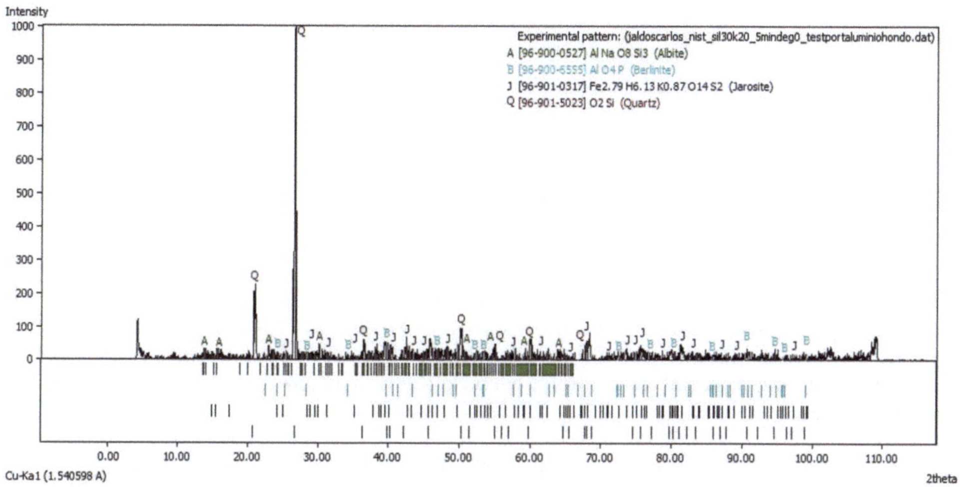

Figure 1. X-ray diffractogram of the material of Dos Carlos tailings, the main mineral species are: quartz (Q), albite (A), berlinite (B) and potassium jarosite (J). Obtained with a Cu Kα1 radiation length of 1.54056 Å, a voltage of 30 kV, a current of 20 mA and a scanning rate of 22°θ / min.

According to these results, we observed the presence of minerals from jarosite family such as potassium jarosite, which according to reports on mining district, these minerals come from mines El Cristo and Paricutin, [3,13]. Similarly, it is known that these jarositic minerals were leached during ore beneficiation and were present in Dos Carlos tailings. Also note that due to the chemical composition of the host rocks such as dacite, trachyte, andesite, basalt and rhyolite,

[14], so common in the rocks of the "Group Pachuca" and where economic mineralization are present in the Mining District Pachuca-Real del Monte, it has been found that these rocks are igneous type (volcanic and clastic vulcanoes), and a low-temperature hydrothermal epigenesis and shallow, also called epitermalism, [14], because the presence of quartz and feldspar minerals such as plagioclase (albite). In the case of minerals like anhydrite and gibbsite, it is suggested that in Dos Carlos tailings, these minerals formed due to weathering in situ. Gibbsite was formed by the weathering of plagioclase, and in the case of anhydrite was due to the chemical reaction between the CaO present in the ore as an additive to maintain the pH during the flotation and acid mine drainage from the hydrolysis of pyrite. Was also observed the presence of argentite, hematite and pyrite. These metallic phases hat could not be leached and left as a residue for subsequent extraction.

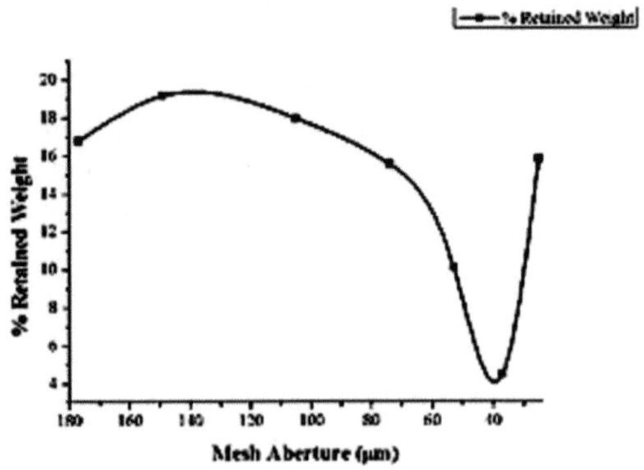

Figure 2. Particle size distribution of the material from Dos Carlos Tailings.

Moreover, it is observed in Figure 2 that the weight mainly is in retained particle size of 149 microns with 19.18%, then presents an abrupt decrease to 37 microns wherein the weight retained minimum of 4.50% is found, according to this, one can determine the possibility of manufacturing a building material, or ceramic, because the raw material presents optimal particle sizes within specifications.

In the case of the block prepared using tailings, 3 composition of cement, coarse aggregate, fine aggregate and water were prepared, using material from Dos Carlos tailings as fine aggregate and as coarse aggregate was used expanded perlite or pumicite, as shown in table I. For the preparation of ceramic composites, a percentage of tailings was used, enough to provide the highest strength after curing in the cementing agent.

Table I. Composition in weight percent, for developing block with material from Dos Carlos tailings.

Composite	Cement	Coarse aggregate	Fine aggregate (Tailing)	Water
1	8.32%	13.51% Perlite	58.21%	19.96%
2	6.49%	38.96% Pumicite	38.96%	15.58%
3	6.49%	46.75% Pumicite	31.17%	15.58%

Figure 3 shows the appearance of the blocks made with the proportions specified in Table I, where the fine aggregate represented by the material of Dos Carlos tailings is joined to the coarse aggregate with the dicalcium silicate, tricalcium silicate and tricalcium aluminate.

Figure 3. Blocks developed with material from Dos Carlos

Table II gives the physical properties obtained for each of the blocks made from material of Carlos Dos tailings. Also, the table shows the dimension and two of the physical properties stipulated in the standard SCT N-CMT-2-01-002/02, the absorption and compression resistance. It is noted that according to the standard, the three types of blocks could be used for internal, external, load or fill walls, taking into account, in the case of block 1, the need to be covered by a waterproof agent if it was used for external walls.

Table II. Physical properties of blocks made with material from Dos Carlos tailings and comparison with the standard

Standard SCT N-CMT-2-01-002/02							
Length (cm)	Width (cm)		Height (cm)	Absorption during 24 hrs. (%)		Compressive Strength (kg/cm^2)	
39.7-40.3	13.8-14.2		11.8-12.2	17-23.5		56.0	
Composites with tailings							
Sample	Length (cm)	Width (cm)	Height (cm)	Volumetric weight (Kg/m^3)	Absorption during 24 hrs. (%)	Permeability (cm^3/s)	Compressive Strength (kg/cm^2)
1	40.5	14.0	12.5	1341.15	35.83	0.048	76.54
2	40.6	12.2	16.0	1622.46	23.47	0.027	81.16
3	40.0	12.0	16.0	1236.59	19.42	0.023	82.5

Conclusions

It is concluded that according to the characterization performed by XRD of material from Dos Carlos tailings, were found as majority phases the following: quartz, albite, berlinite and potassium jarosite. In addition to these mineral phases are present in lower concentrations in the material the following phases: anorthoclase, calcite, anhydrite, gypsum, gibbsite, pyrite, argentite, magnetite, hematite and pyrolusite. Of these, we found that the phases that generate high pozzolanic power and high strength are the majority phases and also the anorthoclase.

In regards to the development of building material in the form of block, according to the standard SCT N-CMT-2-01-002/02, analysis of their properties determined more than stipulated strength. Furthermore, for the case of blocks made with coarse aggregate of pumicita, a value similar to or less than that established by the standard absorption was reached. Hence, these blocks can be used on internal or external walls and filler. Although in the case of the block made by using expanded perlite, which may also have the same use, a waterproofing agent, due to their higher percentage of absorption is necessary.

References

1. CMRMyP, "Resúmen General de presas de Jales," CMRMyP, Pachuca de Soto, (2005).

2. J. Hernández Ávila, E. Salinas Rodríguez , F. Patiño Cardona, I. Rivera Landero, J. Flores Badillo, N. Y. Trápala Pineda, M. Pérez Labra, M. U. Flores Guerrero y I. A. Reyes Domínguez, "Tile Production using Wastes from Mining Industry of the Mining District Pachuca Real del Monte," The Mineral, Metals & Materials Society, (2012), pp. 203-209.

3. J. Hernández Ávila, Cinética de molienda y recuperación de Ag mediante procesos convencionales y no convencionales de las escombreras de la industria minero-metalúrgica del Estado de Hidalgo, Pachuca de Soto, Hidalgo: (Universidad Autónoma del Estado de Hidalgo, 2009).

4. D. E. Ángeles Chávez, L. Lizárraga Mendiola y A. Blanco Piñón, "Análisis Geoquímico-Mineralógico del depósito de jal Dos Carlos, Distrito Minero Pachuca-Real del Monte, Hidalgo," *Actas INAGEQ*, 17, (2011), p. 146.

5. C. D. Jaquays, "Building and other materials containig treadted bauxite tailings and process for making same". US Patent US20050087107A1, 28th. April 2005.

6. G. A. Anicama Acosta, Estudio experimentaldel empleo de materiales de desecho de procesos mineros en aplicaciones prácticas con productos cementicios, Lima, Perú: (Pontificia Universidad Católica del Perú, 2010).

7. E. Montaño Cisneros, J. Robles Camacho, P. Corona Chávez, M. Martínez Medina y Y. R. Ramos Arroyo, "Caracterización mineral y geoquímica de los jales del Distrito Minero El Oro-Tlalpujahua. Reutilización potencial de los desechos mineros," *Memorias del 3er. foro de Ingeniería e Investigación en Materiales*, 3, Morelia, Michoacán: (Universidad Michoacana de San Nicolás de Hidalgo, 2006), pp. 198-203.

8. F. Cardarelli, "Cements, Concrete, Building Stones nd Construction Materials," in *Materials Handbook: A concise desktop reference*, 2nd. ed., London, UK: (Springer, 2008), pp. 967-981.

9. M. Velez, "Concrete and Cement,"*Ceramic and Glass Materials: Structure, Properties and Processing*, New York, USA: (Springer, 2008), pp. 135-151.

10. D. R. Askeland, P. P. Fulay y W. J. Wright, Ciencia e ingeniería de materiales, 6a. Edición ed., México: (CENGAGE Learning, 2011), pp. 705-709.

11. W. D. Callister, Materials Science and Engineering: An Introduction, 7th. ed., New York, USA: (John Wiley & Sons, Inc., 2007), p. 583.

12. SCT, "Standard SCT N-CMT-2-01-002/02". Mexico: Diario Oficial de la Federación (2012).

13. A. R. Geyne, C. J. Fries, K. Segerstrom, R. Black y R. F. Wilson, Geology and mineral deposits of the Pachuca-Real del Monte District, State of Hidalgo, México: (CRM, 1963).

14. CRM, Monografía Geológico-Minera del Estado de Hidalgo, México: (CRM, 1992).

Engineering Solutions for Sustainability: Materials and Resources II
Edited by: Jeffrey W. Fergus, Brajendra Mishra, Dayan Anderson, Emily Allyn Sarver, and Neale R. Neelameggham
TMS (The Minerals, Metals & Materials Society), 2015

INFLUENCE OF THE CONTENT OF DIMENSION STONES SOLID WASTE IN THE PHYSICAL AND MECHANICAL BEHAVIOR OF STRUCTURAL CERAMICS

Alessandra Savazzini dos Reis[1,4], Danilo Marin Fermino[2], Viviana Possamai Della-Sagrillo[3], Francisco Rolando Valenzuela-Diaz[4]

[1]IFES Campus Colatina; Av. Arino Gomes Leal, 1700; Colatina, Espírito Santo,CEP 29700-660, Brasil
[3]IFES Campus Vitória; Av. Vitória, 1729; Vitória, Espírito Santo, CEP 29040-780, Brasil
[2,4]PMT-USP; Av. Prof. Mello Moraes, 2463; Cidade Universitária, São Paulo, CEP 05508-030, Brasil

Keywords: Dimension stones, Structural ceramic, Ceramic body, Waste.

Abstract

The use of wastes in ceramic body is a common practice in structural ceramic industries in Brazil. However, this use is not always followed by the study of the physical and mechanical behavior of the ceramic body. Our aim is to study the behavior of ceramic bodies with the addition of dimension stone waste and to verify the influence of chemical, mineralogical and microstructural composition of the materials contained in ceramic bodies. Samples of clay used in the structural ceramic industry, as well as waste, were obtained from companies in Espírito Santo-Brazil. The materials were characterized, and clayey mixtures with waste of the contents ranging from 10% to 90% (weight) were prepared, burned in industrial furnace and tested for water absorption, apparent porosity and mechanical strength. The behavior of ceramic bodies was found to be influenced by the characteristics of the constituent materials and varies in relation to the content of waste used.

Introduction

Engineering has faced a major challenge all over the word related to waste reuse, in particularly that generated in large-scale industrial processes. Although there are many scientific papers that seek the technical feasibility of using waste, technological innovation has not really happened.
Sustainable development seeks to meet the needs of current and future generations without compromising the possibility of future generations to meet their needs [1]; we see that most of the world production models and present societies follow ways of production and patterns of high consumption that do not abide by the concept of sustainable development.
One of the alternatives for achieving sustainable development is to take advantage of the waste generated in industrial processes to meet technical, financial and environmental feasibility. This procedure may reduce consumption of non-renewable raw materials and the volume of waste disposed in landfills. In this sense, the aim of this research is to study the influence of the addition of dimension stones solid waste in the physical and mechanical behavior of the clayey mass used in a structural ceramics factory located in the state of Espírito Santo – Brazil.
The Brazilian industry for extracting and processing dimension stones has greatly developed. Brazil is among the major global producers of dimension stones, such as China, India and Turkey; and ranks 7th in stone exports [2]. The Brazilian production of dimension stones in 2013 was 10.5 million tons and the quantity exported was 2.7 million tons, 47% of the stones already

processed. More than 60% of the Brazilian production of dimension stones is located in the southeast region, mainly in the states of Espírito Santo (ES) and Minas Gerais (MG) [3] (Figure 1).

Figure 1. Location of ES and MG in the map of Brazil

When cutting blocks of stone into plates, waste is generated, usually consisting of stone powder, steel shot, lime and water. In many cases, the waste in the form of mud, is deposited into sedimentation ponds directly on the ground (Figure 2) or watertight tanks. The mud is air-dried or loses moisture in a filter press, and is then disposed in landfills, at considerable costs for both transportation and disposal; however, there are cases in which the waste is dumped directly into waterways [4] or the soil [5].

Figure 2. Sedimentation pond

After drying, the waste becomes a very fine powder, which can contaminate soil and water; besides being hazardous to health, according to Moreira et al. [6]. In Brazil, the amount of solid waste generated when processing dimension stones in 2013 was 3.7 million tons [3]. In this context, not only the contribution of the stone sector to the Brazilian economy can be confirmed, but also the large amount of waste generated and the damage it may cause. These facts make the studies about reusing waste and about the need of technological innovation for use in industrial processes relevant.

The use of the waste was studied by Menezes et al. [7] in the production of ceramic blocks: mixtures with up to 55% (by mass) of waste met the limit value of water absorption and mechanical strength.

In Silva et al. [8], the waste was used from 10% to 50% (by mass) in clayey mass: the mechanical strength decreased with the increase of the amount of waste, and higher mechanical strength was obtained with 30% waste, the same results found by other authors [9-11].

Manhães, Moreira and Hollanda [12] analyzed the microstructure of clayey mass with granite waste levels in 10% and 20% (by mass). The waste content of up to 20% contributed to greater densification and reduced defects in the microstructure at temperatures above 950ºC. These results corroborate the research by Souza, Pinheiro and Holanda [5], in which the same results were obtained up to 30% of waste.
Based on studies, the waste can be said to contribute to increasing the densification of the clayey mass and to improving the mechanical strength in contents from 30% to 55% (by mass) at temperatures above 950ºC. The paper in question analyzed the influence of the waste content of 10% to 90% (by weight) at a firing temperature of 900ºC used in the structural ceramics industry.

Materials and methods

Samples of clay and dimension stones solid waste were collected in the storage of the structural ceramic industry located in the State of Espírito Santo. After that, the samples were air-dried, broken with the aid of a pounder and homogenized. Part of the material was used at characterization tests, and part was used for preparing the specimens.
The clay and dimension stones solid waste were characterized by X-ray fluorescence (XRF) in Philips PW2400 spectrometer; X-ray diffraction (XRD) with copper source (K α radiation) in Philips X- Pert MPD equipment and scanning electron microscopy (SEM) in Philips XL-30 electron microscope; differential scanning calorimetry (DSC) and thermogravimetric analysis (TG) with a heating rate of 10°C/min in air atmosphere at Shimadzu DTA-50.
The identification of the samples analyzed in XRD was made by comparison with the Crystallographic Open Database – COD standard files.
Molded (spheres) and pressed (bars) specimens were made for the characterization tests after firing.
The bars were made with a mixture of clay and waste (sieved in # 80 of ABNT - Associação Brasileira de Normas Técnicas), measuring 2 cm x 6 cm x 0.5 cm of the mixtures of clay and waste with 10 g dry mass, 10% moisture content, by axial pressing at 250 $kgfcm^{-2}$.
The spheres were made with a mixture of clay and waste (sieved in # 40 of ABNT - Associação Brasileira de Normas Técnicas). The specimens were manually molded with an average diameter of 2 cm, 10 g dry mass, with enough water to allow molding.
The waste content in the clayey mass varied from 0% to 100% (in weight); and twelve bars were molded for each content.
The bars and spheres were dried in a drying oven for 24 hours at 110°C. Mechanical strength was measured in part of the specimens and part of the specimens was burned in an industrial tunnel-type furnace in a structural ceramic industry for 24 hours at a maximum temperature of 900°C. After firing, the following features were evaluated: water absorption, apparent porosity, apparent specific mass, loss on ignition and flexural strength according to the methodology proposed by Souza Santos [13], besides compressive strength. Some spheres and bars burned are shown in Figures 3a and 3b.

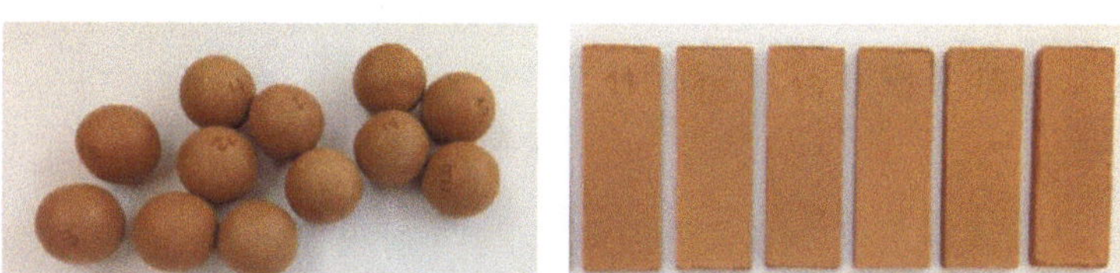

(a) (b)

Figure 3. (a) Spheres; (b) Bars

Results and discussion

The chemical analysis result (Table 1) is shown in terms of percentage through the weight of constituent oxides and loss on ignition. By analyzing the data, the clay was verified to present a typical chemical composition of clayey raw material, with 43.39% silica, 30.16% alumina; and 13.17% loss on ignition obeying ranges of values from 43.20% to 77.60% for SiO_2, from 6.80% to 38.00% for Al_2O_3 and from 6.00% to 15.70% for the loss on ignition, respectively, as specified by Souza Santos [13] for ceramic raw materials.
The percentage of 9.29% iron oxide is responsible for the reddish color after firing. The concentration of Na_2O and K_2O is not high; therefore, even being fluxing oxides, they contribute little to the formation of the glassy phase. Clays with these characteristics (aluminous silica, with high iron content and low content of fluxing oxides) observed in XRD demand high sintering temperature, as reported by Hildebrand et al [14].
As seen in the table, the granite waste presents a high silica content (62.88%) and alumina (14.11%); and low value of loss on ignition (2.97%), compared to clay. The presence of fluxing alkali oxides Na_2O (2.54%) and K_2O (4.36%) may contribute to burning the ceramic bodies, reducing the need of high temperature in the formation of the glassy phase of the clay in question. The presence of Fe_2O_3 is associated with steel shot, and CaO is associated with lime, both used in the mud from which the waste originates.

Table 1. Chemical composition by XRD (wt%)

	SiO_2	Al_2O_3	Fe_2O_3	TiO_2	K_2O	MgO	P_2O_5	Na_2O	CaO	BaO	LOI
Clay	43.39	30.16	9.29	1.39	1.25	0.62	0.12	0.17	0.24	-	13.17
Waste	62.88	14.11	5.72	0.67	4.36	1.02	0.38	2.54	4.42	0.20	2.97

Note: LOI – loss on ignition

Figure 4 shows the X-ray diffraction of the clay and the waste. The XRD analysis of clay indicates that the predominant minerals are kaolinite, quartz, illite and goethite.
The analysis of the X-ray diffractogram of the waste identifies the presence of crystalline phases associated to mica, feldspar and quartz; the latter confirms the high value of SiO_2 (62.88%) in the XRF detected. Calcite and orthoclase were also detected.

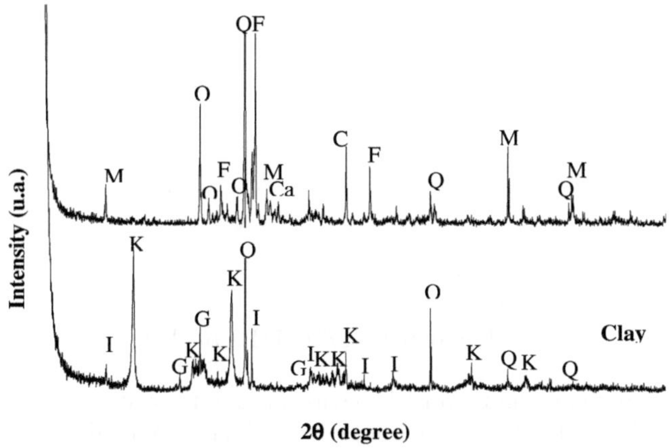

Ca-calcite; F-feldspar; G-goethite; I-illite;
K-kaulinite; M-mica; O-orthoclase; Q-quartz

Figure 4. XRD of materials

The electron microscopy result indicates that the clay (Figure 5a) and the granite waste (Figure 5b) show an irregular shape tending to form lamellar and slender particles, some scattered and others in clumps with sizes ranging from 2 μm to 46 μm.

(a) (b)

Figure 5 – (a) Clay; (b) Granite waste

Figure 6 shows the thermal behavior of the clay. At 110°C, no formation of the endothermic peak occurs because there is no water among the structural layers, which may indicate that the clay is composed of well crystallized kaolinite, according to Souza Santos [13]. An endothermic peak occurs between 200ºC and 300ºC, which probably corresponds to the loss of organic matter. Another endothermic peak of higher intensity occurs at 500°C, characteristic of hydroxyl loss in the kaolinite and illite clay minerals. The thermal event associated with the vitrification that would begin near 900°C has not been detected.

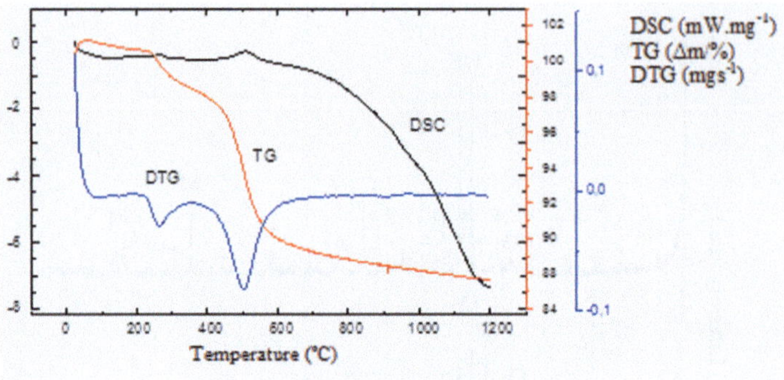

Figure 6. DSC/TG/DTG of clay

In Figure 7, which presents the thermal behavior of granite waste, it is possible to observe an endothermic peak at about 700°C, which may be attributed to the decomposition of carbonate, such as calcite. The exothermic peak above 800°C was not observed, which would represent the formation of a new crystalline phase (silicate).

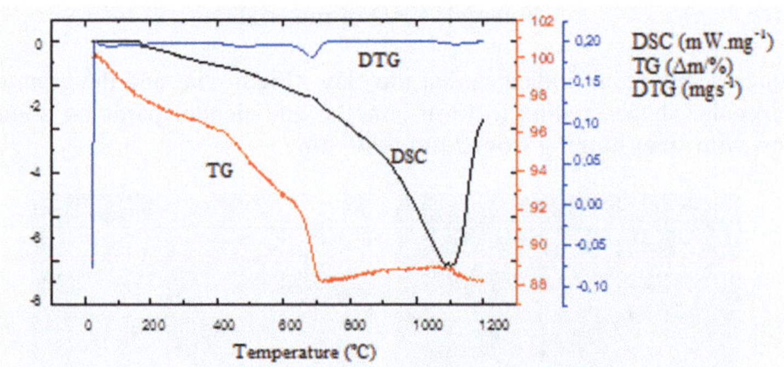

Figure 7. DSC/TG/DTG of granite waste

The results from the spheres made with clay and granite waste are shown in Table 2. The highest value of flexural strength (1.92 MPa) was obtained with the 10% content of waste, which resulted in improvement of this property, since the clay without waste presented a lower value (1.31 MPa). The highest compressive strength value (0.72 MPa) in dry spheres occurred for 20% of waste content.

The values of the compressive strength, loss on ignition and volumetric shrinkage decreased by increasing the waste content.

A gradual reduction is observed in the water absorption and apparent porosity values occurring with the increasing content of waste, obtaining minimum values of these properties at 70% waste content, at which the maximum apparent specific mass (1.56 gcm^{-3}) was also obtained, indicating that this waste content has the highest densification of the clayey mass.

Table 2. Average values of the ceramic properties of spheres

Waste (%)	LOI (%)	WA (%)	AP (%)	ASM (gcm^{-3})	VS (%)	σ_D (MPa)	σ_B (MPa)
0	11.86	25.21	37.41	1.52	3.80	0.62	1.31
10	11.55	24.54	36.19	1.52	4.13	0.55	1.92
20	9.81	23.68	35.94	1.52	3.94	0,72	1.68
30	8.01	23.56	35.89	1.52	3.92	0,43	1.49
40	7.68	23.37	35.59	1.53	3.25	0,39	1.04
50	6.93	23.32	35.46	1.53	3.23	0,33	0.77
60	6.00	23.29	35.33	1.53	1.85	0,24	0.51
70	4.93	21.61	33.62	1.56	1.81	0,20	0.43
80	4.58	22.25	34.42	1.55	1.06	0,16	0.41
90	3.57	24.12	35.34	1.54	0.85	0,15	0.22

Note: LOI= loss on ignition; WA= water absorption; AP= apparent porosity; ASM= apparent specific mass; VS=volumetric shrinkage after burned; σ_D=compression strength after drying; σ_B= compression strength after burning.

In the bars made with clay and granite waste, the results obtained are shown in Table 3, which refer to the limit values prescribed in Souza Santos [13]. Higher values of flexural strength were obtained with 10% residue in dry bars (1.44 MPa) and in the burned bars (3.33 MPa); the latter corroborates the results found in the spheres.

The water absorption and apparent porosity values were also observed to decrease with the increase in waste content, and the minimum value was achieved for 60% of the waste content. From this point onwards, they started to increase with the increase in waste content, showing that as from this content, the waste no longer contributes to the densification of the ceramic body. In the of 60% content, the maximum apparent specific mass was obtained as an consequence of minimum absorption and porosity values measured. The flexural strength, loss on ignition and linear shrinkage values decreased with the increase the waste content. These results also occurred in the spheres. Reis, Della-Sagrillo and Valenzuela-Diaz [15], also allow observing that the spheres and bars, burned and made with clay, showed similar values in the ceramic properties evaluated.

Considering the referential limits described in Table 3, the water absorption and apparent specific mass values met the limits at all the waste contents. The flexural strength in dry bars met the minimum limit in up to 70% waste. On the other hand, the flexural strength in burned bars did not meet the limits for any content.

Table 3. Average values of the ceramic properties of bars

Waste (%)	LOI (%)	WA (%)	AP (%)	ASM (gcm^{-3})	LS (%)	σ_D (MPa)	σ_B (MPa)
0	12.16	24.17	39.33	1.63	1.74	1.36	3.71
10	11.19	23.15	38.85	1.68	1.33	1.44	3.33
20	10.02	23.11	38.40	1.66	1.32	1.34	3.25
30	9.27	23.09	38.28	1.66	0.99	1.29	2.49
40	8.57	22.63	38.13	1.68	0.78	1.26	2.07
50	7.63	22.44	37.76	1.68	0.38	1.25	1.98
60	6.62	21.90	37.23	1.70	0.32	1.01	1.75
70	5.57	24.50	39.71	1.62	0.21	0.56	0.73
80	4.18	24.57	39.71	1.62	0.17	0.40	0.57
90	3.65	24.05	39.21	1.63	0.02	0.46	0.87
Reference**	-	≤ 25.00	≤ 35.00	1.50 a 2.00	2 a 17*	0.50 a 7.00	≥ 6.00

Note: LOI= loss on ignition; WA= water absorption; AP= apparent porosity; ASM= apparent specific mass; LS= linear shrinkage after burned; σ_D= flexural strength after drying; σ_B= flexural strength after burning.
Note:*values for kaolinitic clay; **Souza Santos [13].

Figure 8 was generated from the data of in Tables 2 and 3, showing similar flexural strength behavior of bars and compressive strength of spheres, both with the addition of granite waste. The 10% waste content was confirmed obtain the highest mechanical strength in both types of specimens.

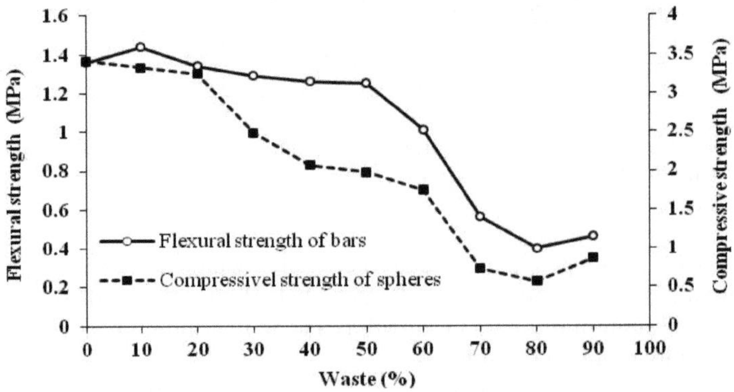

Figure 8. Flexural strength of bars (MPa) and compressive strength of spheres (MPa) x waste (%)

Conclusions

Chemical and mineralogical characterization showed that clay is predominantly kaolinite, containing adequate amounts of quartz and iron. The clay and the granite waste may be considered silica aluminous materials; the waste also contains fluxing oxides that can assist in the process of the clayey mass burning at a temperature of 900°C. The behavior of the properties evaluated [both for spheres and bars] is generally equally influenced by the presence of waste.
In the bars, the values for flexural strength after drying (up to 70% of waste), water absorption and apparent specific mass meet the requirements of the Brazilian standard clays for structural ceramics found in the literature, while apparent porosity and flexural strength after burning does not meet these limits.
It can thus be concluded that the granite waste especially influences the formation of the glassy phase of the ceramic body, reducing water absorption, obtaining a greater densification at 60% of waste in bars and 70% in the spheres, including the comparison of the values obtained from clay without waste. However, the best results in flexural and compressive strength were obtained for burned specimens at 10% residue.
The reuse of granite waste in the clayey mass can contribute to the sustainability of dimension stones and ceramics sectors, reducing raw materials consumption and avoiding waste disposal in landfills.

Acknowledgment

The authors thank IFES, CAPES, FINEP and USP/PMT.

References

[1] SIMON, C.; DEFRIES, R. S. Uma terra, um futuro. Tradução: Maria Claudia Santos Ribeiro Ratto. São Paulo: Makron Books, 189p., 1992.

[2] ASSOCIAÇÃO BRASILEIRA DA INDÚSTRIA DE ROCHAS ORNAMENTAIS (ABIROCHAS). APEX e ABIROCHAS: uma parceria de sucesso. Informe 7/2013. São Paulo. 2013. Disponível em: <http://www.ivolution.com.br/news/upload_pdf/11645/ Informe_07_2013.pdf>. Acesso em: 04 jun. 2013.

[3] ______. Balanço das exportações e importações brasileiras de rochas ornamentais em 2013. Informe 01/2014. São Paulo. 2014. Disponível em: <http://www.ivolution. com.br/mais/fotos/ 6/17/3050/Informe_01_2014.pdf>. Acesso em: 15 ago. 2014.

[4] PONTES, I.; STELLIN JR, A. Utilização de finos de serrarias do Espírito Santo nas Indústrias de construção civil. In: JORNADAS IBERMAC "Caracterización y Normalización de Materiales de Construcción". Programa CYTED, 2001, Madri/Espanha. 2001.

[5] SOUZA, A.J; PINHEIRO, B.C.A; HOLANDA, J.N.F. Efeito da adição de resíduo de rocha ornamental nas propriedades tecnológicas e microestrutura de piso cerâmico vitrificado. Cerâmica, v.57, São Paulo, 2011, p. 212-218.

[6] MOREIRA, J. M. S.; MANHÃES, J. P. V. T.; HOLANDA, J. N. F. Reaproveitamento de resíduo de rocha ornamental proveniente do Noroeste Fluminense em cerâmica vermelha. (Utilization of ornamental rock waste from Northwest Fluminense in red ceramic). Revista Cerâmica. v.51, n.319, p.180-186, São Paulo, Jul./Set. 2005.

[7] MENEZES, R. R.; FERREIRA, H. S.; NEVES, G. A.; LIRA, H. L.; FERREIRA, H. C..Use of granite sawing wastes in the production of ceramic bricks and tiles. Journal of the European Ceramic Society, n. 25, p.1149–1158, 2005.

[8] SILVA, J. B.; HOTZA, D.; SEGADÃES A. M.; ACCHAR, W. Incorporação de lama de mármore e granito em massas argilosas (Incorporation of marble and granite sludge in clay materials). Revista Cerâmica. v. 51, p.325-330, São Paulo, 2005.

[9] SEGADÃES, A. M.; CARVALHO, M. A.; ACCHAR, W. Using marble and granite rejects to enhance the processing of clay products. Applied Clay Science, 2005, 11p.

[10] MOTHÉ FILHO, H. F. Reciclagem: o caso do resíduo sólido das rochas ornamentais. Revista Rochas de Qualidade. São Paulo. Ano XL. Edição 192, p.189-195, jan./fev. 2007.

[11] FERREIRA, L. C. Potencial de utilização de resíduos industriais na formulação de massa de cerâmica vermelha para a fabricação de blocos de vedação. 2012. 76f. Dissertação (Mestrado Urbano e Industrial) - Universidade Federal do Paraná. Curitiba/Paraná. 2012.

[12] MANHÃES, J.P.V.T., MOREIRA, J.M.S., HOLANDA, J.N.F. Variação microestrutural de cerâmica vermelha incorporada com resíduo de rocha ornamental. Revista Cerâmica. v. 55, São Paulo, 2009, p. 371-378.

[13] SOUZA SANTOS, P. Tecnologia das argilas. v.1-2. SP: Edgard Blucher, 1975.

[14] HILDEBRANDO, E.A.; SOUZA, J.A.S.; ANGELICA, R.S.; NEVES, R.F. Application of bauxite waste from Amazon region in the heavy clay industry. *Materials Research.* 2013; v.16, n.6: 1418-1422.

[15] REIS, A.S; DELLA-SAGRILO, V.P.;OLIVEIRA, J.N.; VALENZUELA-DIAZ, F.R. Caracterização de argila usada em massa de cerâmica vermelha. In: 58° Congresso Brasileiro de Cerâmica, Bento Gonçalves, RS, 2014. Anais... RS, 2014, p.118-129.

III

Materials Selection

Engineering Solutions for Sustainability: Materials and Resources II
Edited by: Jeffrey W. Fergus, Brajendra Mishra, Dayan Anderson, Emily Allyn Sarver, and Neale R. Neelameggham
TMS (The Minerals, Metals & Materials Society), 2015

AN INTERACTIVE AND VISUAL TOOL FOR SUSTAINABLE USE OF MATERIALS IN ENGINEERING DESIGN

Claes Fredriksson[1]

[1]Granta Design Ltd; 62 Clifton Rd, Cambridge CB1 7EG, UK

Keywords: Materials, Properties, Database, Selection, Eco, Carbon Footprint

Abstract

Sustainability in engineering is an abstract and complex issue entailing both material and energy flows, alongside societal and economic aspects. We describe how a software resource with a large dataset of material properties can be used in engineering teaching and industry to enable a rational and well informed response to eco challenges. The CES EduPack and its relative, CES Selector, can aid the selection of materials from comprehensive databases based, *e.g.*, on mechanical performance manufacturability cost and eco properties, such as embodied energy or carbon footprint, using interactive visualization and selection tools. Life-cycle properties can be estimated already at the design stage using a so called Eco Audit which also delivers guidelines for improved environmental performance. Case studies and examples demonstrate the technical utility. The Sustainability database of CES EduPack is used to show how to support social and resource aspects in teaching and in Corporate Social Responsibility reporting.

1 Introduction and Background to the Software

The global engineering community has a key role to play in the essential and inevitable transformation into a more sustainable way of living. The latest IPCC-report reinforces the empirical evidence for the anthropogenic contributions to global warming [1]. These arise mainly from fossil energy use and a significant part can be attributed to material production and utilization. Embodied energy in materials results from extraction of ores and minerals, as well as from processes associated with preparing these materials. Prime examples are steel making or cement/concrete production, where CO_2-emissions are inherent to the processes. Moreover, most industrial polymers are produced literally from fossil fuels, such as oil or natural gas.

Materials-related data, such as the *Annual world production*, illustrates the enormous quantities of materials used for construction, industrial purposes and consumer products, see Figure 1. Charts, such as this one are a visual starting point for discussions in the engineering community around sustainability issues. This is also a good starting point in Engineering and Design Education for well-informed and fact-based teaching of such topics.

The bar chart is based on data from CES EduPack (hereafter referred to as *the software*). This commercially available software [2] is used to support teaching in Engineering, Materials Science or Design, and was originally developed at the Engineering Department of Cambridge University (UK) for education. It has evolved over the past 20 years and is now part of a family of tools used for materials-related University teaching as well as for applications in industry and research (CES Selector and Granta MI) [2]. This platform therefore serves as a bridge linking Students and Academics with industrial R&D. Here, we mainly consider the *Sustainable Development Edition* of CES EduPack, which has tools and data particularly suited for the topic of this paper. Many of the features are, however, also available in the Standard or other specialized editions (Aerospace, Polymer *etc.*) and in CES Selector.

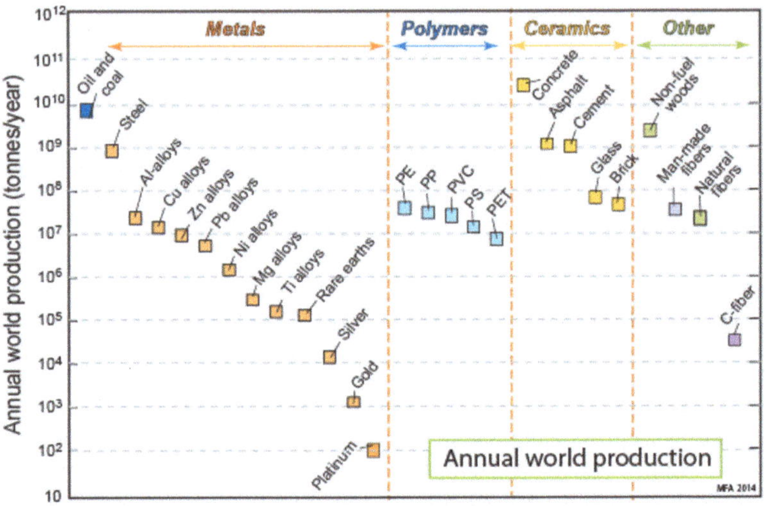

Figure 1. Annual World Production of some relevant engineering materials from CES EduPack.

The software is built around databases that contain information on material properties and manufacturing processes useful to several areas of engineering, see Table I. It also has interactive visualization of material-, process- and environmental (Eco) properties in charts, which are used to facilitate understanding in the educational context, as well as selection, decision-making and visual communication. An important feature is the built-in *Eco Audit* tool, which is used to analyze *Carbon footprints*, *Embodied energies* and recently also *Cost* information, already during the design process. This is particularly useful in product development, re-design or life-cycle engineering.

Table I. Summary of properties in the record for Aluminum 7075 T6 at Level3 of CES EduPack

Property Types	**Number of Attributes**	**Examples of Data (some have discreet values)**
General info, Composition overview, Price and Physical properties	11 + Composition	UNS number, Typical uses, % of components, estimated price per kg and Density
Mechanical Properties	16 + 4 diagrams	Poisson's ratio, Strengths *vs* temp
Impact and Fracture Properties	1	Fracture toughness
Thermal Properties	7	Melting point, Min/max service temperature, Thermal conductivity
Electrical, Optical, Magnetic Properties	4	Resistivity, Galvanic potential, Magnetic type, Transparency
Processing Properties	5	Metal casting, deep drawing etc.
Durability	10	Resistance to acids, water etc.
Material Eco properties	3	Embodied Energy, CO_2-Footprint, and Water use
Processing Eco properties	23	Energy, CO_2-Footprint, and Water use for various processes
Recycling and End of life	8	Recyclability, Bio-degradability

As can be seen from the example in Table I, around half of the attributes in the data record have relevance to sustainability (Durability, Eco properties and Recycling). Each material data record is also linked to other parts of the database that give further information, such as data on the Elements included in the material composition. In addition to traditional *Periodic table* data, there is Geo-economic data containing the *Annual world production* for the Element, the *Abundance* in the earth's crust or in seawater, the estimated global *Reserves* and the *Main mining areas* (Countries) in the world for that Element. These links will be further discussed in the next section.

The standard set of educational databases has three "Levels" – the database described in this paper is specialised for exploring Sustainability and is at Level 3, the advanced level. This database contains nearly 4000 materials from the main *material families*, including metal alloys, polymer blends, hybrids and composites, as well as some 240 manufacturing processes, which enables realistic projects to be carried out. Approximate cost information and Eco properties are included to facilitate comparisons and qualitative discussions in classrooms and around product development projects, for instance. To facilitate self-learning, Science Notes, which are interactive *on-demand* features, are available for every property included in the database. These give definitions and background to properties. In addition to Science Notes, there are also generic (folder level) records that give supporting information on the contents of the data record folders. An example for *Wrought aluminum alloys* is shown in Figure 2, together with a Science Note on the property *Abundance in Earth's crust and oceans*.

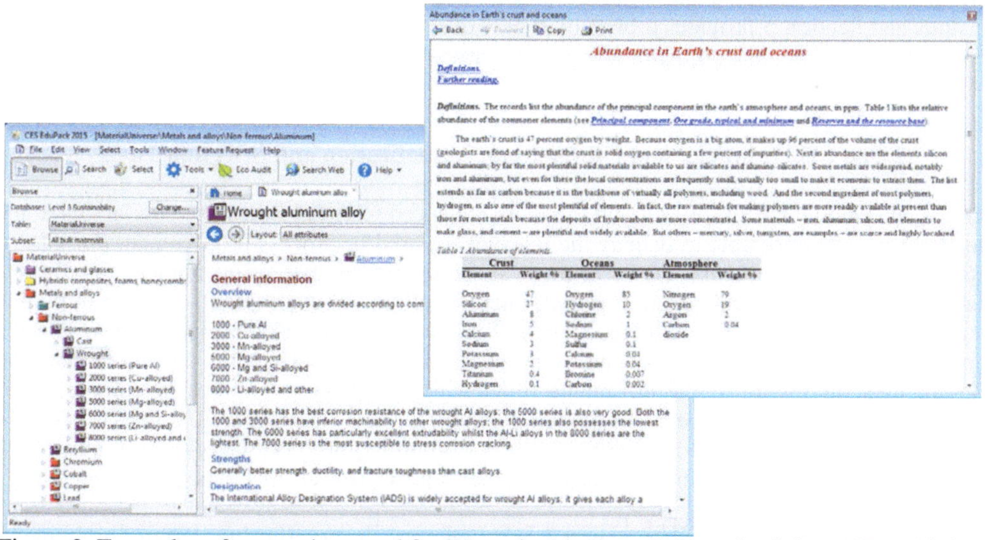

Figure 2. Examples of a generic record for Wrought Al alloys (left) and a Science Note (right).

This was a brief introduction to the relevant software content. The database structure of the sustainability database is schematically shown in Figure 3, below. It shows the individual data tables and how they are linked to each other for easy access to information and for coupled selection or analysis using criteria from a combination of these.

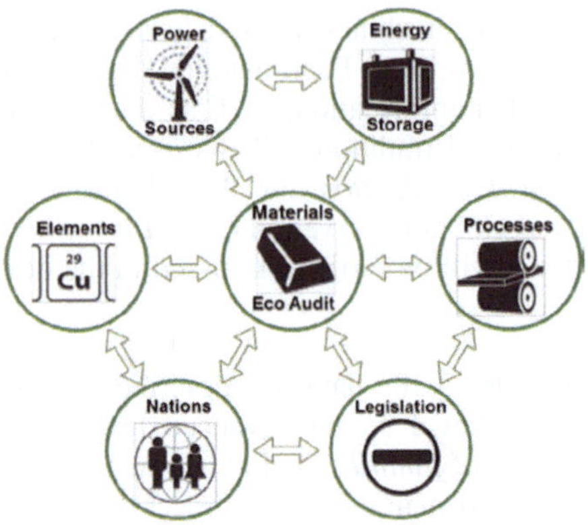

Linked Data Tables:

MaterialUniverse

ProcessUniverse

Legislation and Regulations

Power Systems (Storage, Sources)

Nations of the World

Elements (Reserves, Criticality etc.)

Manufacturing Processes Eco data

Figure 3. Structure of the sustainability database for CES EduPack and names of data tables.

This combination of traditional material properties, Eco properties as well as fact based sustainability-related information, such as *Annual world production* for constituent Elements, *Abundance* and estimated global *Reserves*, clearly facilitates teaching, discussions or decision-making relating to materials and sustainability in engineering. This Paper will introduce some of the interactive and visual tools in Section 2 and give examples of how the database can be used for eco-selection. Section 3 describes Property Audits which is demonstrated for Carbon footprint and Cost. Section 4 shows how criticality and geo-economic data can support materials decisions and Section 5 provides a brief overview of the Sustainability database with a property chart application. The general methodologies are discussed in greater detail in several textbooks [3-4].

2 The CES EduPack and CES Selector Visualization and Selection Tools

Both the educational CES EduPack and the more advanced CES Selector have similar Chart and Selection tools. It is possible to display the relevant material or process properties in *Bar Charts* or *Bubble Charts* (see Figure 4). Screening can be applied, either using numeric or discrete values in a property list (called Limit stage), interactively in the chart (Box select) or through a filter in the tree structure of data (Tree stage). This produces a visual representation of selected property data that provides meaning and overview that could never be achieved using traditional tables of data. In Figure 4, the advantage of viewing all material families in comparison is shown. The relationship between the material families can be emphasized using color-coded envelopes, moreover, combined properties such as Carbon footprint [kg CO_2/kg] multiplied with the Density [kg/m^3] can easily be generated for a new entity [kg CO_2/m^3] to be plotted.

In order to systematically select a material according certain criteria, the design requirements should be translated into constraints (for screening) and objectives to be minimized or maximized (via ranking). A methodology for systematic visual, yet objective, and interactive selection using material performance indices directly in the charts can be used [3-4]. The material performance index (material index) represents the relevant metric for the particular conditions of the selection and can be plotted as a straight line in the property chart (see Figure 4) and used for automatic ranking of the candidates. CES Selector has more efficient and advanced tools for selection and materials substitutions, needed for industrial applications.

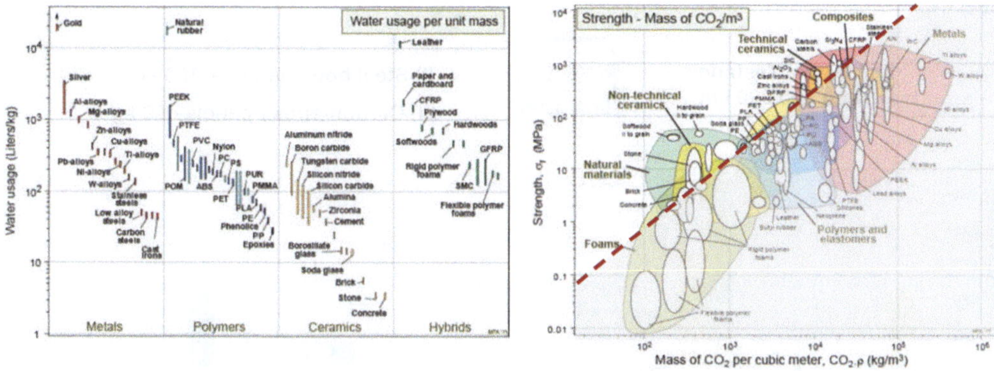

Figure 4. Examples of property charts (bar and bubble charts) to investigate sustainability of materials and that might inspire re-engineering in the framework of Design for Sustainability. The dashed selection line to the right indicates combined High-strength – Low CO_2-footprint.

3 How to Use Property Audits

In modern Engineering, it is natural and inevitable to consider the whole life-cycle, from *Cradle* to *Grave*, (or even *Cradle* to *Cradle*) of products that are being conceived, designed, produced and operated. Although the software does not carry out a Life-Cycle Assessment (LCA), life-cycle thinking is integrated in the database. It contains information on the typical Carbon footprints of materials, energy use of manufacturing processes for different sources of energy and recycling properties of materials, for example. A schematic of the life-cycle phases considered in the *Eco Audit* is shown in Figure 5

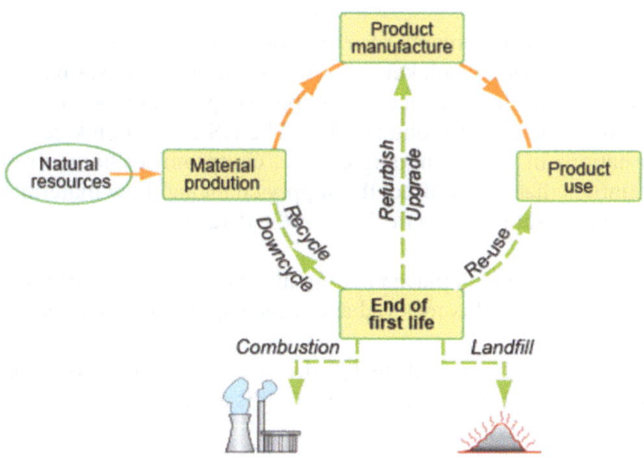

Figure 5. Phases considered in the Eco Audit tool

To illustrate the use of the Property Audits, two scenarios for a family car are considered. One has a typical bill of materials and one is a light-weight option where 100 kg of steel has been replaced by 60 kg of Carbon-Epoxy composite panels. Carbon footprints and cost information of these scenarios in the advanced Eco Audit tool are shown in Figure 6.

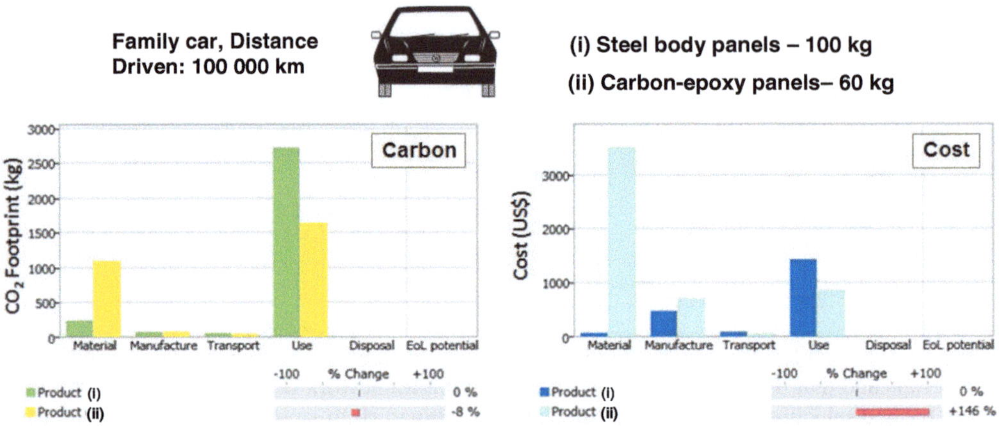

Figure 6. Carbon footprints (left) and Cost info (right) for alternatives (i) and (ii) above.

As can be seen, CO_2-emissions are reduced in the use phase based on a 100 000 km life but increase by almost as much in the material production phase for the composite case, the total reduction is 8%. The Cost Audit reveals that, although the reduced weight saves cost in the use phase, the increased materials cost cannot be justified. The total cost increases by 146% over the life. A similar audit can be produced for embodied energy and, in general, guidelines for reducing the main contributing phase (revealed by the audit) are given interactively by clicking on the bars.

4 How to Use Criticality and Geo-Economic Data

In 2011, the price of rare earth elements rose by more than a factor of 10 when China imposed restrictions on export. The price of nickel is set to rise after Indonesia banned exports of metal ores in 2014. Increasingly, ethical concerns inhibit the purchase of minerals from nations in which the proceeds might be used to fund conflict. The US Dodd-Frank act, for instance, creates trade barriers with nations in which ongoing conflict compromises human rights. Many of these are mineral-rich countries that might affect the supply chain and how materials can be sourced. A less dramatic case can illustrate the practical use of the database.

A stainless steel pipework is to be replaced and, in this case, Hastelloy W is considered. We want to know the *Carbon footprint* of this material and if the supply-chain can be trusted.

Data for Hastelloy W can be found using the search function on the tool bar, in the MaterialUniverse data table at Level 3, see Figure 6. It is a Ni-Mo-Cr alloy.

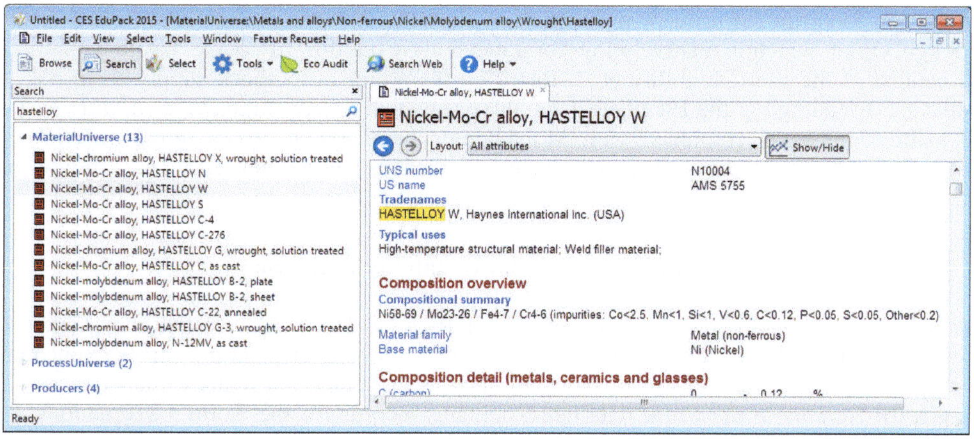

Figure 7. Search result from the investigation into sustainability properties of Hastelloy W.

Using the *Browse* function in the Elements data table or the links provided at the bottom of the data record for Hastalloy W, the main components, Nickel (58-69%) and Molybdenum (23-26%) can be found easily. These data records show that Nickel has a medium *Abundance risk level* and very low *Sourcing and geopolitical risk* level. It is not on the US or EU Critical lists but it has very high *Price volatility* risk. The data for Molybdenum, shown in Table 2, show high *Price volatility risk* and, moreover, Mo is on the EU Critical list.

Table II. Excerpt, showing some relevant sustainability Mo data (notes and references excluded)

42. Molybdenum				
Geo-economic data				
Typical exploited ore grade	0.096	-	0.106	%
Minimum economic ore grade	0.002	-	0.2	%
Abundance in the Earth's crust	0.8	-	1.5	ppm
Abundance in seawater	0.01			ppm
Annual world production	2.7e5			tonne/yr
World reserves	1.1e7			tonne
Eco properties				
Embodied energy, primary production	107	-	118	MJ/kg
CO2 footprint, primary production	7.3	-	8.04	kg/kg
Water usage, pure element	341	-	377	l/kg
Critical materials information				
In EU Critical list?	True			
In US Critical list?	False			
Abundance risk level	Medium			
Environmental country risk Herfindahl-Hirschman Index (HHI)	1			
Environmental country risk level	Low			
Sourcing and geopolitical risk Herfindahl-Hirschman Index (HHI	1.27			
Sourcing and geopolitical risk level	Low			
Price volatility	343			%
Price volatility risk	High			
Principal uses and substitutes				
35% used in constructional engineering steels for oil and gas exploration (piping and refineries), heavy machinery, Transportation (automotive for powertrain systems, off-road vehicles and ships), power generation, and petrochemical machinery and tanks. Alternative: molybdenum-free constructional engineering steels.				
Quality: adequate				

The carbon footprint for primary production of Hastelloy W (not shown) is estimated to 10.7-11.8 kg CO_2/kg, which is relatively low for a metal alloy but this, of course, has to be put in the full context of the use (Durability etc.). Based on the premises to check supply-chain risk, the results are discouraging: the Elements data table show high or very high *Price volatility risk* for the main components combined with Molybdenum being on the EU Critical materials list. Molybdenum-free stainless steels should be investigated, as advised in the data on *Principal uses and substitutes*, above.

5 How to Use the Sustainability Database

The Sustainability database enhances the CES EduPack Level 3 to include five major additions. The first is a data table of *Regulation*, listing legislation, regulations and incentives to encourage or restrict the use of materials or of practices such as recycling that relate to material use. The second, the *Nations of the world* data table, contains records for the world's 210 nations, with data for population, governance, economic development, energy use and engagement with human rights, together with information that may impact security of supply and the ethical sourcing of materials. The new data tables are linked to the pre-existing MaterialUniverse (Materials) and ProcessUniverse (Processes) Level 3 data tables, allowing individual materials to be linked to legislation that affects them and to the nations from which they are sourced. The materials records themselves have been extended to include a ranked list of the *Countries of origin* and a *Critical material status*. The *Materials* data table has additional links to two further data tables relating to energy: one with records for *Electric-Power generating systems*, the other for *Energy Storage systems*. Figure 8 shows the structure.

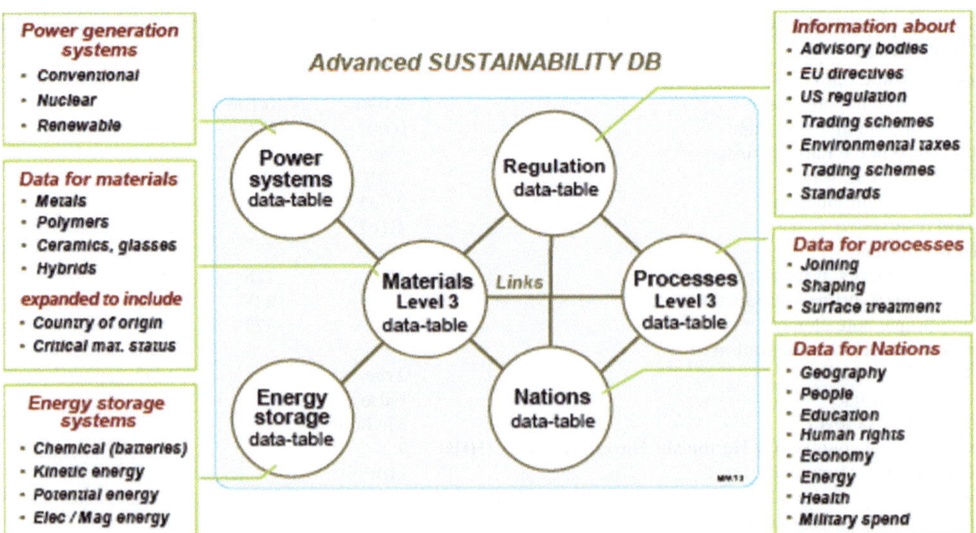

Figure 8. Detailed description of the data tables enhancing the Standard Level 3 database.

Environmental properties include information on water usage, carbon footprint, embodied energy, and end of life information. It is supplemented by data on the material's country of origin and whether it is considered a critical material. This links to data in *Nations of the world* and to the *Elements* data table (not shown above).

The data was chosen by the Granta Design team, in collaboration with users, to enable a wide range of project-based learning exercises to be supported. The aim was to facilitate students' understanding of the complexities of sustainable development and gaining skills in information analysis and debate, ready to take part in policy decisions as well as technical decisions. The Sustainability database can be used for assessing articulations of sustainable technology and the place of materials in them. It is comprehensive and allows in-depth studies using a 5-step methodology that has been tested and is described elsewhere [5]. However, it can also be the basis of more general critical discussions, using the *Nations of the world* data table, as can be seen in Figure 9. Other interesting Charts that can be produced are Satisfaction with life or UN Human development Index vs GDP/Capita.

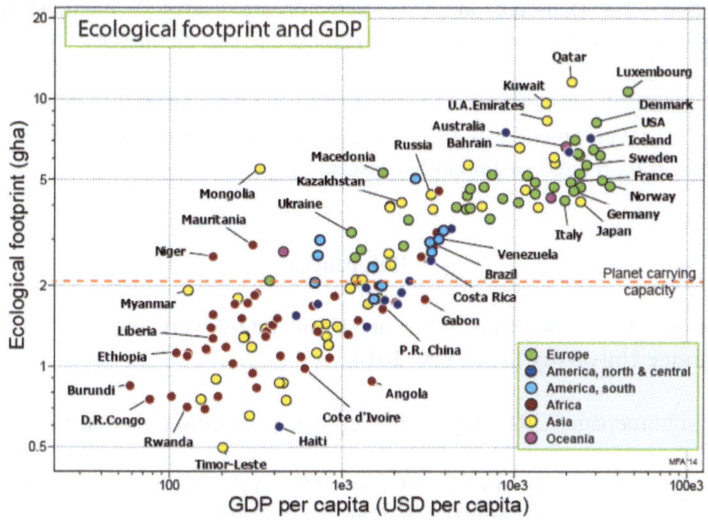

Figure 9. Example of Chart made possible by the Nations of the world data table.

6 Quality of the Data

The database is compiled, maintained, up-dated and expanded by Granta Design on an on-going basis. Records of the provenance of the data and the reasoning used in selecting it for inclusion are also stored. The whole MaterialUniverse is very large – various sub-sets of it are tailored for specific needs, such as the Sustainability database discussed here. The Elements Eco data is individually referenced and traceable. The model behind the Eco Audit Tool is spelled out in detail in the *Help* menu of the software so that it can be discussed with students. A general warning as to the uncertainty of environmental data is also given in the notes accessible through the *Help* menu. The measurable materials data is generally not individually traceable but references are given to the general sources used, such as the American Society for Materials, ASM, or other credible sources. Estimated values are clearly indicated with an asterix (*) for clarity and great effort has been taken to ensure that values are reliable and comparable throughout the materials in each data table.

7 Summary and Conclusions

By examples, we have demonstrated the technical utility of the software and the Sustainability database has been used to show how to support investigations into social and resource issues relating to materials.

In the paper, we have described the interactive and visual tools in the software and given examples of how the database can be used for eco-selection, how Property Audits can aid life-cycle thinking and how criticality and geo-economic data can support material decisions.

We conclude that CES EduPack and CES Selector are useful in engineering teaching and industry to enable a rational and well informed response to eco challenges.

By using CES EduPack, students and researchers can gain a more holistic view of a product already during its development using a database that is designed to be relevant for material selection and materials-related sustainability issues.

References

1. Climate Change 2013 – The Physical Science Basis, Intergovernmental Panel on Climate Change (Cambridge University Press, UK, 2014)

2. Granta Design homepage, URL: http://www.grantadesign.com (Accessed February 1, 2015)

3. Mike Ashby, Hugh Shercliff, and David Cebon, *Materials: engineering, science, processing and design,* 3rd edition. (Butterworth Heinemann, Oxford, 2014)

4. Mike Ashby, Materials Selection in Mechanical Design, 4th edition (Butterworth Heinemann, Oxford, 2011)

5. Mike. Ashby, Didac Ferrer Balas, and Jordi Segalas Coral, *Materials and Sustainable Development,* (Butterworth Heinemann, Oxford, 2015)

Engineering Solutions for Sustainability: Materials and Resources II
Edited by: Jeffrey W. Fergus, Brajendra Mishra, Dayan Anderson, Emily Allyn Sarver, and Neale R. Neelameggham
TMS (The Minerals, Metals & Materials Society), 2015

MELT CHARACTERISTICS OF POLY-LACTIDE (PLA) NATURAL FIBRE–COMPOSITES

[1]Akpan, E. I., [2]Adeosun S. O., [3]Lawal.G. I., [4]Balogun S. A.

[1]Department of Materials and Production Engineering, Ambrose Alli University, Nigeria
[2,3]Department of Metallurgical and Materials Engineering, University of Lagos, Nigeria
[4]Department of Mechanical and Biomedical Engineering, Bells University of Technology, Ota, Nigeria

Keywords: Polymer composites, biodegradable, glass transition temperature, specific mechanical energy (SME)

Abstract

Melt processing characteristics and thermal properties of cellulosic fibre-polylactide (PLA) composites have been investigated. A twin screw mixer is used for composites compounding of PLA with five agro-wastes natural fibres obtained through two process paths. Distribution of torque, temperature and energy of the mixing process are generated directly from the twin screw mixer. Specific mechanical energy is calculated from the torque-time relationship while thermal properties (glass transition, melting and crystallization temperatures) are deduced from results of Differential scanning calorimetry test. Results show that ample time is required for rheological changes to take place in composites while changes in neat PLA commenced almost immediately. It is found that lower mechanical energy is consumed in compounding composites than neat PLA and this result in improved melt processing characteristics of composites compared to neat PLA. Analyses of thermal properties show that addition of fillers promotes lowering of inter molecular forces at the fibre-matrix interface giving rise to higher chain mobility. Thus, lower glass transition temperature with increase melting results.

Introduction

The use of non-renewable, petroleum-based chemicals for the synthesis and manufacture of structural polymers, and the environmental concerns generated by their disposal, pose major challenges to the polymer industry. The main strategies to address these problems are the utilization of polymeric materials from renewable sources, and development of biodegradable polymeric materials. Poly lactic acid (PLA) is the most important biodegradable polymer, and has been applied in more and more industries due to its lower energy consumption and is non-toxic to the environment [1 - 2]. PLA has attracted an increasing interest in various markets, such as packaging, textile [3 – 4], and automotive industries [5 - 6, 1]. It can be processed using injection-moulding, compression moulding, extrusion, and thermoforming etc.

However, there exists much weakness such as slow crystallization rate, low crystalline degree and poor heat resistance etc. Other drawbacks which tend to limit its widespread applicability include; high cost, brittleness, and low heat distortion temperature. Thus, in order to broaden the applications of PLA, material properties and process-ability has to improve. Various additives such as plasticizers, toughening agents, reinforcing fillers and compatibilizers have been incorporated into PLA [7 – 9]. Although these fillers are able to impact good properties, they are mostly synthetic and pose environmental problems in some cases. Some of these fillers are also expensive and not readily available. The use of renewable and biodegradable fillers is desirable to provide cost-competitive polymer composites [10]. Natural fibres have attractive technical, economic and environmental advantages and this includes; low cost, low abrasion, medium mechanical and thermal properties, high disposability, high renewability and biodegradability, low energy consumption and generation of neutral carbon dioxide [11 - 13].

Natural fibres present some disadvantages such as high moisture sorption, poor dimensional stability, low thermal resistance, anisotropic impact resistance, variability and poor adhesion to the surface of base polymers [14, 15]. The high content of non-cellulosic components in natural fibres (hemicelluloses, lignin, pectin and waxes) and impurities are the major influence on fibre properties. To improve the properties of fibres for composite applications, individual methods have been used in processing/treatment of the fibres and this includes; steam explosion, alkaline treatment, thermal alkaline degradation, shear alkaline peroxidation, silane treatment, acid hydrolysis and high pressure homogenization. These methods prove to achieve improved fibre mechanical and chemical properties. However, the adhesion property does not significantly improve. Considering the fact that the adhesion properties of fibres depend basically on the amount of non-cellulosic materials present in the fibre structure [16] we recently developed and tested two processing paths for processing agro-wastes fibres to facilitate the removal the non-cellulosic materials. Characterization of the resulting fibres shows that the processed fibres possess excellent morphological characteristics, thermal properties, topography, crystallinity and thermal degradation which indicate that improved polymer composites properties will be obtained if used as reinforcement.

In processing polymer composites, the melt behaviour, thermal and especially melt properties are very important. The melt rheology of polymers and polymer based composites is temperature-dependent. The processing window (the range of temperature which the composite can be processed without damage to the various components) is a crucial parameter to control nucleation and cell growth and to stabilize the overall polymer and composite network [17 – 18]. Typical biopolymers are generally sensitive to thermo-mechanical stress during processing. These often exhibit a small processing window and a significant property loss due to thermal degradation. The investigation of melt properties under processing conditions as well as the broadening of the thermoplastic processing window without thermal degradation of the biopolymer are essential factors to increase the application of biopolymers in composite application. Knowledge of the melt characteristics of the polymer/natural fibre systems is as important as the melt behaviour of the polymer itself [19].

Energy input during extrusion is an important parameter as it relates to the physical and chemical transformations in the resulting product [20 – 21]. Specific mechanical energy (SME) is the amount of mechanical energy (work) dissipated as heat inside the material. It is an important process parameter influencing the final product characteristics such as solubility, density, expansion index and hardness [22 – 24, 20]. Its value is a good measure of the extent of molecular breakdown or degradation the material undergoes during the extrusion process. For this reason specific mechanical energy (SME) is necessary to interpret the observed phenomena regarding the processing dependent macroscopic fibre dispersion [20].

Five natural fibres earlier processed using two novel batch processing paths, developed by the authors to process agro-wastes fibres for composite applications are used as fillers in PLA to make composites. These composites are compounded using the Twin Screw mixer and characterized for their SME, glass transition, melting and crystallization temperatures. Thus, this article discusses effect of fibre type and treatment on the melt processing characteristics and thermal properties of natural fibre filled PLA composites. SME, melt behaviour, glass transition, melting and crystallization temperatures are used as process characteristics and are correlated with the processing methods and fibre types.

Methodology

Materials

Fibres are obtained from five agro – wastes sources namely; Palm fruit stalk, palm fruit bunch, rice husk, groundnut shell, and coconut shell. These wastes materials are collected from farm fields in Ekpoma, a city located in southern Nigeria. The unwanted parts are removed and the fibrous sheaths dried in air for two weeks. All samples are designated as shown in Table 1. Each agro-waste material is subjected to two processing paths of treatment designated M1 and M2 at National Institute of Science Laboratory Technologist, Ibadan, Nigeria. Poly lactic acid pellets are obtained from Nature Works LLC supplier in China.

Table 1: Sample designation

Sample	Designation
GM1	Groundnut shell fibre processed via M1
GM2	Groundnut shell fibre processed via M2
CM1	Coconut husk fibre processed via M1
CM2	Coconut husk fibre processed via M2
RM1	Rice husk fibre processed via M1
RM2	Rice husk fibre processed via M2
PBM1	Palm fruit bunch fibre processed via M1
PBM2	Palm fruit bunch fibre processed via M2
PSM1	Palm fruit stalk fibre processed via M1
PSM2	Palm fruit stalk fibre processed via M2
PLAP	As-received PLA

Fibre Processing

The study plant wastes are collected, washed, sunlight dried, cut into small pieces, ground to pass a screen of 10 mm in a mechanical crusher and divided into two. The first set (M1) is subjected to steam explosion at a temperature of 175^{0}C and a pressure of 1bar in an autoclave. Alkaline hydrolysis is conducted on the resulting fibre in 2 % solution of NaOH overnight, neutralized in acetic acid and bleached with 8 % solution of hydrogen peroxide. Further acid hydrolysis is undertaken with a mixture of 10 % (w/w) nitric acid and 10 % (w/w) chromic acid at a temperature of 60^{0}C for 15 minutes. The average particle size of fibres is found to be 347 micrometre (Particle size analysis is done using Laser size analysis equipment). The second set (M2) of fibres is first de-waxed with benzene – ethanol treatment in a Soxhlet extractor, washed to neutrality and oven dried at a temperature of 45^{0}C. Resulting fibres are subjected to steam explosion at a temperature of 175^{0}C and a pressure of 1bar in an autoclave for an explosive reaction on the fibres. Enzymatic hydrolysis is carried out on the resulting fibre using crude celulase suspended in 50 mL of 50 mM Na acetate buffer (pH 4.8) at 50^{0}C for two hours. This is followed by alkaline hydrolysis, bleaching and acid hydrolysis as done in the first set. All treated fibres are then washed in water, centrifuged in ethanol and dried in still air. The average particle size of fibres is found to be 347 micrometre (Particle size analysis was done using Laser size analysis equipment).

Melt Mixing

Compounding is conducted on a co-rotating, twin-screw mixer/extruder (HAAKE POLYLAB Rheomix 600 OS). The screw speeds is fixed at 90 rpm for feeding and melt mixing, process time is 10 minutes at 180^{0}C heating temperature. The temperature of compounding is chosen because the melting and crystallization temperatures of PLA have been found to range between

160 – 190 ^{0}C [25 – 26], and thus, the composite is processed at a temperature within this range to achieve melt processing in the range of crystallization. The screw speed and compounding time are chosen in line with previous successful works to allow for comparison [25 – 26]. Each time test is done the mixer cavity is cleaned with polypropylene. In each run 7.5 wt. % of fibre is added to PLA to make up 70 g of total charge. All samples are tagged to follow the designation of the fibres contained in it. Mass temperature and torque are measured during melt processing. The specific mechanical energy input (SME) in Jg^{-1} is calculated from the torque-time graphs according to Eq. (1) [27].

$$SME = \frac{\omega}{m}\int_0^{t_{max}} C(t)\,dt \ldots\ldots\ldots\ldots\ldots\ldots\ldots\ldots\ldots\ldots\ldots\ldots\ldots\ldots [1]$$

Where ω is the rotor speed, m is the sample mass, C(t) is the torque at time t and t_{max} is the mixing time.

Differential Scanning Calorimetry

Thermal properties of the composites and neat PLA are determined using differential scanning calorimeter (DSC) model DSC Q200 equipped with a cooling attachment, and operated under a nitrogen atmosphere. About six milligram (6mg) of each sample is placed in covered aluminium sample pans and then placed into the DSC sample holder. The sample is then heated at 10ºC/min, in the 40-250ºC temperature range. Two heating cycles (up to 250^{0}C) and one cooling cycle (down to 40^{0}C) are conducted with composite samples with a heating and cooling rate of 10^{0}C min^{-1}. The heat flow changes of the samples in the sealed aluminium pans are recorded with reference to an empty aluminium pan. The data received after the test is used to determine glass transition, melting and crystallization temperatures of composites.

Results and Discussion

Effect of Fibre and Processing Conditions on Melt Processing Characteristics

The melt processing behaviour of composites and virgin PLA are shown in Figures 1 – 3. These Figures are truncated to allow for a clearer view of the changes in the behaviour of the composites before they attain a constant value. Figure 1 shows the variation of torque and temperature of extrusion with time. The torque evolution is characterized by a lag phase for modified PLA followed by exponential increase and later a continuous decrease. The reduction in torque is attributed to shear thinning behaviour of the composite material in the melt state and is related to the reduction in matrix viscosity during the extrusion process [28]. Temperature evolution is characterised by an initial decrease in temperature with time and subsequent increase following sigmoid shaped curve reaching a maximum and continues with constant temperature for the remaining compounding time. The initial decrease in temperature is due to the admittance effect of cold pellets into the chamber which cause an overall decrease in the temperature. The presence of fibres led to a higher starting torque for the composites but a lower starting torque for virgin PLA. This indicates that at the beginning of the mixing process, cohesive, plastic and very sticky dough is formed with stronger dough for both composites but a less stronger dough for virgin PLA. Composite filled with M2 treated fibre has a higher starting torque followed by that filled with M1 treated fibre and virgin PLA. The highest torque is attained by the composite with M2 treated fibre followed by virgin PLA. This confirm that M2 treated fibres possess a strong adhesion to the PLA than the M1 treated fibres. Viscous heat dissipation is important and leads to an increase in temperature of the composites to a maximum and continues constantly till end of process time. Rheological changes occur in GM2 earlier than GM1 while changes in virgin PLA started almost immediately. These differences in torque-time behaviour can be attributed to a higher shear stress needed to initiate transformation in GM2 than in GM1 and virgin PLA. Virgin PLA shows the highest specific mechanical energy (SME) followed by GM2 and GM1 has the lowest (see Table 2). This shows the viscosity of virgin PLA may be higher than both GM2 and GM1 [20]. It has also been established [28] that increase in melt SME leads to increase in melt yield stress to a maximum before it remains constant. This thus, indicates that GM2 will have higher yield stress compared to GM1 which is in close correlation with the fact that GM2 has a strong adhesion with the fibres leading to improvement in properties.

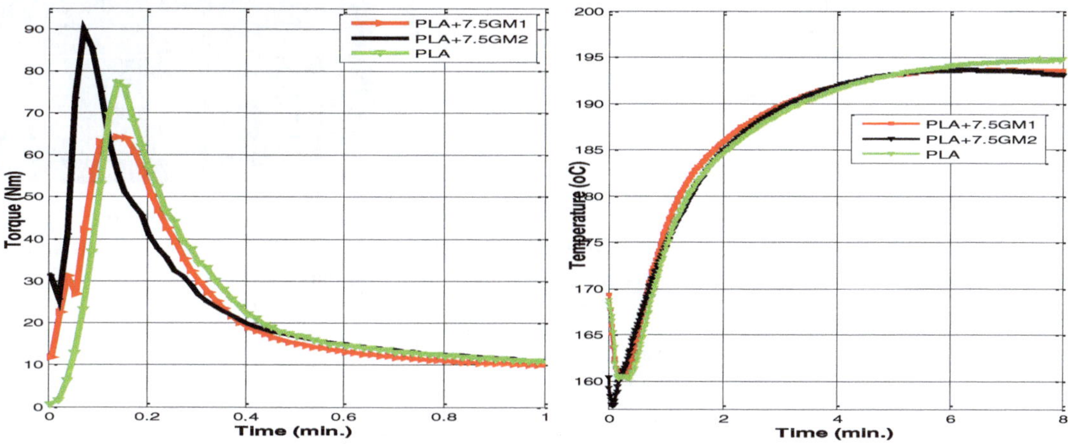

Figure 1: Torque and temperature of groundnut fibre-PLA composite (truncated for clarity).

The torque and temperature/time evolutions in the twin screw melt mixer during coconut fibre filled PLA composite processing are shown in Figure 2 (truncated for clarity). The torque-time curve increase exponentially and later decrease continuously to a constant value. The starting torque is highest for CM2 PLA composite followed by CM1 PLA composite with virgin PLA having the lowest starting torque. The highest processing torque is recorded for virgin PLA followed by CM2 composite with CM1 as the lowest. This shows that CM1 displays the lowest shear and thermo-mechanical stress during mixing and will most likely have less mechanically induced thermal degradation. It is assumed that addition of CM1 fillers led to lower inter molecular forces at the fibre matrix interface giving rise to higher chain mobility and lubricity resulting in reduce torque. The chain mobility is lower for CM2 and virgin PLA. The temperature-time curve show that the highest melt temperature (194.8°C) is recorded for virgin PLA while CM1 and CM2 composites have similar mass temperature. This is an indication that there is reductions in heat dissipation during the composites melt processing attributed to decline in both shear and thermo-mechanical stress induced in the composites [18]. Specific mechanical energy measurements show that virgin PLA has the highest value followed by CM1 with CM2 as the least (514.69 J/g). Thus, low mechanical energy is expended in compounding composites than virgin PLA. This is an indication of low shear and thermo-mechanical stress and consequently less induced thermal degradation. It has been shown by Domenech et al. [28] that higher specific strength means higher melt yield stress which points to improve dispersion of filler fibres, showing that the fibres cause little distortion in PLA molecular structure. The SME results also show that PLA has the highest melt viscosity.

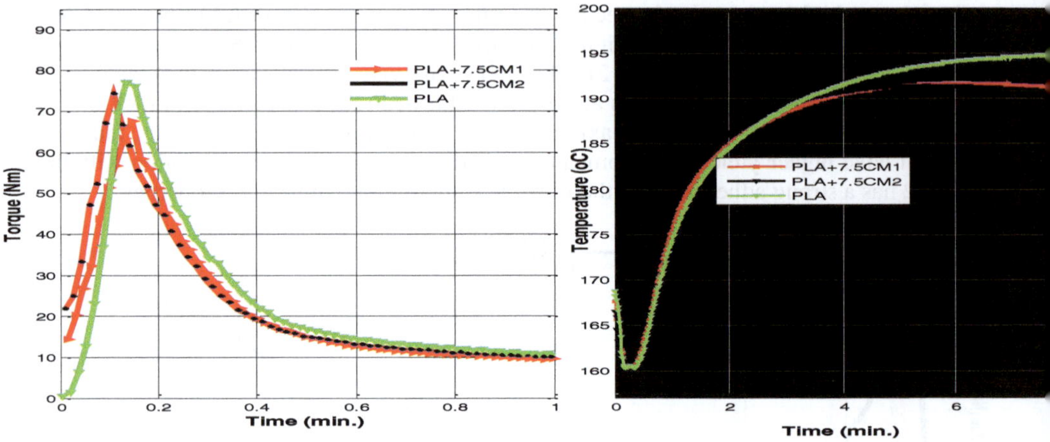

Figure 2: Variation of torque and temperature of coconut husk fibre PLA composite with time.

Table 2: Mixing conditions and mixing parameters

Sample	SME (J/g)	Peak Torque (Nm)	Final Torque (Nm)
PLA+GM1	589.83	64.20	3.8
PLA+GM2	622.15	89.80	3.6
PLA+CM1	546.29	67.80	3.0
PLA+CM2	514.69	74.30	2.2
PLA+RM1	593.24	64.00	3.1
PLA+RM2	570.42	73.20	2.9
PLA+PBM1	559.50	65.00	2.9
PLA+PBM2	550.94	61.00	2.7
PLA+PSM1	505.85	55.20	1.8
PLA+PSM2	483.52	61.30	1.7
PLA	631.34	77.20	4.3

Figure 3 shows the variation of torque and temperature of melt with time and the rheological transformations taking place immediately at the onset of mix process. The torque evolution is characterized by a steady exponential rise to a maximum value. This is followed by a continuous decline to a constant value for all samples except for RM1 composite which has a local peak before further rise to a maximum peak and finally followed by a continuous decrease. Similar to coconut and groundnut fibres composites, the starting torque is higher than that of virgin PLA. Virgin/unreinforced PLA has highest melt torque followed by RM2 and the RM1 composites in decline order. This is an indication of reduction in shearing, heat dissipation, and overall thermo-mechanical stress during mixing. This give rise to low thermal and degradation temperature with improve melt processing. The SME results show that unreinforced PLA is least with consumption of high mechanical energy (see Table 2) than the composites. Thus, shear and thermo-mechanical stress are minimal during the compounding of the composites and consequently less induced thermal degradation involvement. The SME results also reveal that the presence of fibres increases the mobility of polymer chains at the vicinity of the fibre-matrix interface in the composites. The variation of mass temperature with time shows that the highest melt temperature is in PLA followed by RM1 composite. This confirms composites have improved melt processing than unreinforced PLA [18].

A similar torque-time relationship is obtained for palm fruit bunch filled composites showing that the starting torque is higher for PBM1 and PBM2 composites than unreinforced PLA. The highest processing torque occurred in unreinforced PLA followed by PBM1 and PBM2 in decline order. Thus, PBM2 has lowest shear and thermo-mechanical stress in melt mixing and obviously will show low mechanically induced thermal degradation. The temperature-time curve indicates that there is reduction in heat dissipation during the melt mixing of the composites, which is attributed to reduction in both shear and thermo-mechanical stress induced in the composites [13]. Calculated specific mechanical energy shows that unreinforced PLA (631.34 J/g) has the highest value followed by PBM2 (559.50 J/g) and PBM1 (550.94 J/g).

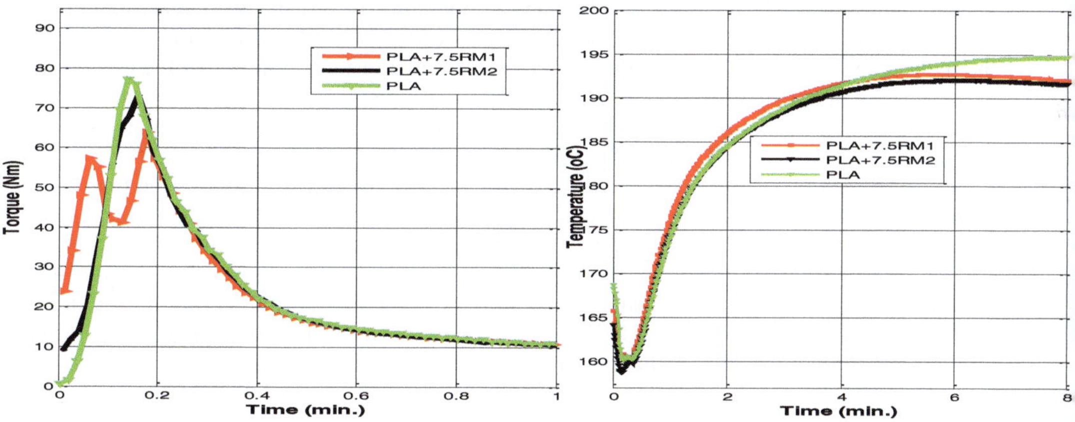

Figure 3: Torque and temperature of rice husk fibre-PLA composite as a function of mixing time

Similarly, melt processing characteristics of composites with palm fruit stalk fibres show that composite with M2-fibre possess higher starting torque followed by M1-fibre composite and unreinforced PLA. The highest torque is shown by M2-fibre composite followed by unreinforced PLA. This confirms M2-fibre has strong adhesion to PLA matrix than M1-fibres. Viscous heat dissipation is important and leads to an increase in temperature of the composites to a maximum and continues constantly till the maximum time of mixing. These differences in torque-time behaviour can be attributed to higher shear stress needed to initiate transformation in PLA than in PSM1 and PSM2. Unreinforced PLA records the highest specific mechanical energy (631.31 J/g) followed by PSM1 (505.85 J/g) and PSM2 (483.52 J/g) (see Table 2). Thus, the viscosity of unreinforced PLA is higher than that of PSM1 and PSM2. Therefore, increase in melt SME leads to increase in melt yield stress to a maximum before it remains constant [18]. This indicates that PSM1 will possess a higher yield stress compared to PSM2 which agrees with the fact that PSM1 fibre has a strong adhesion with the matrix resulting in improved properties.

Effect of Fibre and Processing Conditions on Thermal Characteristics

For semi-crystalline PLAs, both the glass transition (Tg) and melting (Tm) temperatures are important for determining the use and processing temperatures across various applications. The glass transition, melting and crystallization temperatures of the composites are determined from the data collected from the DSC test. Table 3 shows the glass transition temperature of the various polymer composites examined in this study. The glass transition temperatures show that there is no significant change in the transition temperature of composites against as received PLA pellets. However, the addition of fillers promotes slight decrease in the glass transition temperature with the highest effect shown by PSM2 composites. These results are consistent with the works of Silverajah et al. [29] and Mustapa et al. [30]. Results of the crystallization temperature show that the introduction of fibres into the PLA matrix led to a decrease of the Tcc for all the composites compared to unreinforced PLA (see Table 3). Decrease in cold-crystallization temperature (Tcc) of composites signifies fibres hindrance of the migration and diffusion of PLA molecular chains to the surface of the nucleus in the composites [31 – 32]. Reduction in Tcc is also associated with crystallization and increase in viscosity of the bio-composite mixture [31].

The crystallisation temperature of as-received PLA (without melt mixing) is higher than that of all the composites. The lowest crystallisation temperature is shown by PLA+PSM1 (107.9^0C) followed by PLA+GM1 (108.3^0C). The difference between the glass transition temperature and the melting temperature of the composites show the range of temperature that the composite can be processed in the melt state without damage. It is observed from Table 3 that this range increased for all composites against that of melt processed PLA. The range is highest for PLA+CM2 followed by PLA+PSM1. This increase in melt processing shows that the fibres improve the melt processing characteristics of PLA. Moreover, it is an indication that the fibres possess good surface adhesion with the polymer matrix leading to tougher composite dough that withstands temperature deterioration. Double melting endotherms are detected at low temperature (Tm1) and higher melting peaks (Tm2) for all composites (see Figure 4). The unreinforced PLA shows one melting temperature (154.8^0C) against all composites which show two distinct melting temperatures. This is attributed to the absence of recrystallization during the melt mixing of the PLA. It is noted that the presence of fibres created sites for recrystallization at the fibre matrix interface so that the composites show two melting endotherms.

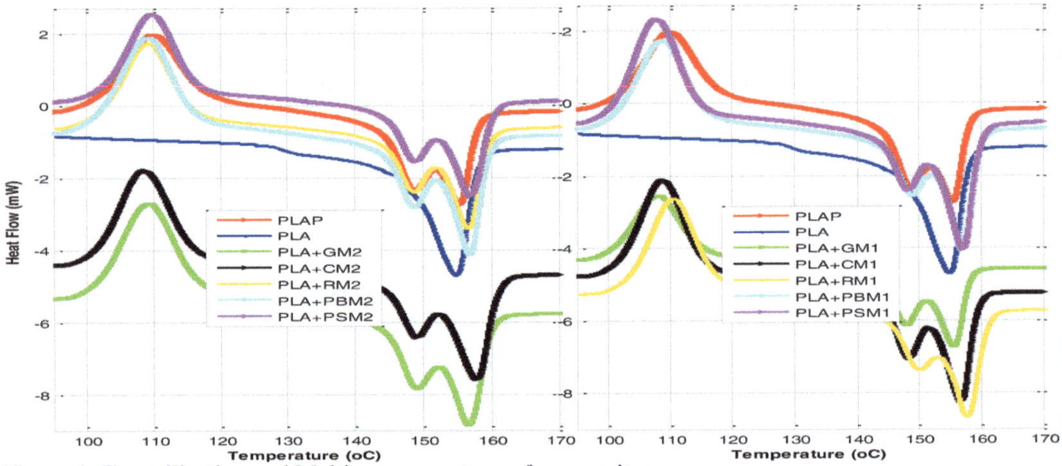

Figure 4: Crystallization and Melting temperature of composites

Table 3: Thermal properties

Sample	$T_g(^0C)$*	$T_{m1}(^0C)$*	$T_{m2}(^0C)$*	$T_c(^0C)$	$\Delta T(^0C)$
PLA+GM1	57.8	148.0	155.6	108.3	97.8
PLA+GM2	57.4	149.1	156.7	109.3	99.3
PLA+CM1	57.4	148.4	156.7	108.7	99.3
PLA+CM2	56.6	149.2	157.8	108.6	101.2
PLA+RM1	57.9	150.0	157.4	110.4	99.5
PLA+RM2	57.8	148.6	156.5	109	98.7

PLA+PBM1	57.9	149.0	156.7	108.9	98.6
PLA+PBM2	58.5	148.7	156.8	109.3	98.3
PLA+PSM1	56.4	148.4	156.9	107.9	100.4
PLA+PSM2	58.7	148.9	156.8	110.0	98.1
PLA (as received pellets)	62.3	-	154.8	-	92.5
PLAP (processed PLA)	58.3	148.9	155.4	110.3	

*T_g is the glass transition temperature and T_{m2}(°C) as the melting temperature.$\Delta T$is the difference between melting temperature and glass transition temperature

Conclusion

Studies on the thermal and melt processing characteristics of cellulosic fibre- PLA composites show that; the presence of fibres imparts a higher starting torque for the composites to that of unreinforced PLA. Lower mechanical energy is expended in melt mixing of composites than unreinforced PLA. This indicates low shear and thermo-mechanical stress and consequently less induced thermal degradation. Unreinforced PLA have highest specific mechanical energy (631.31 J/g) followed by PLA+PSM1 (505.85 J/g) and PLA+PSM2 (483.52 J/g). The SME results also show that PLA has the highest melt viscosity.

References

[1.] E.Y. Hassan, H.J. Wei and M. Yu, "Dynamic Mechanical Properties and Thermal Stability of Poly(lactic acid) and Poly(butylene succinate) Blends Composites," *Journal of Fibre Bioengineering and Informatics*, 6 (2013), 85-94.

[2.] Y. Tadashi and Y. Masayuki, "Structure and properties for biomass-based polyester blends of PLA and PBS," *European Polymer Journal*, 44 (2008), 677-685.

[3.] B. Gupta, N. Revagade and J. Hilborn, "Poly(lactic acid) fibre: An overview," *Prog Polym Sci*, 32(2007), 455-482.

[4.] B. Gupta et al., "Preparation of poly(lactic acid) fiber by dry-jet-wet-spinning. I. Infuence of draw ratio on fiber properties," *J Appl Polym Sci*, 100 (2006), 1239-1246.

[5.] A.S.T. Singha and K. Vijay, "Fabrication of Hibiscus Sabdaria fibre reinforced polymer composites," *Iranian Polym J.*, 17 (2008), 541-54.

[6.] Hoydonckx H., Renewable Furan Resins in Composite Applications. Composites Innovation, 2007 Conference Proceedings, www.netcomposites.com.

[7.] H. Liu and J. Zhang, "Research progress in toughening modification of poly(lactic acid)," *J Polym Sci Part B: Polym Physi*, 49(2011), 1051-1083.

[8.] H. Wang, X. Sun and P. Seib, "Strengthening blends of poly(lactic acid) and starch with methylenediphenyldiisocyanate," *J Appl Polym Sci*, 82 (2001), 1761-1767.

[9.] L. Yu, K. Dean and L. Li, "Polymer blends and composites from renewable resources" *Prog Polym Sci.,* 31(2006), 576-602.

[10.] V.L. Finkenstadt et al., "Poly(lactic acid) green composites using oilseed co-products as fillers," *Industrial Crops and Products,* 26 (2007), 36–43

[11.] P. Wambua, J. Ivens and I. Verpoest, "Natural fibres: can they replace glass in fibre reinforced plastics?" *Comput. Sci. Technol.,* 63(2003), 1259–1264.

[12.] P.J. Herrera-Franco, A. Valadez-Gonzalez, "Mechanical properties of continuous natural fibre-reinforced polymer composites," *Compos. Part. A: Appl. Sci.*, 35 (2004), 339–345.

[13.] S.J. Eichhorn and R.J. Young, "Composite micromechanics of hemp fibres and epoxy resin microdroplets," *Comput. Sci. Technol.*, 64 (2004), 767– 772.

[14.] A. Stamboulis and C.A. "Baillie Environmental durability of flax fibres and their composites based on polypropylene matrix," *Appl Comp Mater*, 7(2000), 273–94,

[15.] A. Bessadok et al., M., "Effect of chemical treatments of Alfa (Stipatenacissima) fibres on water-sorption properties," *Composites Science and Technology*, 67(2007), 685–697.
[16.] S.O. Adeosun et al., "Review of Green Polymer Nanocomposites," *Journal of Minerals & Materials Characterization & Engineering*, 11, (2012) (4), 483-514, 2012
[17.] M. Mihai, M.A. Huneault and B.D. Favis, "Rheology and Extrusion Foaming of Chain-Branched Poly(lactic acid)," *Polym. Eng. Sci.*, 50(2010) 629-642.
[18.] S. Zepnik et al., Cellulose Acetate for Thermoplastic Foam Extrusion, In Biomass Conversion, ed. J. Kadla (InTech, DOI: 10.5772/56215, 2013), 17 – 44.
[19.] T. Villmow, B. Kretzschmar and P. Pötschke, "Influence of screw configuration, residence time, and specific mechanical energy in twin-screw extrusion of polycaprolactone/multi-walled carbon nanotube composites," *Composites Science and Technology,* 70 (2010), 2045-2055.
[20.] S. Godavarti and M.V. Karwe, "Determination of Specific Mechanical Energy Distribution on a Twin-Screw Extruder," *J. Agric. Engng Res.,* 67 (1997), 277-287.
[21.] J. Tayeb and G.D. Valle, "Simulation of transport phenomena in twin-screw extruders," *Food Extrusion Science and Technology*, ed. J. Kokini, C.T. Ho, and M.V. Karwe (Marcel Dekker, New York, 1992), 41 - 70
[22.] J.M. Harper, "Food Extruders and their applications," *Extrusion Cooking*, ed. C. Mercier, P. Linko, J.M. Harper (American Association of Cereal Chemists Minnesota, 1989), 1 - 15.
[23.] J.M. Harper, Extrusion of Foods, vol. I (CRC Press, Boca Raton, 1981), 21 – 25.
[24.] B. Van Lengrich and F. Meuser, "Determination of the viscosity of starch during its extrusion cooking with a co-rotating twin-screw extruder," *Trends in Food Processing*, ed. A.H. Ghee, (Singapore Institute of Food Science and Technology, Singapore, 1989), pp. 19 - 24.
[25.] A. Redl et al., "Rheological Properties of Gluten Plasticized with Glycerol: Dependence on Temperature, Glycerol Content and Mixing Conditions," *Rheol. Acta.,* 38 (1999), 311-320.
[26.] T. Domenech, P. Edith and V. Bruno, "The importance of specific mechanical energy during twin screw extrusion of organoclay based polypropylene nanocomposites," *Composites Science and Technology*, 75 (2013), 7–14.
[27.] Y. Cai, "Crystallization and Melting Behavior of Biodegradable Poly(L-lactic acid)/Talc Composites," *E-Journal of Chemistry*, 9 (2012), 1569-1574
[28.] M. S. Huda, A. K. Mohanty, L. T. Drzal, E. Schut, M. Misra "Green" composites from recycled cellulose and poly(lactic acid): Physico-mechanical and morphological properties evaluation, Journal of Materials Science,40 (2005) (16), 4221-4229
[29.] V. S. Silverajah et al., Mechanical, Thermal and Morphological Properties of Poly(lactic acid)/Epoxidized Palm Olein Blend, *Molecules*, 17 (2012), 11729-11747.
[30.] I. R. Mustapa, A. S. Robert and K. Ing, Melting behaviour and Dynamic Mechanical Properties of Poly(Lactic acid)-Hemp-Nanosilica Composites, *Asian Transactions on Basic and Applied Sciences*, 3 (2013) (2), 29– 37.
[31.] H. Cheung et al., "A potential material for tissue engineering: Silkworm silk/PLA biocomposites," *Composites Part B: Engineering,* 39 (2008), 1026-1033.
[32.] M.S. Huda et al., "Effect of fiber surface-treatments on the properties of laminated biocomposites from poly(lactic acid) (PLA) and kenaf fibers," *Compos Sci Technol*, 68 (2008), 424–32.

Engineering Solutions for Sustainability: Materials and Resources II
Edited by: Jeffrey W. Fergus, Brajendra Mishra, Dayan Anderson, Emily Allyn Sarver, and Neale R. Neelameggham
TMS (The Minerals, Metals & Materials Society), 2015

INVESTIGATING *SOLANUM AETHIOPICUM* LEAF-EXTRACT AND SODIUM-DICHROMATE EFFECTS ON STEEL-REBAR CORROSION IN SALINE/MARINE SIMULATING-ENVIRONMENT: IMPLICATIONS ON SUSTAINABLE ALTERNATIVE FOR ENVIRONMENTALLY-HAZARDOUS INHIBITOR

Joshua Olusegun OKENIYI[1,*] Adebanji Samuel OGBIYE[2], Olubanke Olujoke OGUNLANA [3], Elizabeth Toyin OKENIYI[4], Oluseyi Ebenezer OGUNLANA [5]

[1]Mechanical Engineering Department, Covenant University, Ota 112001, Nigeria
[2]Civil Engineering Department, Covenant University, Ota 112001, Nigeria
[3]Biochemistry Programme, Department of Biological Sciences, Covenant University, Ota 112001, Nigeria
[4]Petroleum Engineering Department, Covenant University, Ota 112001, Nigeria
[5]Biochemistry Programme, Department of Biological Sciences, Crawford University, Igbesa 112001, Nigeria

Keywords: Sustainable corrosion-protection admixture, natural plant-extract, sodium-dichromate, saline/marine test-environment, electrochemical test-measurements analyses, Inhibition efficiency

Abstract

This paper investigates *Solanum aethiopicum* leaf-extract and the well-known but environmentally-hazardous sodium-dichromate inhibitor effects on concrete steel-rebar corrosion in 3.5% NaCl medium (simulating saline/marine environment). Different equal-concentration models (wt% cement) of the natural-plant leaf-extract and of sodium-dichromate were admixed in steel-reinforced concrete slabs from which electrochemical test-measurements were obtained for comparing admixture performance. Test-results, analyzed as per ASTM G16-95 R04, showed that only the 0.083% sodium-dichromate admixture outperformed the 0.083% *Solanum aethiopicum* leaf-extract in corrosion-inhibition effectiveness. The other natural-plant leaf-extract exhibited better inhibition-efficiency performance than their equal-concentration models of sodium-dichromate. The 0.25% *Solanum aethiopicum* leaf-extract exhibited optimal performance, η = 98.28%, at inhibiting steel-rebar corrosion among the also effective different concentrations of the plant-extract and of sodium-dichromate admixtures employed. These and phytochemical test-results bare indications that *Solanum aethiopicum* leaf-extract is a suitable, sustainable and eco-friendly alternative for the environmentally-hazardous sodium-dichromate inhibitor of steel-rebar corrosion in concrete designed for saline/marine environments.

Introduction

Chloride contaminated environments from natural marine, or from artificial de-icing salts constitute highly aggressive medium, that induce corrosion degradation of concrete steel-rebar and affect durability and sustainability of steel-reinforced concrete buildings and infrastructures [1-3]. Severity of chloride induced corrosion degradation of steel-reinforcement in concrete is usually high because the chloride ion inducing the corrosion is not usually consumed and is always available for further corrosion attacks on the reinforcing steel in the concrete immersed in chloride medium [4]. If not addressed in time, usually through costly maintenance and/or rehabilitations, chloride induced rebar corrosion could lead to volume expansive by-products that culminate in cracks, spalling, delamination and general loss of structural integrity of steel-reinforced concrete member [5-7].

A generally accepted method, among many other techniques, for mitigating chloride-induced corrosion attacks on concrete steel-reinforcements includes the use of corrosion inhibitors, especially, as admixtures in steel-reinforced concrete [8]. The use of inhibitors as admixtures in concrete finds acceptability among researchers and construction stakeholders due to its comparative advantages of low cost, ease of application, and effectiveness at inhibiting steel-reinforcement corrosion [5-6,8–10]. For this, compounds of chromates have been well-known as effective inhibitor of concrete steel-reinforcement corrosion in chloride contaminated environment [5,9,11]. However, in spite of their effectiveness, the use of compounds of chromates for addressing steel-reinforcement corrosion in aggressive environments is being restricted in many countries due to the toxicity and hazardousness of these compounds to the environmental eco-system [9,12]. Apart from environmentally-friendliness, other factors militating against sustainable usage of the compounds of chromates as corrosion inhibitors include ease of availability and cost effectiveness [13-14]. These constitute reasons research focus are shifting towards search for sustainable alternatives that can combine environmentally-friendliness with ready availability, cost effectiveness, and acceptable inhibition effectiveness on steel-rebar corrosion in aggressive environment [8,13-14].

Solanum aethiopicum, also known as African eggplant or garden egg, is a crop species readily available in the tropics with edible parts, i.e. fruits, leaves, stems, and roots, which are eaten raw or cooked as vegetables or for traditional medicine in the East and the West African sub region [15-17]. Also, the reported study by Chinedu et al [17] on the fruit of *Solanum* species indicated that *Solanum aethiopicum* is constituted of important phytochemicals such as tannins, saponins, glycosides, flavonoids, and more especially low level, i.e. as low as 14% of values considered toxic, of glycoalkaloids. Another study by Rahim et al [18] indicated that tannins from natural plant are potent inhibitors of pre-rusted steel corrosion in 3.5% NaCl medium. These engender motivations for investigating effects of the leaf-extract from this natural and medicinal plant on concrete steel-reinforcement corrosion in saline/marine environment. No reported work in literature has deliberated on this form of corrosion inhibition study using *Solanum aethiopicum* leaf-extract. Based on these, the performance of the natural plant extract on steel-rebar corrosion in the chloride contaminated medium is being compared with that of sodium dichromate that has been a well known inhibitor of steel-rebar corrosion in the medium. Therefore, the objective of this study is to investigate inhibition of *Solanum aethiopicum* leaf-extract and sodium dichromate on the corrosion of steel-rebar in concrete immersed in 3.5% NaCl medium, for simulating saline/marine environment.

Experimental Methods

Experimental Materials

Fresh leaves of *Solanum aethiopicum* (*S. aethiopicum*) *Solanaceae* were collected from a compound in Soka Area, Iwo Road Expressway, Ibadan, Nigeria. The leaves were identified at the Forestry Herbarium Ibadan (FHI), Nigeria, where a sample was also deposited with the voucher FHI. No. 109498. The leaves were dried under cover in a well aerated room maintained at 20 °C, blended to powder, and extract obtained from it (using methanol as the solvent in a Soxhlet extractor) were employed as concrete admixtures in the present case [9,13].

Sodium-dichromate ($Na_2Cr_2O_7$) used in the experiment is of Analar grade chemical and was obtained from Eurostar® Scientific Ltd. Four variations of different concentrations of this chemical was employed in equal-mass model with the leaf-extract of *S. aethiopicum*, for aiding performance comparisons, as admixtures in 100 mm × 100 mm × 200 mm steel reinforced concrete specimens. The variations of admixed concentrations in the specimens steel-reinforced concrete slabs employed for the chemical and natural plant extract include 0.083, 0.167, 0.25, and 0.333 wt% cement for the concrete mixing formulation. These constitute four specimens admixed with the different concentrations of $Na_2Cr_2O_7$ chemical and four specimens admixed with the different concentrations of *S. aethiopicum* leaf-extract. As a control specimen, a steel-reinforced concrete sample was also cast which contained neither of the $Na_2Cr_2O_7$ nor of the *S. aethiopicum* leaf-extract admixture (i.e. 0% admixture) such that the number of steel-reinforced concrete samples for this experimental study totalled 9 specimens.

The reinforcing steel employed is of 12 mm diameter and it has elemental composition in % of: 0.27 C, 0.40 Si, 0.78 Mn, 0.04 P, 0.04 S, 0.14 Cr, 0.11 Ni, 0.02 Mo, 0.24 Cu, 0.01 Co, 0.01 Nb, 0.01 Sn and the remainder Fe. This material was cut into rods of specimens, each of which are 190 mm long. Surface preparation for each of the steel rods were done uniformly according to standard procedure prescribed in ASTM G109-99a [19] and described in reported studies [6,13]. Each rod was centrally embedded in their respective specimens of concrete during casting such that 40 mm of the steel was protruding leaving the remaining 150 mm embedded in the concrete. The 40 mm protrusion, which finds usefulness as connector for electrochemical test-measurements, was painted with glossy paint after casting of each concrete specimen. For each slab of concrete specimen, mixing formulation employed includes 300.0 kg/m^3 cement, 890.6 kg/m^3 river sand, 1106.3 kg/m^3 granite stones and 149.7 kg/m^3 water; this indicates water/cement ratio = 0.499 [5-6,13,20].

Experimental Setup and Test-Data Measurements

For the experimental setup, each sample of steel-reinforced concrete specimen was immersed partially along the 200 mm length in a plastic bowl containing 3.5% NaCl test-solution, for simulating saline/marine environment [20-21]. This test-solution was made up in the bowl to just below the protruding reinforcing steel, in that longitudinal arrangement, but without touching the rebar.

From each steel-reinforced concrete sample, electrochemical test-measurements, of half-cell potential (*HCP*) versus Copper/copper sulphate electrode, CSE, (Tinker & Razor®), and of corrosion rate (*CR*) through 3-electrode LPR Data Logger (Metal Samples®) [5,22], were obtained. Connection mode for the *HCP* followed standard procedure prescribed in ASTM C876-91 R99 [23], and this and connection mode for the *CR* followed description in reported works [5-6,11,24] The electrochemical tests were measured from each concrete sample in five days interval for the first forty-days, then in seven days interval for the following four weeks, which gave 13-point measurements in an experimental period totalling sixty-eight days.

Experimental Test-Data Modelling and Analyses

From the steel-reinforced concrete specimens, measured test-data of each electrochemical test-monitoring method were subjected to the statistical analyses, as prescribed in ASTM G16-95 R04, of the Normal and of the Weibull distributions. These have respective cumulative distribution functions (cdf) given by [6,24]:

$$F_N(x) = \frac{1}{\sigma\sqrt{2\pi}} \int_{-\infty}^{x} \exp\left[-\frac{(x-\mu_N)^2}{2\sigma^2} \right] dx \tag{1}$$

$$F_W(x) = 1 - \exp\left\{ -\left(\frac{x}{c}\right)^k \right\}; \quad c > 0,\ k > 0 \tag{2}$$

Where: x is measured test-data of corrosion rate (mm/y). Also, μ_N is Normal mean/location parameter (mm/y); σ is Normal standard deviation/scale parameter (mm/y); k is dimensionless Weibull shape parameter and c is Weibull scale parameter (mm/y); these were estimated as described in [9]. The Weibull parameters find usefulness for computing the Weibull mean model, μ_W, from:

$$\mu_W = c\Gamma\left(1 + \frac{1}{k}\right) \tag{3}$$

Compatibilities of the scatter of each datasets, of electrochemical measurements per specimen, to the Normal and to the Weibull distributions were also studied using the Kolmogorov-Smirnov (K-S) goodness-of-fit (GoF) statistics at $\alpha = 0.05$ significant level [13,24-25]. This was done to heed the prescriptions from [26-27] for avoiding grossly erroneous conclusion that could be inherent from the use of distribution model other than that which a particular corrosion test-data distributed like for describing that corrosion test-data. Thus, the establishment of the scattering of corrosion rate test-data like any of the distributions model facilitate use of the corrosion rate mean model of that distribution for estimating inhibition efficiency, $\eta(\%)$, of each admixture concentration in the concrete samples from:

$$\eta(\%) = \frac{\mu_{dist,CR\_control\ sample} - \mu_{dist,CR\_admixed\ sample}}{\mu_{dist,CR\_control\ sample}} \times 100 \tag{4}$$

Results and Discussion

Results of Test-Data Modelling and Analyses

The plots of the K-S GoF analyses of the corrosion test-data, from the steel-reinforced concrete samples are presented in Figure 1; in which linear plot of $\alpha = 0.05$ was also included for interpreting distribution followed (or not followed) by each test-dataset directly from the plots. This showed that *CR* datasets from 3 samples, the 8 g $Na_2Cr_2O_7$, the 4 g and the 8 g *S. aethiopicum* admixed samples, were not distributed like the Normal distribution; while *HCP* dataset from the 2 g $Na_2Cr_2O_7$ admixed sample was not distributed like the Weibull distribution. Rather, all *HCP* datasets in the study followed the Normal distribution, and all the *CR* datasets followed the Weibull distribution. This form of selective probability density distribution of corrosion test-variables followed that which was encountered in [9].

Based on this, plots of the Normal mean models of *HCP* are presented in Figure 2 for all the steel-reinforced concrete samples. The figure also includes linear plots for interpreting probability of corrosion risk, as per ASTM C876-91 R99 [23] criteria, directly from the figure. This showed that the steel-reinforced concrete samples admixed with *S. aethiopicum* leaf-extract generally exhibited lower probability of corrosion risks than the control sample and the samples admixed with $Na_2Cr_2O_7$.

Similarly, plots of the Weibull mean models of corrosion rate, overlaid with the plots of resultant inhibition efficiency by the equal-mass admixture models of $Na_2Cr_2O_7$ and the *S. aethiopicum* leaf-extract, are presented in Figure 3 for the steel-reinforced concrete samples. This figure shows that though the $Na_2Cr_2O_7$, for which inhibition efficiency $\eta > 90\%$ for all the concentrations studied, was confirmed as highly effective at inhibiting steel-rebar corrosion, the toxic chemical was outperformed in effectiveness by most of the *S. aethipicum* leaf-extract admixtures. More specifically, only the 0.083% *S. aethiopicum* was outperformed by the $Na_2Cr_2O_7$ among the other equal-mass models of steel-reinforced concrete admixtures by the plant extract and the dichromate chemical being compared for corrosion inhibition effectiveness in the study. Inhibition efficiency $\eta = 91.95\%$ by the 0.083% *S. aethiopicum* compared to $\eta = 93.92\%$ by the 0.083% $Na_2Cr_2O_7$ admixture.

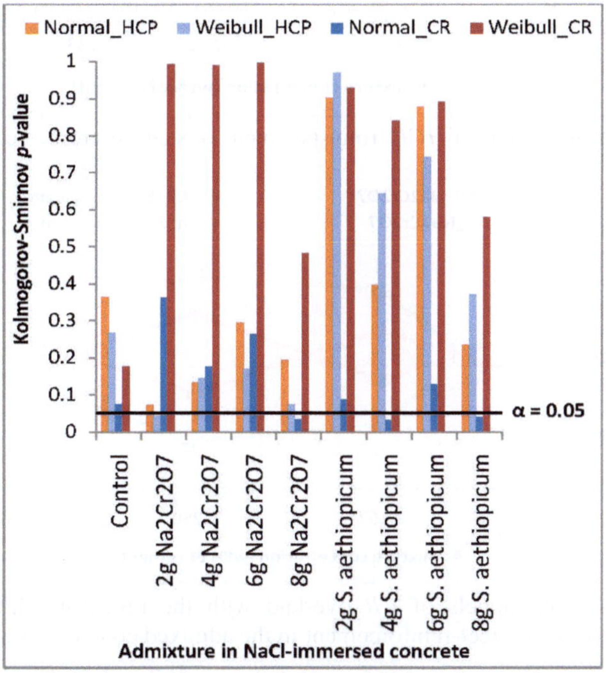

Figure 1. Kolmogorov-Smirnov goodness-of-fit test-statistics of corrosion test-data.

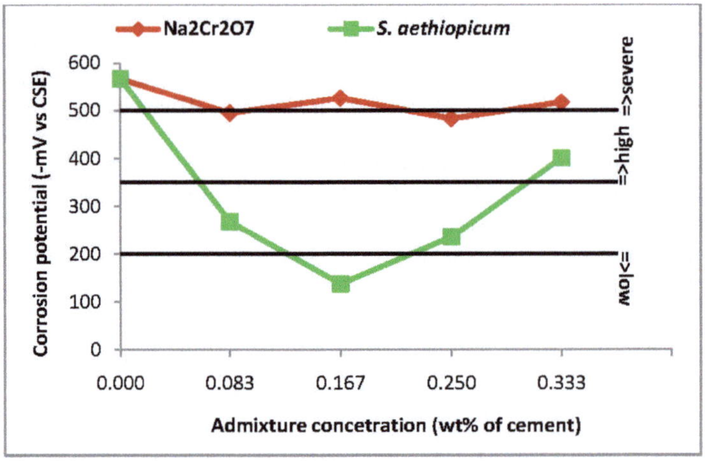

Figure 2. Normal mean models of *HCP* from specimens of steel-reinforced concrete samples.

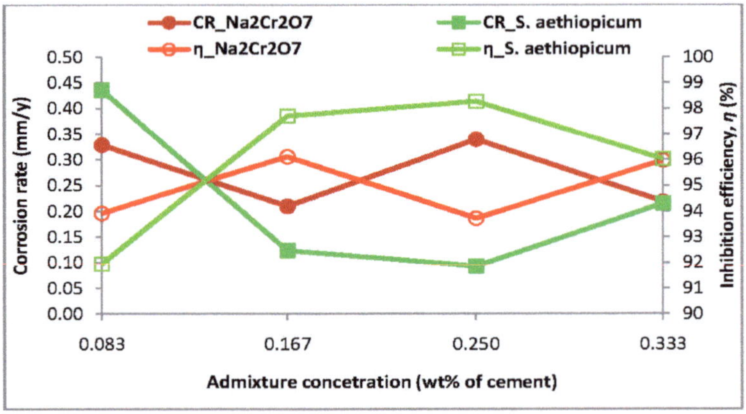

Figure 3. Weibull mean models of *CR* overlaid with the resultant inhibition efficiency of admixtures on the corrosion steel-reinforcement in the admixed concrete samples.

However, the other concentrations of *S. aethiopicum* exhibited higher inhibition efficiency performance than their counterparts of equal-mass model of $Na_2Cr_2O_7$ admixture. By this, the 0.250% *S. aethiopicum* exhibited optimal effectiveness performance of η = 98.28%, among the other also effective admixtures in the study, at inhibiting steel-rebar corrosion in the concrete immersed in the saline/marine simulating-environment. In comparison, the 0.167% $Na_2Cr_2O_7$ was the most effective, η = 96.12%, among the admixture concentrations of the $Na_2Cr_2O_7$ chemical, on the corrosion inhibition of the studied concrete steel-rebar. This supports the use of *S. aethiopicum* as a suitable substitute for sodium-dichromate inhibitor of concrete steel-reinforcement corrosion in chloride contaminated environment.

Phytochemical test-results of *S. aethiopicum* leaf-extract

The phytochemical analyses of *S. aethiopicum* leaf-extract was carried out in this study, using procedures that were prescribed in [28], because phytochemical analyses from other research works [16-17] had been deliberated on the fruit-extract of this natural-plant. Table 1 shows the results of the phytochemical analyses of the *S. aethiopicum* leaf-extract used in this study. This confirmed that *S. aethiopicum* leaf-extract is constituted not only of tannins that had been found to be useful at inhibiting metallic corrosion but also of saponins that could hydrolyse glycoside and consequently prevent toxicity that might be inherent in the intact molecule [17]. These and the effective inhibition performance of *S. aethiopicum* leaf-extract bare indications that it is a suitable, sustainable, and eco-friendly alternative for the environmentally-hazardous sodium-dichromate inhibitor of steel-rebar corrosion in saline/marine immersed-concretes.

Table 1. Phytochemical test-results of *S. aethiopicum* leaf-extract*

Tannins	Phlobatannins	Saponins	Glycosides	Flavonoids	Stenoids	Terpenoids	Alkaloids
+	+	+	+	+	+	+	+

* Present ≡ +; Absent ≡ –

Conclusions

The conclusions that could be drawn from this study include:

- In spite of the high inhibition effectiveness, $\eta > 90\%$, performance exhibited by the sodium-dichromate admixtures, only 0.083% $Na_2Cr_2O_7$, $\eta = 93.92\%$, outperformed its equal-mass model of 0.083% *S. aethiopicum* , $\eta = 91.95\%$, while other concentrations of the natural plant extracts outperformed their equal-mass counterparts of the $Na_2Cr_2O_7$ at inhibiting steel-reinforcement corrosion in the NaCl-immersed concrete samples;
- The 0.250% *S. aethiopicum* exhibited optimal performance, $\eta = 98.28\%$, which still outperformed the 0.167% $Na_2Cr_2O_7$ that exhibited the best effectiveness, $\eta = 96.12\%$, among the toxic chemical admixtures, at inhibiting steel-rebar corrosion in concrete immersed in the saline/marine simulating environment;
- Phytochemical analyses showed that *S. aethiopicum* is, among others, constituted of tannins that has been known to inhibit metallic corrosion as well as saponins that is potent at averting associated toxicity in the plant's constituents;
- These corrosion-inhibition and phytochemical test-results bare indications of *S. aethiopicum* as a suitable, sustainable, and eco-friendly alternative for the environmentally-hazardous sodium-dichromate inhibitor of steel-rebar corrosion in concrete designed for the saline/marine environment.

References

1. B. Pradhan, "Corrosion behavior of steel reinforcement in concrete exposed to composite chloride–sulfate environment," *Construction and Building Materials*, 72 (2014), 398–410.

2. E. Güneyisi, M. Gesoğlu, F. Karaboğa, K. Mermerdas, "Corrosion behavior of reinforcing steel embedded in chloride contaminated concretes with and without metakaolin," *Composites: Part B*, 45 (2013), 1288–1295.

3. D. Chen, S. Mahadevan, "Chloride-induced reinforcement corrosion and concrete cracking simulation," *Cement & Concrete Composites*, 30 (2008), 227–238.

4. J. Wei, J.H. Dong, W. Ke, "Corrosion evolution of scaled rebar in concrete under dry/wet cyclic condition in 3.5% NaCl solution," *International Journal of Electrochemical Science*, 8 (2013), 2536–2550.

5. J.O. Okeniyi, I.O. Oladele, I.J. Ambrose, S.O. Okpala, O.M. Omoniyi, C.A. Loto, A.P.I. Popoola, "Analysis of inhibition of concrete steel-rebar corrosion by $Na_2Cr_2O_7$ concentrations: Implications for conflicting reports on inhibitor effectiveness," *Journal of Central South University*, 20 (12) (2013), 3697–3714.

6. J.O. Okeniyi, O.M. Omoniyi, S.O. Okpala, C.A. Loto, A.P.I. Popoola, "Effect of ethylenediaminetetraacetic disodium dihydrate and sodium nitrite admixtures on steel-rebar corrosion in concrete," *European Journal of Environmental and Civil Engineering*, 17 (5) (2013), 398-416.

7. C.Q. Li, J.J. Zheng, W. Lawanwisut, R.E. Melchers, "Concrete delamination caused by steel reinforcement corrosion," *Journal of Materials in Civil Engineering*, 19 (7) (2007), 591–600.

8. J.J. Shi, W. Sun, "Effects of phosphate on the chloride-induced corrosion behaviour of reinforcing steel in mortars," *Cement & Concrete Composites*, 45 (2014), 166–175.

9. J.O. Okeniyi, C.A. Loto, A.P.I. Popopla "Electrochemical performance of *Anthocleista djalonensis* on steel-reinforcement corrosion in concrete immersed in saline/marine simulating-environment," *Transactions of the Indian Institute of Metals*, 67 (6) (2014), 959–969.

10. X. Zhou, H. Yang, F. Wang, "Investigation on the inhibition behavior of a pentaerythritol glycoside for carbon steel in 3.5% NaCl saturated $Ca(OH)_2$ solution," *Corrosion Science*, 54 (2012), 193–200.

11. J.O. Okeniyi, O.A. Omotosho, O.O. Ajayi, and C.A. Loto, "Effect of potassium-chromate and sodium-nitrite on concrete steel-rebar degradation in sulphate and saline media," *Construction and Building Materials*, 50 (2014), 448–456.

12. R. Kahraman, H. Saricimen, M. Al-Zahrani, S. Al-Dulaijan, "Effect of inhibitor treatment on corrosion of steel in a salt solution," *Journal of Materials Engineering and Performance*, 12 (5) (2003), 524–528.

13. J.O. Okeniyi, C.A. Loto, A.P.I. Popoola, "Electrochemical performance of *Phyllanthus Muellerianus* on the corrosion of concrete steel-reinforcement in industrial/microbial simulating-environment," *Portugaliae Electrochimica Acta*, 32 (3) (2014), 199-211.

14. N.O. Eddy, P.O. Ameh, M.Y. Gwarzo, I.J. Okop and S.N. Dodo, "Physicochemical study and corrosion inhibition potential of *Ficus tricopoda* for aluminium in acidic medium," *Portugaliae Electrochimica Acta*, 31 (2) (2013), 79–93.

15. T.K. Lim, *Edible Medicinal And Non-Medicinal Plants, Volume 6, Fruits* (Springer Netherlands, 2013), 310–317.

16. S.O. Eze, C.Q. Kanu, "Phytochemical and nutritive composition analysis of *Solanum aethopicum* L.," *Journal of Pharmaceutical and Scientific Innovation*, 3 (4) (2014), 358–362.

17. S.N. Chinedu, A.C. Olasumbo, O.K. Eboji, O.C. Emiloju, O.K. Arinola, D.I. Dania, "Proximate and Phytochemical Analyses of *Solanum aethiopicum* L. and *Solanum macrocarpon* L. Fruits," *Research Journal of Chemical Sciences*, 1 (3) (2011), 63–71.

18. A.A. Rahim, E. Rocca, J. Steinmetz, M.J. Kassim, "Inhibitive action of mangrove tannins and phosphoric acid on pre-rusted steel via electrochemical methods," *Corrosion Science*, 50 (2008), 1546–1550.

19. ASTM G109-99a, *Standard Test Method for Determining the Effects of Chemical Admixtures on the Corrosion of Embedded Steel Reinforcement in Concrete Exposed to Chloride Environments* (West Conshohocken, PA: ASTM International, 2005).

20. M. Ormellese, M. Berra, F. Bolzoni, T. Pastore, "Corrosion inhibitors for chlorides induced corrosion in reinforced concrete structures," *Cement and Concrete Research*, 36 (2006), 536–547.

21. M.M. Mennucci, E.P. Banczek, P.R.P. Rodrigues, I. Costa, "Evaluation of benzotriazole as corrosion inhibitor for carbon steel in simulated pore solution." *Cement & Concrete Composites*, 31 (2009), 418–424.

22. V.S. Sastri, *Green corrosion inhibitors: theory and practice* (Hoboken, New Jersey: John Wiley & Sons, Inc., 2011).

23. ASTM C876–91 R99, *Standard test method for half-cell potentials of uncoated reinforcing steel in concrete* (West Conshohocken, PA: ASTM International, 2005).

24. J.O. Okeniyi, I.J. Ambrose, S.O. Okpala, O.M. Omoniyi, I.O. Oladele, C.A. Loto, and P.A.I. Popoola " Probability density fittings of corrosion test-data: Implications on $C_6H_{15}NO_3$ effectiveness on concrete steel-rebar corrosion," *Sadhana - Academy Proceedings in Engineering Science,* 39 (3) (2014), 731–764.

25. J.O. Okeniyi, E.T. Okeniyi, "Implementation of Kolmogorov–Smirnov p-value computation in visual basic®: Implication for microsoft excel® library function," *Journal of Statistical Computation and Simulation*, 82 (2012), 1727–1741.

26. ASTM G16–95 R04, *Standard guide for applying statistics to analysis of corrosion data* (West Conshohocken, PA: ASTM International, 2005).

27. P.R. Roberge, "Statistical interpretation of corrosion test results," *ASM handbook, Vol 13A – Corrosion: fundamentals, testing, and protection*, ed. S.D. Cramer and B.S. Covino Jr., (Materials Park, OH: ASM International, 2003), 425–429.

28. H.O. Edeoga, D.E. Okwu, B.O. Mbaebie, "Phytochemical constituents of some Nigerian medicinal plants," *African Journal of Biotechnology*, 4 (7) (2005), 685–688.

Engineering Solutions for Sustainability: Materials and Resources II
Edited by: Jeffrey W. Fergus, Brajendra Mishra, Dayan Anderson, Emily Allyn Sarver, and Neale R. Neelameggham
TMS (The Minerals, Metals & Materials Society), 2015

SUSTAINABILITY METRICS FOR EFFICIENT AND INNOVATIVE RESIDENTIAL BUILDING WALL SYSTEMS

Ryan L. Solnosky[1] and Ali M. Memari[1]

[1] The Pennsylvania State University
104 Engineering Unit A University Park, Pa 16802, USA

Keywords: Residential Construction, Sustainability, Wall Systems, Design Metrics

Abstract

With sustainability becoming more recognized and desirable by homeowners, designers need to have better metrics to make solutions possible. One of the most effective building components that can be optimized for sustainability purposes is the wall system. To select the proper wall system that is efficient and at times innovative, there is a need to better understand how metrics relate to different wall configurations and materials. This paper expands upon prior works to identify the key sustainability metrics for residential wall systems that designers, builders and homeowners can use to assess and select the best wall system. The metrics highlighted include: embodied energy, material-specific, and impact on the environment. These metrics will be examined for wall systems made up of: Insulated Concrete Form, Wood-Frame, Steel Stud, Structural Insulated Panel, Concrete Masonry Unit, Autoclaved Aerated Concrete, and Precast Concrete.

Introduction

Building construction poses highly negative impacts to the environment [1], including: consuming 40% of natural resources, generating over 45 % of waste disposal, as well as emitting large amounts of CO_2 through building material production, construction, renovation, and demolition [2, 3, 4]. The Environmental Protection Agency (EPA) [5] states that in the U.S., the environmental impacts of buildings over their life cycle include 39% of total CO2 emissions, 72% of total electricity consumption, 38.9% of total primary energy use, 13% of potable water consumption, 2.3 billion acres use, and 40% of landfill material generated (254 million tons annually). Moreover, construction materials generate annually millions of tons of waste around the world and consume large amount of embodied energy. Lomite [6] estimates that about 3.5 billion metric tons of cement will be produced by the end of 2050, which with embodied energy for cement (7.8 MJ/Kg), will result in a doubling of CO_2 emissions. A majority of this natural resource energy is the energy consumed during the building while it is in use. Here, heating, cooling and lighting make up the major fraction of energy usage by commercial and residential buildings [7]. A large percentage can be directly related to an individual's household [8]. Therefore, residential designs play a significant role in reducing environmental impacts.

In new design and construction across the building industry from residential to commercial, a prominent area being considered in the design of building is lifecycle sustainability. Sustainability is defined by a life-cycle and material analysis of the product/project. For building projects, sustainable design encompasses a multitude of technical, environmental, economic, social, organizational, and innovation-related issues [9]. Achieving sustainability depends on how well these issues are handled during the lifecycle of the project. Designs that meet sustainability refer to solutions that have a minimal negative impact on the environment, including: reducing environmental impacts in site planning, safeguarding water and water efficiency, energy efficiency and renewable energy, conservation of materials and resources, indoor environmental quality, user satisfaction, etc. [10].

In the U.S., the government, non-profit nongovernmental organizations, such as the U.S. Green Building Council (USGBC) and the National Institute of Building Sciences (NIBS) have commissioned analyses and reports for a sustainable built environment [11-13]. Based on the Energy Policy Act of 2005, the federal, state, and local governments have provided incentives for high performance sustainable buildings [13-16].

One of the challenges in the sustainable development movement has been defining metrics that more effectively aid in making decisions and how they affect the environment [17]. In looking at residential construction (where impacts are highest), sustainability can have a multitude of facets to what it encompasses. Presently, sustainability can look at energy efficiency, embodied carbon and energy, recycling of materials, LEED, and high performance design, to name a few desirable attributes [1, 12, 21]. It is the responsibility of the design professional to ensure that the solutions are in accordance with applicable codes and regulations regardless of the sustainability. However, this often can lead to conflict in design decisions due to poor understanding of metric relevance between domains [3, 20]. From an environmental standpoint, designers should choose more energy efficient materials and systems in their designs to reduce energy consumption [1]. The concept of sustainability, whether defined as the total energy consumed during a product's life-cycle or the resultant environmental impacts, remains up for interpretation [18-19].

The absence of a comprehensive list of criteria for sustainable materials is a major problem facing designers and architects when selecting sustainable building materials [1]. This leads to the need for more studies into how wall systems behave, so that we can make better informed design decisions. For this paper's scope, the understanding of more rigorous design metrics can provide better solutions that directly impact the economy and the environment. As part of this preliminary investigation, the results summarize the appropriateness of different wall systems for single-family dwellings in a simple rating form across metrics that can allow professionals and builders to make more informed decisions about these wall systems.

Metrics of Interest

Different wall systems employ different materials and methods of construction, which lead to different metrics for selection consideration as part of design. In these wall systems, predominantly a single material type carries the structural load, whereas for thermal protection, it is the combination of materials that can provide superior performance. Due to conflicting/competing requirements across performance types, what is best for one may not be best for another one [22-23]. Building certifications that presently define sustainability around the world disregard many positive aspects to designs and have limited objective metrics not reflected on the points and/or rating systems by these agencies [24-28].

A literature study conducted has identified a wide range of metrics with different classifications of wall systems that can be considered [1, 29]. Here, this study identified the relevant wall metrics of importance from the literature and organized them into categories proposed by Memari et al. [29]. Table 1 identifies the metrics of interest across Environmental, Design, and Serviceability categories. Environmental driven metrics are the measurements that directly impact the environment, whether it is through limiting resources, toxic contaminations, or reusing materials.

Design driven metrics are indirect values in that their influence can lead to decisions that will impact sustainable features of the wall system. Serviceability, like design, is an indirect measure; here however, these metrics measure the effects the system has on its occupants.

Table 1: Metrics Organized by Category

Environmental	Design	Serviceability
Recyclable	Aesthetics	Durability
Embodied Energy	Indoor quality	Service life
Natural Resources	Reusability	Thermal comfort
Non-toxic Materials	Flexibility	Acoustic comfort
Low VOC	Lifecycle cost	
Biodegradability	Construction time	
	Labor skill	
	constructability	

Walls of Interest

Although there exists many different viable load bearing wall systems that can be used for residential construction [30], each can be classified according to their broad structural systems [31-32]. Furthermore, residential building trends vary depending on where the home is located. While most homes have similarities across regions, there are locations globally that have unique residential systems suitable for those regions, which were not included in this study; instead, only the U.S. market was considered. For this study, only the following seven wall types that are more widely used, were investigated (Figure 1): Conventional wood-frame system (WF), Light Gage Steel Stud (SS), Structural Insulated Panels (SIP), Precast Concrete Products (PCP), Concrete Masonry Unit (CMU), Autoclaved Aerated Concrete (AAC), and Insulated Concrete form (ICF).

Wood-Frame (by author) | Light Gage Steel Stud (by author) | Concrete Masonry Unit (by author) | Precast Concrete Products (Image courtesy of Dukane Precast)

Structural Insulated Panels (image courtesy of Murus SIPs) | Autoclaved Aerated Concrete (image courtesy of PCA) | Insulated Concrete Form (image courtesy of Quad-Lock Building Systems Ltd.)

Figure 1: Wall Systems on Interest

Environmental Metrics

Of the identified metric categories (Table 1), the environmental class is directly correlated to sustainability. This study identified 7 metrics within the environmental class that are applicable to residential wall systems. They are: Recyclable, Biodegradability, Natural Resources, Non-toxic Materials, Low Volatile Organic Compounds (VOCs), and Embodied Energy. This section discusses the highlights regarding the applicability of these metrics for the identified wall systems previously mentioned.

Recyclable is a metric of the material's capacity to be used as a resource in the creation of new products after its existing use is completed [33]. Materials can have varying levels of being recyclable. WF and SIP walls are the most recyclable in terms of not needing to convert the material back into a base material or reused as long as they have not been damaged in such a manner that they become ill-capable of functioning as building elements. CMU, AAC, ICF, and PCP can all be recycled by taking the concrete portions and grinding the concrete back into aggregate for future applications as fill for sites or in future concrete mixes. SS walls and the reinforcing in the concrete walls are recyclable from the standpoint of being melted but the energy necessary is demanding, thus not ideal.

The biodegradability metric considers the material's potential to naturally decompose when discarded [33]. Of the wall systems listed, only the WF system meets biodegradability as wood will naturally decompose back into organic material. If WF walls are left to biodegrade, then all non-wood based products (nails, drywall, insulation and finishes) must be removed [1]. There are certain types of biodegradable foam plastics available that could be used in ICFs but they are not presently used due to the exposure the foam witnesses during its usage, which advances decay causing problems during building's usage.

A broader metric is natural resources. This identifies materials found in nature [34]. Out of the seven wall systems under investigation, WF walls are the most in line with natural resources as the primary material is natural growing wood [35]. Concrete based wall system (CMU, ICF and PCP) can claim some natural resources in that the aggregates are naturally occurring materials with minimal modifications needed. The same can be said about SS except that most steel utilized in building applications is recycled and no longer a natural resource.

In comparing the natural resources, biodegradability, and recyclable metrics, a matrix (Table 2) was generated that rates each of the wall systems against one another in these similar but distinct areas. This matrix can be considered to offer guidance for selecting alternative systems. The values assigned in the matrix are based on available literature by looking at the subcomponents for each system.

Table 2. Wall Systems Rating Based on Biodegradability, Natural Resources, and Recyclable Criteria

Wall Type	Biodegradability	Natural Resources	Recyclable
WF	III	III	II
SS	I	I	III
SIP	II	II	II
CMU	I	II	III
AAC	I	I	III
ICF	I	II	II
PCP	I	II	II

Note: 3-level rating where: I=fails the definition, II=meets some of the definition and III=meets or most all of the definition.

One of the major health risks in homes, particularly in the recent past is the toxicity of materials [36]. This leads the way to a non-toxic material metric. This metric is objective in that it is easily met or exceeded based on predefined acceptable limits for different situations. According to the

ERC [37], toxic materials are those with a lethal oral dose LD50 at or below 300 mg/kg, a lethal dermal dose LD50 at or below 1,000 mg/kg, or a lethal aerial dose LD50 at or below 4 mg/l. Material finishes and coatings are perhaps the most common sources for this metric in the wall system. For other similar health issues, metrics are grouped in this category such as allergens and insects/pests [38]. Proliferation or inherent abundance of these conditions can cause health issues, particularly in children [39] and the elderly [40].

The VOCs metric relates back to sustainability, and different classes of sustainable materials have different levels of VOCs [41, 42]. Different areas within the wall have varied amounts of VOCs due to when the section emits the VOCs, they are basically airborne and may remain suspended there. However, with any air and/or vapor infiltration/leakage across the wall section under differential pressure, these VOCs can move with that infiltration and be transported to the building's interior and stay air-borne for some time. In the building enclosure, sources can include: manufactured wood products, adhesives, sealants, caulks, paints, coatings, surfacing, cladding, insulation, air barriers, and even wall board [43]. Most interior coating products, including paints, primers, stains, and wood finishes, emit unhealthy VOCs. Conventional caulking materials or sealants and construction adhesives also off gas VOCs. The off gassing that occurs during application can be substantial for days, weeks and even months afterward, and can potentially affect the health and comfort of people living and working in the home [43].

For non-toxic and VOCs metrics, a matrix (Table 3) was generated similar to the prior matrix. The VOC and Non-toxic material comparison looks at the primary materials the make up the building enclosure. Paints and finishes are critical but are not compared here due the seemingly unlimited number of choices in material types but also in their low impact on the overall building performance of the wall system.

Table 3. Wall Systems Rating Based on Nontoxic Material and Low VOC Criteria

Wall Type	Non-toxic Materials	Low VOC
WF	III	I-II
SS	III	III
SIP	II	I-II
CMU	III	III
AAC	III	III
ICF	II	I-II
PCP	III	II-III

Note: 3-level rating for non-toxic where: I=fails the definition, II=meets some of the definition, and III=meets all or most of the definition.
3-level rating for VOC where: I=High VOC, II = Some possible VOC, and III=No significant VOC.

Analysis of the wall systems based on the embodied energy or the total energy needed for a product (human resources, materials, required power, etc.) provides a standard to gauge the overall energy efficiency [12, 44-45]. Previous research by Hsu [46-47] has shown that the embodied energy plays a critical role when operating energy can no longer be optimized. Limited studies are available that focus on properties related to embodied energy for wall systems. A study done by the authors [29] looked to examine through simulations the embodied energy performance of different walls, both with and without windows. This study looked at accounting for embodied energy within the lifecycle during the following stages: extraction/harvest of raw materials, refining, manufacturing of building materials and products, construction activities, renovations and alterations, demolition and disposal, and transportation.

Upon completion of computer simulations, it was evident that the embodied energy of each wall system, whether comprised of concrete, polymer foams, steel, or traditional wood, would vary with great discrepancy from 90% to 161% from a WF baseline. Both the PCP and SIP homes require a considerable amount of energy in order to manufacture and implement, while the wood and steel stud homes demonstrate lesser amounts of required energy by 18%. Construction is nearly the same for wood and steel studs, including SIPs (45 GJ), but for PCPs the energy is 80 GJ, a 43% increase. In looking at SS and SIP systems, they are at a comparative level to be beneficial to greenhouse gas emissions. Renewable wood stud walls reduce carbon dioxide emissions by nearly 38% when compared to a PCP.

CMUs require the additional process of forming the concrete into symmetric units, and therefore consume the second most fossil fuel in their production. AAC contains a large amount of air, and thus requires the least amount of fossil fuel because it contains the lowest percentage of cement per unit area. The ICF system, in addition to a large amount of concrete, also requires synthetic insulation in large quantities causing this system to require the most fossil fuel. The PCP systems are thought to have the greatest levels of fluctuation for construction, maintenance, and end of life embodied energies, as they roughly require 24% more embodied energy than the wood stud home.

The geographic location contributes substantially to embodied energy due to the availability of local resources, transportation distance, and material efficiency for a climate. A residential home was compared in two locations: Pittsburgh, PA, USA and Los Angeles, CA, USA [29]. When comparing the two locations, it was observed that CMU wall systems have a significantly higher fossil fuel consumption for the Los Angeles location at 1.9 GJ compare to 1.18 GJ in PA. While CMU and AAC homes have slightly higher values overall (1.33 time larger for both locations), Adobe and ICF have relatively the same embodied energy as the WF (90% and 95% of wood in CA) based on the tradeoffs between maintenance, manufacturing, and materials. Similarly, we have AAC with a 101% of baseline in CA and 133% in PA. ICFs fall in the middle of embodied energy consumption (171 GJ below the average of all walls at 2,500 GJ), which can be attributed to the embodied energy from the production of the cements and the insulation materials. WF walls are one of the best systems (15-24% energy reduction) from sustainability and embodied energy standpoint. On comparison, WF wall systems require 15-16% less total embodied energy than thermally comparable homes employing alternative steel or concrete based systems [49]. A matrix (Table 4) was developed to suggest a rating of the performance of the systems in a side-by-side manner. The values assigned in the matrices are somewhat subjective but do follow the overall trends of the simulation results.

Concluding Remarks

The main objective this paper was to discuss the primary sustainability metric classes and categories that need to be considered. In particular, this paper focused on providing a comparison between different wall systems for different metrics within the environmental category of sustainability. The metrics for the different wall systems indicate that not all wall system types are ideal for all situations, thus a clear "winner" in wall types cannot be easily selected. The matrices within this paper provide a starting point for designers to help in the selection of the optimum wall system. It is recommended that the embodied energy be looked at first followed by the other environmental metrics for they have less overall impact compared to embodied energy. To advance this study, further analytical and experimental studies are needed to provide quantifiable data necessary to make more objective decision-making aids, particularly, looking at the tradeoffs between the metrics within a class and across sustainable classes.

Table 4. Rating of the Wall Systems Based on Embodied Energy Criterion

Wall Type	**PA**	**CA**
WF	III	III
SS	II-III	II-III
SIP	II	II
CMU	I	I
AAC	II	II
ICF	I-II	I-II
PCP	I	I

Note: 3-level embodied energy rating where: I=poor, II=average, and III=good

References

1. S.M. Baharetha, A.A. Al-Hammad, and H.M. Alshuwaikhat, "Towards a Unified Set of Sustainable Building Materials Criteria," *ICSDEC*, (2012), 732-740.

2. Department of Energy, "U.S. Energy Information Administration Annual Energy Review," U.S. Department of Energy, Washington, DC, (2009).

3. Hermreck, C., and W. K. Chong, "Embodied Energy of CDW Recycling and Technical Metabolism Due to Regional Differences and Building Designs." *2009 Construction Research Congress*, (2009), 273-281.

4. B. Milani, "Building Materials in a Green Economy: Community-based Strategies for Dematerialization," University of Toronto, (2005).

5. Environmental Protection Agency (EPA). "Buildings and their Impact on the Environment: A Statistical Summary." (2009). Available at http://www.epa.gov/greenbuilding/pubs /gbstats.pdf. Accessed 03/20/2012.

6. F.R. Lopez, and G.F. Sanchez, "Challenges for Sustainability Assessment by Indicators." *Leadership and Management in Engineering,* October (2011), 321-325.

7. S. Mulki and A. Hinge, "Green Investment Horizons: Effects of Policy on the Market for Building Energy Efficiency Technologies," *World Resources Institute*, November (2010).

8. J. Carpenter and J. Zhou, "Life Cycle Analysis of a St. Louis Flat Roof Residential Retrofit for Improved Energy Efficiency," *ICSDEC*, (2012), 20-28.

9. F.H. Halicioglu, D. Arditi, and S. Gunhan, "Towards Advanced Solutions for Achieving Sustainability Goals in Buildings: Lessons Learned from Case Studies," *ICSDEC*, (2012), 835-843.

10. World Commission on Environment and Development (WCED), *Our common future*. Oxford: Oxford University Press, (1987). 43 pp.

11. U.S. Green Building Council (USGBC), "Building a greener future. Special advertising section in partnership with Fortune." *Fortune*, March, 20, (2006), 2-14.

12. U.S. Green Building Council (USGBC), "Leadership in energy and environmental design (LEED)." http://www.usgbc.org/ [accessed 03/16/2012].

13. National Institute of Building Sciences (NIBS), "High performance buildings," *NIBS,* (2012).

14. U.S. Department of Energy (U.S. DOE), "Federal energy management program (FEMP) - High performance Federal buildings." http://femp.buildinggreen.com/ [accessed 03/16/2012].

15. U.S. General Services Administration (U.S. GSA), "Office of Federal high performance green buildings." http://www.gsa.gov/portal/category/101107 [accessed 03/16/2012].

16. Whole Building Design Guide. (WBDG), "Federal high performance and sustainable buildings." http://www.wbdg.org/references/fhpsb.php [accessed 03/16/2012].

17. E. Stek, D. DeLong, T. McDonnell, and J. Rodriguez, "Life cycle assessment using ATHENA® impact estimator for buildings: A case study," *Structures Congress* (2011) 483-494.

18. R. Haynes, "Embodied Energy Calculations within Life Cycle Analysis of Residential Buildings," (2010), [Online], http://etool.net.au/wp-content/uploads/2012/10/Embodied -Energy-Paper-Richard - Haynes.pdf.

19. L. Klotz, and M. Horman, "Counterfactual Analysis of Sustainable Project Delivery Processes." *Journal of Construction Engineering and Management*, 136(5), (2010), 595-605.

20. T. Graedel, and B. Allenby, *Industrial Ecology and Sustainable Engineering*, Prentice Hall, Upper Saddle River, NJ, (2010).

21. B.V. Reddy and K.S. Jagadish, "Embodied energy of common and alternative building materials and technologies, Energy and Buildings," 35, (2003), 129-137.

22. P. Wu and S. P. Low, "Project Management and Green Buildings: Lessons from the Rating Systems." *Journal of Professional Issues in Engineering Education and Practice*, 136(2), (2010), 64-70.

23. Chong, W. K., S. Kumar, C. T. Haas, S. M. Beheiry, L. Coplen, and M. Oey. "Understanding and Interpreting Baseline Perceptions of Sustainability in Construction among Civil Engineers in the United States." *ASCE Journal of Management in Engineering*, 25 (3), (2009), 143-154.

24. R. Cole, "Building environmental assessment methods: Clarifying intentions." *Building Research and Information, 27* (4/5), (1999), 230–246.

25. G. Ding, "Developing a multi-criteria approach for the measurement of sustainable performance." *Building Research and Information, 33* (1), (2005), 3–16.

26. J. San-Jose, R. Losada, J. Cuadrado, and I. Garrucho, "Approach to the quantification of the sustainable value in industrial buildings," *Building and Environment, 42* (11), (2007), 3916–3923.

27. R.C. Retzlaff, "Green Building Assessment Systems A Framework and Comparison for Planners," *Journal of the American Planning Association*, 74(4), (2008), 505-519.

28. K.Y. G. Kwok, C. Statz, B. Wade, K. Wai, and O. Chong, "Carbon Emission Modeling for Green Building: A Comprehensive Study of Calculations,"

29. A. Memari, R. Solnosky, J. Tufano, and M. Dillen, "Comparative Study on Multi-hazard Resistance and Embodied Energy of Different Residential Building Wall Systems," J. Civil Eng. Architect. Res. 1(6), (2014), pp. 367-387

30. SWA. *Residential Panels Benchmark Requirements*, Steven Winter Associates (SWA), Prepared for U.S. Department of Housing and Urban Development, office of Policy Development and Research, Washington, DC, (2004).

31. M.G. Salvadori and R.A. Heller, *Structure in Architecture*, Prentice-Hall, (1975).

32. M. L. Obiso, "Analysis of means and methods of construction improvement in single family housing in mid-Atlantic rural university towns." *M.Sc. Thesis*, Virginia Polytechnic Institute and State University, (1997).

33. J.J. Kim, and B. Rigdon, "Sustainable Architecture module: Qualities, use, and Examples of Sustainable Building Materials," *National Pollution Prevention Center for Higher Education*, University of Michigan, Ann Arbor, MI, (1998).

34. E. Hubbs, "If Walls Could Talk." *Nursing Homes/Long Term Care Management*, 5(6), (2003), 1-3.

35. D.M. Kestner, J. Goupil, and E. Lorenz, "Sustainability guidelines for the structural engineer," American Society of Civil Engineers (ASCE), Reston, VA, (2010), 1-315.

36. D. E. Jacobs et al., "The prevalence of lead-based paint hazards in U.S. housing." *Environ. Health Perspect*, 110(10), (2002), A599–A606.

37. ERC, "ERC web." [Online], http://www.ercweb.com/resources/viewreg.aspx?id=7102.

38. H. A. Denmark and H. L. Cromroy, "Featured creatures: House dust mites." Publication No. EENY-59, (2007). [Online Jun. 30, 2013], http://entnemdept.ufl.edu/ creatures/urban/house_dust_mite.htmæ.

39. J. Krieger and D.L. Higgins, "Housing and health: Time again for public health action." Am. J. Public Health, 92(5), (2002), 758–768.

40. K. Yeatts et al. "A brief targeted review of susceptibility factors, environmental exposures, asthma incidence, and recommendations for future asthma incidence research." Environ. Health Perspect., 114(4), (2006), 634–640.

41. Green Building Home Page (GHP), "Green Building," [Online] http://www.ciwmb. ca.gov/GreenBuilding/ .(accessed in 2011).

42. EPA. "History of RCRA." [Online Jun. 21, 2013] http://www.epa.gov/wastes/laws-regs/ rcrahistory.html.

43. P.J. Arsebault and P. Bates, "Designed, Sealed, Delivered - IAQ and the Building Enclosure" McGraw Hill Construction Continuing Education, (2013), [Online] http://continuingeducation .construction.com/article_print.php?C=1140&L=376

44. A.M. Memari, "Comparative study of multi-hazard performance of different wall systems used in single-family dwelling construction," *6th Congress on Forensic Engineering*, (2012), 1-10.

45. W. Werner, and J.G. Burns, "Quantification and optimization of structural embodied energy and carbon," *ASCE Structures Congress*, (2012), 929-940.

46. S.L. Hsu, *Life cycle assessment of materials and construction in commercial structures: Variability and limitations*, Massachusetts Institute of Technology, Cambridge, MA, 2010.

47. C. Jayasinghe, "Embodied energy of alternative building materials and their impact on life cycle cost parameters," ICSECM 201, (2011), 1-20.

48. C. Reardon, G. Milne, C. McGee, and P. Downton, "Your Home Technical Manual – Australia's guide to environmentally sustainable homes," 4th Edition (Reprinted with minor updates) (2010), [Online], http://www.environment.nsw.gov.au/resources/

49. P. Pierquet, J.L. Bowyer, and P. Huelman, "Thermal performance and embodied energy of cold climate wall systems," *Forest Products Journal*, 48 (6) (1998) 53-60.

50. T. Ariosto, and A.M. Memari, "Comparative study of energy efficiency of glazing systems for residential and commercial buildings," *AEI*, (2013) 408-417.

Engineering Solutions for Sustainability: Materials and Resources II
Edited by: Jeffrey W. Fergus, Brajendra Mishra, Dayan Anderson, Emily Allyn Sarver, and Neale R. Neelameggham
TMS (The Minerals, Metals & Materials Society), 2015

WET CHEMICAL METALLIZATION OF AEROSPACE COMPOSITES AS A LIGHTNING PROTECTION STRATEGY

Rajesh P S M[1], Xavier Cauchy[1], Martin Gagne[2], Jolanta E Klemberg-Sapieha[2], Frederic Sirois[3], Daniel Therriault[1]

[1]Mechanical Engineering Department, [2]Engineering Physics Department, [3]Electrical Engineering Department
Ecole Polytechnique de Montreal, C.P. 6079, Succursale Centre-Ville
Montreal, QC H3C 3A7, Canada

Keywords: Polymer composite, Lightning protection, Aircraft safety, Electroless plating, Silver coating.

Abstract

Polymer composite aircraft presently utilize metallic (mostly copper) meshes/foils for lightning strike protection (LSP). However, metallic meshes/foils require additional material for their integration with composites. In this paper, we explore wet chemical metallization of composites via electroless plating of silver. We investigate and report: (i) the effect of the reaction bath on integrity of the composite studied via mechanical analysis, (ii) the effect of bath composition on coating resistivity studied via 4-probe measurements, and (iii) the adhesion characteristics of the coating studied via tape adhesion tests. Although silver is denser than copper, it is less resistive and noble. Wet metallization also offers a quick, effortless means to achieve a continuous surface coating conforming to various geometries. This reduces integration complications and weight, both key benefits during manufacture, operation and repair. It is thus envisaged that with further research, the proposed engineering solution could potentially serve to improve air-travel safety and sustainability.

Introduction

Increased oil prices coupled with the requirements for greener aircraft are driving aircraft manufacturers to adopt composite materials; today's Boeing 787 and the Airbus A350XWB being leading examples. The main advantage of using fiber-reinforced polymer composites (FRPC) in lieu of the conventional aluminum alloys for aircraft structures comes from their high specific stiffness and strength that makes aircraft much lighter and thus more fuel efficient. However, like every technology composite materials also have their disadvantages. One of the major issues with composites is their electrical conductivity, which is far inferior to that of aluminum alloys. However, high conductivity is required to ensure protection again lightning strikes by providing a conducting path for the lightning current. Lightning strike protection (LSP) is thus a necessary feature for aircrafts with composite structures.

The most widely employed LSP strategy is the use of metallic meshes and foils that are placed over the composite. However, in addition to the weight of the mesh, there is also the weight of excess resin used for its binding. Additionally, repair and maintenance scenarios present an even bigger challenge of ensuring electrical conductivity between the new patch of mesh with that existing on the remaining aircraft. For these reasons, aircraft manufacturers are looking at newer technologies such as lightweight, conductive coatings and films that can be externally applied to aircraft structures. With this line of thought, we investigated the possibility of using electroless coating of silver directly onto the composite surface. Such a technique would have the advantage of providing a simple and reliable means for achieving a continuous coating conforming to different shapes.

Experimental Details

Sample preparation: The coating process [1, 2] was carried out room temperature and atmospheric pressure. Three solutions were first prepared for this purpose.

Sensitizing solution: Tin chloride ($SnCl_2.2H_2O$, Sigma Aldrich) was added to water to make 0.05 M solution. HCl was added until the ingredients completely dissolved.

Tollen's reagent: Silver nitrate ($AgNO_3$, Alfa Aesar) was dissolved in DI water to make a 0.5 M solution. 1 M potassium hydroxide (KOH, Anachemia) was added to the above (≤ 1:20 volume ratio); this generated a reddish brown precipitate. 28% ammonia solution (H_5NO, Alfa Aesar) was gradually added to the above solution until the precipitate was completely dissolved.

Reducing agent: 0.2 wt./wt. dextrose ($C_6H_{12}O_6$, Sigma Aldrich) solution was used to reduce the silver ions to silver. This quantity of dextrose solution was chosen so that it was not a limiting factor in the reaction.

The coating was realized on carbon fiber-epoxy composites through a three-step process as discussed below:

Sensitization: The samples were left undisturbed in the sensitization solution where the tin ions adhere to the surface. After 15 minutes, the samples were thoroughly rinsed in DI water. They were kept with the smoother face up in a container.

Tollen's reaction: Immediately after the previous step, the Tollen's reagent and the dextrose solution were quickly mixed and poured into the container with the composite samples.

Plating: The container was gently rocked to keep its constituents agitated until little after a clean silvery layer appeared on the samples. The samples were then rinsed thoroughly with DI water and were left to dry overnight.

X-ray diffraction (XRD) studies: A Philips X'pert diffractometer was utilized with copper Kα radiation to obtain the diffractograms of the coating in the range of 5 - 90° at $0.5°\cdot min^{-1}$.

Visual characterization: In addition to photography, optical microscopy was carried out to study the quality and continuity of the coating using a Zeiss ScopeA1 AXIO microscope with OCView OptixCam image capture software from Microscope Store LLC.

Mechanical and adhesion characterization: 3-point bending tests were conducted on an MTS Insight universal testing machine with a 1000 N load cell in accordance with ASTM D790 standard [3] albeit using a constant span length of 2.5 in. Adhesion characterization was carried out via the tape adhesion method as per ASTM D3359, procedure B [4].

Electrical characterization: 4-probe electrical measurements were carried out to estimate the coating resistivity. For this purpose, a Keithley 220 programmable current source and a Hewlett

Packard 34401A multimeter (for voltage measurement) were used in conjunction. Ten different currents were injected and the resulting voltage was measured at three different locations. A linear fit of these 10 voltage measurements is made and the resulting resistance at these 3 locations was averaged. A correction factor (CF) based on sample/probe geometry is used to calculate sheet resistance. Specific resistivity was calculated by multiplying sheet resistance with volume density (of pure silver) and coating thickness (assumed as 5 μm).

Results and Discussion

A high resolution photograph of the coated composite is shown in fig. 1. Bare sample is also shown for comparison. A smooth coating with no visually discernible defects was achieved.

Fig. 1: Photograph of a coated (bottom) and uncoated (top) composite

The coating was further subjected to optical microscopy. Fig. 2(a) shows the face of the coating while fig. 2(b) shows it's cross section. The latter was obtained by sandwiching the coating between the existing epoxy of the substrate (left) and phenocure resin (right). The images indicate a continuous coating with a thickness of less than 10 μm.

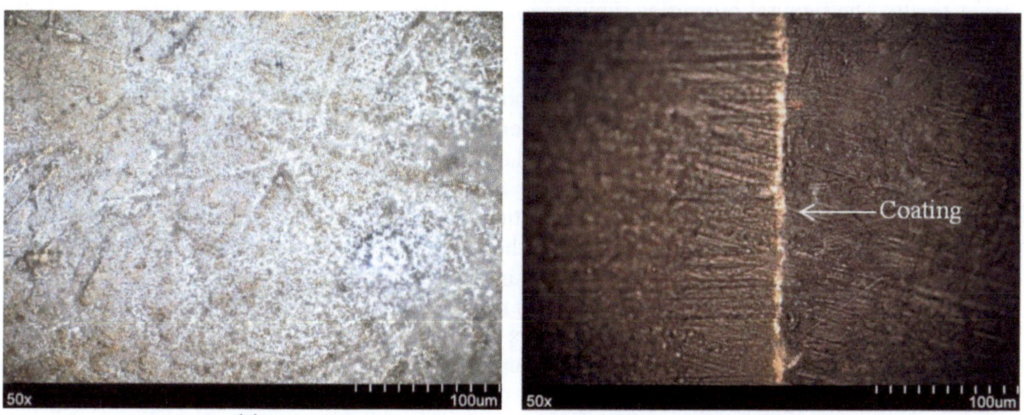

(a) (b)

Fig. 2: Optical images of (a) the face, and (b) cross-section of the coating

The coated composites were also subjected to X-ray diffraction studies (fig. 3). Peaks observed at approximately 39°, 45.5°, 65.4° and 78.5° associated with the (111), (200), (220) and (311)

planes are an indicative of the presence of silver, although the peaks are shifted to the higher angle side by about 1° compared to the XRD observations published in the literature [1, 5].

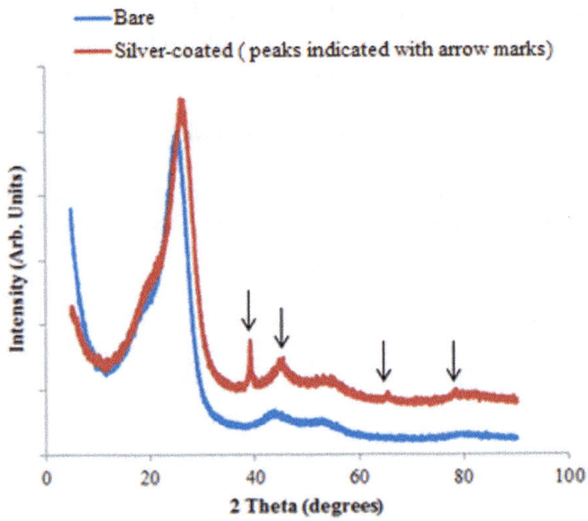

Fig. 3: X-ray diffractograms of coated and uncoated composites

In order to confirm that neither the sensitization nor the reaction bath poses any issues to the sample integrity, 3-point bending tests were conducted; the mechanical characteristics of the coated samples were compared to that of bare samples.

Three samples each were chosen for: (i) sensitization + coating, and (ii) bare composite samples. The average flexural strength and modulus of the coated samples were 100.7% and 102.6% that of the bare samples, the order of the actual values being many 100s of MPa and 10s of GPa, respectively. The flexural properties show no major differences implying that neither of the baths affected the composites adversely.

The adhesion of the silver coating to the composite surface was assessed via the peel-adhesion test. As seen in fig. 4, no flaking occurred after peeling the tape. This intactness of the region between the squares indicates superior adhesion and can be classified as grade 5B.

In the 4-probe technique, a set of four identical, collinear probes were pressed against the sample. Electrical current was passed through the outer probes while the inner probes were used to measure the voltage. An average specific resistivity of 3.695×10^{-3} $\Omega\cdot g\cdot cm^{-2}$ was obtained with a standard deviation of 0.424×10^{-3} (about 10%) for the coated composites. This value of specific resistivity is far less good as it is two orders of magnitude higher than that of pure silver. Attempts are underway to investigate the source of this difference with the two immediate considerations being the accurate measurement of coating thickness and its purity. Additionally, coating uniformity and continuity may also be responsible and will be studied.

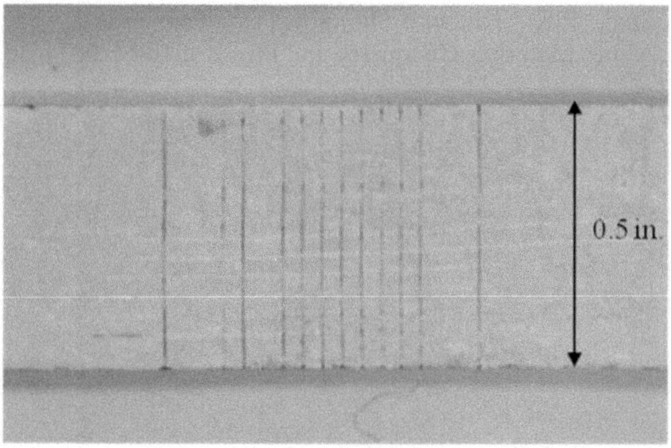

Fig. 4: Cross-scratched surface of the sample after peel adhesion test.

Conclusion

The preliminary results associated with our attempts to directly coat silver onto composites have been presented. Although a successful coating was achieved, the conductivity of the coating was lower than that of pure silver. An examination of this matter is underway. On the other hand, the fact that the chemical processes do not affect composites' mechanical properties coupled with good adhesion characteristics of the coating encourage us to pursue further research on the proposed LSP technology. The prospects of such an engineering technology are that both the environment and the aerospace sector will benefit from improved fuel economy and simpler maintenance and repair protocols in the long run.

Acknowledgements

The authors sincerely thank J. Qian and R. Vernhes of Engineering Physics department for their assistance in optical microscopy and S. Bodkhe of Mechanical Engineering department for carrying out the XRD runs. The composite panels were provided by Bombardier (Aerospace) Inc. This work was carried out as a part of the NSERC CRIAQ COMP-502 project.

References

1. Majid Montazer et al.; In situ synthesis of nano silver on cotton using Tollens' reagent; Carbohydrate Polymers, 87 (2012), 1706– 1712.

2. Supin Sangsuk; Preparation of high surface area silver powder via Tollens process under sonication; Materials Letters, 64 (2010), 775–777.

3. ASTM D790-10, Standard test methods for flexural properties of unreinforced and reinforced plastics and electrical insulating materials, ASTM International, West Conshohocken, PA, 2010.

4. ASTM D3359-09e2; Standard test methods for measuring adhesion by tape test; ASTM International, West Conshohocken, PA, 2009.

5. Narendra Singh and P.K. Khanna; In situ synthesis of silver nano-particles in polymethylmethacrylate; Materials Chemistry and Physics, 104 (2007), 367–372.

Engineering Solutions for Sustainability: Materials and Resources II
Edited by: Jeffrey W. Fergus, Brajendra Mishra, Dayan Anderson, Emily Allyn Sarver, and Neale R. Neelameggham
TMS (The Minerals, Metals & Materials Society), 2015

COMPARATIVE LCA STUDY OF LIGHTWEIGHT AUTO PARTS OF MMLV MACH-I VEHICLE AS PER ISO 14040/44 LCA STANDARDS AND CSA GROUP 2014 LCA GUIDANCE DOCUMENT FOR AUTO PARTS

Lindita Bushi[1], Tim Skszek[2], and David Wagner[3]

[1]Life Cycle Assessment Consulting; 8 Tollgate Mews, Toronto, ON, M1M 3X5, Canada
[2]Magna International; 750 Tower Drive; Troy, Michigan, 48098, USA
[3]Ford Motor Company; 2101 Village Road; Dearborn, Michigan, 48121, USA

Keywords: LCA, Mach-I design, 2013 Ford Fusion, lightweight, ISO 14040/44, CSA Group LCA Guidance

Abstract

Weight reduction in automotive design and manufacture is a priority across the industry, as strict new regulations push for greater vehicle efficiency and CO_2 reduction in the US, Europe and Asia.
In response to the global automotive market changes, the US Department of Energy (DOE) co-funded Magna International and Ford Motor Company to engineer, prototype and test a new lightweight aluminum-intensive passenger vehicle, facilitating extensive use of advanced lightweight and high-strength materials, resulting in significant environmental and fuel economy benefits. The Multi Material Lightweight Vehicle (MMLV) Mach-I design achieved an overall 364 kg (23%) full vehicle mass reduction, enabling engine downsizing. The life cycle assessment (LCA) calculations estimate the magnitude of mass-induced fuel savings due to the lightweight auto parts and the down-sized engine to be 962 gallons over the useful vehicle life of 250,000 km (155,343 miles), which implies the potential of achieving a combined fuel economy of 34 mpg (6.9 l/100 km) associated with the Mach-I design compared to 28 mpg (8.4 l/100 km) for the 2013 Ford Fusion. Additionally, the LCA calculations estimate a 16% reduction in Global Warming Potential (GWP) from 68,446 kg CO2-eq for the Fusion to 57,629 kg CO2-eq for the MMLV Mach-I. A similar 16% reduction in Total Primary Energy (TPE) from 986,090 MJ for the Fusion to 829,893 MJ for the MMVL Mach-I is estimated over the useful life of the vehicles. This paper is aimed at communicating the results of a comparative life cycle assessment (LCA) study of the MMLV Mach-I auto parts [1] under contract of Magna International, in accordance with International Organization for Standardization (ISO) standards 14040/44 and CSA Group 2014 LCA Guidance document for auto parts [2, 3, 4]. ISO 14040/44 provide general over sectorial rules and requirements for conducting LCA study, which cover but are not specifically tailored for the auto sector. To complement ISO 14040/44 rules, the CSA Group LCA Guidance establishes auto sector specific technical parameters and LCA rules and requirements to calculate the life cycle environmental impact of auto parts/vehicles in North America. A variety of LCA tools, softwares, databases and protocols can be used to assess the life cycle environmental impact of auto parts/vehicles. The Mach-I LCA model was created using SimaPro v.8.03 2014, the world's leading LCA software chosen by industry, research institutes, and consultants in more than 80 countries. Input data associated with the LCA calculation was provided by Ford Motor Company and Magna International, as part of work supported by the Department of Energy National Energy Technology Laboratory under Award Number No. DE-EE0005574.

Introduction

Regulation requirements such as the 2020 CAFE (corporate average fuel economy) standard, growing public demand, and increased fuel prices are pushing auto manufacturers worldwide to increase fuel economy through incorporation of lightweight materials in newly-designed vehicle structures.

In response to the global automotive market changes, the US DOE, co-funded Magna International and Ford Motor Company to engineer, prototype and test the MMLV Mach-I design. The Mach-I design achieved an overall 364 kg (23%) full vehicle mass reduction as compared to the 2013 Ford Fusion. Mach-I weight reduction (in absolute and percent basis) per each vehicle sub-system is presented in Table I. The breakdown of the vehicle system in constituent vehicle's sub-systems, components, and auto parts (sub-assemblies) follows the automotive system cost modeling (ASCM) for the purpose of this LCA [5]. 2013 Ford Fusion and Mach-I design are classified as "Advanced High Strength Steels (AHSS)" and "aluminum" intensive vehicles. Table II details the main material composition per vehicle architecture.

This paper presents the main results of the LCA study "Comparative LCA Study of Lightweight Auto parts of MMLV Mach-I Vehicle as per ISO 14040/44 LCA Standards and CSA Group 2014 LCA Guidance Document for Auto Parts" which was initiated and commissioned by Magna International [1]. The LCA study following the ISO 14040/44 LCA Standards and CSA Group LCA Guidance compares the lightweight auto parts of the new MMLV Mach-I (1.0l I3) vehicle design to the more conventional auto parts of the baseline 2013 Ford Fusion (1.6l I4), both built and driven for 250,000 km in North America.

Table I: Mach-I Weight Reduction on a Vehicle Sub-system Basis

Vehicle systems and sub-systems	2013 Ford Fusion weight (kg)	Mach-I weight (kg)	Mach-I curb weight (%)	Mach-I weight reduction (kg)	Mach-I weight reduction (%)
1. Body	525.0	400.4	33.5%	-124.6	-23.7%
2. Interior	260.4	202.7	17.0%	-57.7	-22.2%
3. Chassis	355.0	260.0	21.8%	-95.0	-26.8%
4. Powertrain	337.0	263.1	22.0%	-73.9	-21.9%
5. Electrical	57.0	49.5	4.1%	-7.5	-13.1%
A. Assembly	25.0	19.5	1.6%	-5.5	-22.0%
Total vehicle	1559.4	1195.2	100%	-364.2	-23.4%

Table II: Main Material Composition per Vehicle Architecture

Material	2013 Ford Fusion weight (kg)	2013 Ford Fusion weight (%)	Mach-I weight (kg)	Mach-I weight (%)
AHSS	417.5	27%	66.9	6%
Conventional steel	413.7	27%	289.8	24%
Cast iron	50.0	3%	19.6	2%
Forged iron	16.0	1%	10.0	1%
Stainless steel	19.1	1%	9.7	1%
Die-cast aluminum.	146.4	9%	147.7	12%
Cold-rolled aluminum	12.8	1%	143.8	12%
Extruded aluminum	15.6	1%	66.9	6%

Material	2013 Ford Fusion weight (kg)	2013 Ford Fusion weight (%)	Mach-I weight (kg)	Mach-I weight (%)
Forged aluminum	0.0	0%	9.8	1%
Magnesium	2.3	0%	16.0	1%
Copper	33.7	2%	29.3	2%
Titanium	0.0	0%	3.3	0%
CFRP	0.0	0%	54.2	5%
GFRP	0.0	0%	3.4	0%
Plastic	235.4	15%	177.1	15%
Rubber	72.6	5%	52.0	4%
Glass	37.5	2%	26.4	2%
Ceramics	0.8	0%	0.8	0%
Batteries	14.0	1%	8.0	1%
Paint	8.0	1%	7.7	1%
Fluid, adhesive and other	64.1	4%	52.8	4%
Total vehicle	1559.4	100%	1195.2	100%

The LCA of the auto parts has been conducted in accordance with ISO 14040/44 standards and follows the specific rules and guidance provided in the Canadian Standard Association (CSA Group) 2014 LCA Guidance for Auto Parts document [2, 3, 4]. ISO 14040/44 provide general over sectorial rules and requirements for conducting LCA study, which cover but are not specifically tailored for the auto sector. To complement ISO 14040/44 rules, the CSA Group LCA Guidance establishes auto sector specific technical parameters and LCA rules and requirements to calculate the life cycle environmental impact of auto parts/vehicles in North America. The LCA study assesses the potential environmental impacts of the auto parts throughout their cradle-to-grave life cycle, with a focus on weight differences between design options due to material composition, manufacturing technology, or part geometry. The LCA report is peer reviewed for conformance with ISO 14040/44 requirements and the specific rules and guidance provided in the CSA Group document.

Goal definition

The goals of the Mach-I LCA study are to:

- Compare the lightweight auto parts of the MMLV Mach-I (1.0l I3) vehicle design to the more conventional auto parts of the baseline 2013 Ford Fusion (1.6l I4), both built and driven for 250,000 km in North America;
- Calculate the total net change in “cradle-to-grave” environmental performance of the lightweight Mach-I auto parts incorporating weight changes due to material composition, manufacturing technology, or part geometry;
- Determine the relative magnitudes of the selected life cycle stage impacts, and identify the stages where future modification and improvement are most appropriate; and
- Calculate the relative contribution of the main vehicle sub-systems and auto parts to the total net change of the “cradle-to-grave” environmental profile of the vehicle system, and identify the sub-systems and auto parts where future R&D and improvement are most appropriate.

The primary intended application is to report the improved environmental performance associated with the MMLV Mach-I midsize vehicle mass reduction and resized powertrain. The LCA study also serves as a baseline for future benchmarking of automotive design works.
Primary interested parties are the US DOE, Province of Ontario, Ford Motor Company and Magna International Inc. Potential interested parties include auto part OEMs and suppliers, auto manufacturers, industry associations, governmental organizations, policy makers, LCA practitioners and other stakeholders who desire reliable information on MMLV Mach-I auto parts.
The results of the LCA study are intended to support external communication; therefore, to be compliant with ISO 14044 and CSA Group LCA Guidance for Auto Parts, a critical review of the study was conducted. An ISO 14044 conformant external critical review has been conducted by Simone Ehrenberger, researcher with broad experience in LCA of automotive material and transport, based at the German Aerospace Center, Institute of Vehicle Concepts, Vehicle Systems and Technology Assessment Department, Stuttgart, Germany.

Scope definition

The LCA study has been conducted for auto parts of Mach-I (the MMLV design) and the 2013 Ford Fusion (the baseline design), both internal combustion engine vehicles (gasoline fuelled, direct injection, turbo charged), built and driven for 250,000 km in North America. Both vehicles fall under the US Environmental Protection Agency (US EPA) size class "midsize sedan" with interior passenger and cargo volumes (PCV) of $110 \leq PCV \leq 119$ cu ft.
The functional unit is defined as the transportation service of auto parts that:

a) Have undergone mass changes in the Mach-I, enabling engine downsizing, relative to the 2013 Ford Fusion, due to material composition, manufacturing technology, or part geometry, while maintaining performance and vehicle configuration;

b) Are manufactured and intended for use in North America for 250,000 km;

c) Are engineered to meet National Highway Traffic Safety Administration (NHTSA) and Insurance Institute for Highway Safety (IIHS) 5-star safety criteria and match stiffness, noise, vibration, and harshness (NVH) performance, and durability of the 2013 Ford Fusion baseline vehicle.

For the comparative LCA study, the functional unit is identical between the MMLV Mach-I and 2013 Ford Fusion baseline auto parts. The reference flow of the auto parts is defined based on the auto part replacement factor (F_R) calculations. The F_R is calculated by dividing the baseline vehicle's lifetime driving distance ($LTDD_V$) of 250,000 km by the auto part lifetime driving distance ($LTDD_A$). This way, the $F_R=1$ means that the auto part has a $LTDD_A$ of 250,000 km and no replacement is conducted during the vehicle lifetime other than the first auto part installation in the vehicle architecture. With the exception of two auto parts (tires and energy storage), the reference flow of both 2013 Ford Fusion and Mach-I auto parts is defined as 1 auto part ($F_R=1$). In addition, the reference flow of tires is defined as 4 auto parts ($F_R=4$) for both vehicle designs. For purposes of this LCA, the reference flow of the energy storage is defined as 3 ($F_R=3$) and 2 ($F_R=2$) auto parts, for 2013 Ford Fusion and Mach-I, respectively.
The study is a cradle-to-grave LCA and includes the following three main life cycle stages: (1) production stage, (2) use stage and (3) end-of-life (EOL) stage. The system boundaries follow these two principles: (1) The "modularity principle": Where processes influence the product's environmental performance during its life cycle, they shall be assigned to the module of the life cycle where they occur (e.g. production, use or EOL); all environmental aspects and impacts are declared in the life cycle stage where they appear; and (2) The "polluter pays principle":

Processes of waste processing including waste water treatment shall be assigned to the product system that generates the waste until an end-of-waste state is reached.
The following processes are excluded from the system boundary: (a) Human energy inputs to processes and/or preprocessing (e.g. hand assembly rather than by machinery); (b) Employee commute (e.g. employees to and from their normal place of work); (c) Capital infrastructure (e.g. factories, roads, trains, ships, production facilities & machinery); (d) Production overhead (e.g. HVAC, lighting, offices); and (e) Auto parts maintenance and repair processes.
The allocation rules and cut-off criteria considered within the system boundary conform to ISO 14044 and CSA Group LCA Guidance [3, 4]. The cradle-to-grave environmental profile of the Mach-I auto parts conform to CSA Group LCA Guidance Clauses 7.3.4, 7.3.5 and 7.3.6, and include all cradle-to-grave consecutive unit processes [4]. The activity and LCI data used to create the inventory model of both vehicle systems are as representative, complete, consistent, reproducible and transparent as possible with regards to the goal and scope of the study under given time and budget. A variety of LCA tools, softwares, databases and protocols can be used to assess the life cycle environmental impact of auto parts. The Mach-I LCA model was created using SimaPro v.8.03 2014, the world's leading LCA software chosen by industry, research institutes, and consultants in more than 80 countries [6]. The US EPA Tool for the Reduction and Assessment of Chemical and Other Environmental Impacts (TRACI), version 2.1, 2012 impact categories are used as they provide a North American context for the mandatory category indicators to be included in Mach-I LCA study [7]. The Mach-I study includes a comprehensive set of LCA (both LCIA and LCI) indicators.

Life Cycle Inventory

Data Collection

Detailed data on the auto part name, number of constituent parts per vehicle sub-system, assembly and sub-assembly, mass per auto part in kilograms, material composition, fabrication process, sleeves and fasteners, and adhesives were provided by Ford Motor Company. The US EPA combined fuel economy of the 2013 Ford Fusion was obtained from the www.fueleconomy.gov website.
This comparative LCA study uses the best available LCI data provided by:

- North American and global metals and other material industry associations;
- North American auto manufacturers; and
- North American and global industry-supported public and commercial LCI databases such as the US LCI database and ecoinvent 2.2 (both included in the LCA software SimaPro v.8.03 2014), GREET-1 2013, and GREET-2 2012 [6, 8, 9].

The applied LCI datasets for the two selected vehicle systems, the transportation modes and distances for all material per type of activity, main auto part fabrication yields (in %) and the typical net amount of EOL scrap and EOL recycling yield values per material are all documented and detailed in the Mach-I LCA study.
As 2013 Ford Fusion and Mach-I design are classified as "AHSS" and "aluminum" intensive vehicles, "cradle-to-gate" LCI datasets of both North American steel and aluminum products are of the high importance for the Mach-I LCA results.
Table III presents the input scrap and carbon dioxide (CO_2) emissions for the selected North American steel and aluminum products. The critically reviewed North American cradle-to-gate LCI datasets for steel and aluminum products below were endorsed for the first time in 2013, by the World Steel Association (worldsteel) in cooperation with US Steel Recycling Institute and the US Aluminum Association, respectively [10], [11].

Table III: Input Scrap and CO_2 Emissions for the Selected North American Steel and Aluminum Products (Cradle-to-Gate LCI Data)

Aluminum product (cradle-to-gate)	Input scrap (kg/kg Al product)	Carbon dioxide[1)] (kg/kg Al product)	Steel product (cradle-to-gate)	Input scrap (kg/kg steel product)	Carbon dioxide[1)] (kg/kg steel product)
Al casting	0.883	2.520	HDG	0.439	2.054
Al extrusion	0.426	5.854	PHRC	0.198	2.111
Al CRC	0.649	4.792	CRC	0.177	2.076
Al primary	0	7.875[4)]	EG[2)]	0.064	2.428
Al recycling ingot (100% scrap)	1.045	0.634	ES[2)]	1.011	0.676
Secondary ingot (primary metal and alloy added)	0.978	1.134	Steel primary[2)3)] (BOF slab) (theoretical value)	0	1.92
			Steel secondary[2)3)] (EAF slab)	1.092	0.386

Notes:
[1)] CO_2 emissions should not be mistaken for the GWP indicator (in kg CO_2 eq), which is calculated based on the potency of greenhouse gases (including carbon dioxide, methane, nitrous oxide, fluorinated gases etc.) relative to carbon dioxide (CO_2).
[2)] World average cradle-to-gate LCI data; North American cradle-to-gate LCI data are not available yet.
[3)] To avoid any data misuse, steel primary and secondary LCI profiles are not made available to LCA practitioners. All interested parties are recommended to use the rolled-up North American LCI profiles of steel products including the "value of steel scrap" as provided by worldsteel. To ensure data comparability, LCA practitioners should also use the rolled-up North American LCI profiles of aluminum products as provided by US Aluminum Association.
[4)] Based on the IAI annual energy survey which covers all aluminum smelting facilities in North America, the 2010 production year power mix consists of 75% hydro (clean energy), coal (24%) and natural gas & nuclear (1%), which has contributed to an overall improved environmental profile of North American aluminum products- see Table 11, Section 5.1.1.4.3, The Aluminum Association 2013, LCA report [11].

Data Calculation

Production Stage. Since auto part replacements are applicable in this study, the total net change in the production stage environmental profile of the Mach-I auto parts, with powertrain adaptation ($\Delta E_{P,v_{r,a}}$), is calculated as follows:

$$\Delta E_{P,v_{r,a}} = \sum_{k=1}^{n} (E_{P,m_k} \times F_{R,m_k} - E_{P,b_k} \times F_{R,b_k}) \quad (1)$$

Where,
$E_{P,m}$= environmental profile of the production stage of the auto part exhibiting the mass change (MC)

$E_{P,b}$= environmental profile of the production stage of the baseline auto part
$F_{R,m}$= the auto part replacement factor of the auto part exhibiting the MC
$F_{R,b}$= the auto part replacement factor of the baseline auto part
n= total number of auto parts exhibiting the MCs (58 auto parts)
It should be noted that 23.4 kg of baseline auto parts have not undergone any mass changes and the total net change in the production stage of these auto parts is "null".

The environmental impact of process scrap recycling (PSR) and/or disposal is reported as part of the production stage. The PSR covers the environmental impact associated with the collection of the process scrap, sorting, melting, refining and avoided burden of primary material production. The environmental impact of PSR (E_{PSR}) is calculated as follows:

$$E_{PSR} = W_f \times Y_f \times (E_{sec} - E_{prim}) \tag{2}$$

Where,
W_f= mass input of process scrap to secondary production
E_{sec}= environmental profile per unit of mass of material due to secondary production, e.g. EAF slab
E_{prim}= environmental profile per unit of mass of material due to primary production, e.g. BOF slab
Y_f= yield for the recycling of process scrap from manufacturing of material that will displace primary production
Total amount of process scrap to secondary production is assumed to be generated during the auto part fabrication processes. This analysis assumes that no process scrap is generated during the vehicle assembly process.

Use stage. The maximum total life cycle mass-induced fuel savings of the Mach-I design ($C_{A,v}$), with powertrain adaptation, is calculated as follows:

$$C_{A,v} = \sum_{k=1}^{n} (m_{m_k} - m_{b_k}) \times F_{CP} \times LTDD_V \tag{3}$$

= (1,171.8 – 1,536.0) kg × 0.40 L/(100 km×100 kg) × 250,000 km
= - 3,642 L or - 962 gal.

Where,
$C_{A,v}$= the maximum total life cycle mass-induced fuel savings (decrease) of the auto parts exhibiting the MCs, in litres
m_m= mass in kg of the auto parts exhibiting the MCs, total of 1,171.8 kg
m_b= mass in kg of baseline auto parts, total of 1,536.0 kg
n= total number of auto parts exhibiting the MCs, 58 auto parts
F_{CP}= mass-induced fuel change potential value, with adaptation, 0.40 L/(100 km×100 kg) [4, 12]
$LTDD_V$= baseline vehicle lifetime driving distance, 250,000 km
Note: negative values (-) represent a decrease in fuel consumption (fuel savings) and positive values (+) represent increase in fuel consumption.

Total net change in the use stage environmental profile of the Mach-I auto parts ($\Delta E_{Use,v_{r,a}}$), with power train adaptation, is calculated as follows:

$$\Delta E_{Use,v_{r,a}} = C_{A,v} \times (E_{FP} + E_{FC}) \quad (4)$$

Where,
$C_{A,v}$= - 3,642 L or 962 gallons
E_{FP}= environmental profile of producing 1 L or gal. of gasoline (WTP – well-to-pump)
E_{FC} = environmental profile of combusting 1 L or gal. of fuel gasoline (vehicle operation)
The well-to-wheel (WTW) LCI profile (both production and combustion) of the gasoline is generated with GREET-1 2013 Software, version 1.1.0.9745 [8]. The total net change in the use stage of "not changed" auto parts is "null".

The US EPA combined fuel economy of the 2013 Ford Fusion (CFE_F) with curb weight (C_{WF}) of 1,559 kg is reported by Ford as 8.4 L/100 km or 28 mpg [13].
The US EPA CFE of the Mach-I (CFE_M) with curb weight (C_{WM}) of 1,195 kg is calculated as follows [14]:

$$\begin{aligned} CFE_M &= CFE_F - (C_{WF} - C_{WM}) \times F_{CP} \times 0.01 \\ &= 8.4 - (1{,}559.4 - 1{,}195.2) \times 0.40 \times 0.01 \\ &= 6.94 \text{ L/100 km or 34 mpg} \end{aligned} \quad (5)$$

Where,
F_{CP}= mass-induced fuel change potential value, with adaptation, 0.40 L/(100 km×100 kg) [4, 12].
For the baseline $LTDD_V$ of 250,000 km, the total life cycle fuel consumptions of the 2013 Ford Fusion and Mach-I, are calculated to be 21,000 L (= 8.4 L/100 km × 250,000 km/100) and 17,358 L (= 6.94 L/100 km × 250,000 km/100), respectively. The total life cycle mass-induced fuel savings of the MMLV Mach-I design equals to - 3,642 L (=17,358 - 21,000) or - 962 gal.
EOL Stage. Since auto part replacements are applicable, the total net change in the EOL stage environmental profile of the Mach-I auto parts, with power train adaptation ($\Delta E_{EOL,v_{r,a}}$), is calculated as follows:

$$\Delta E_{EOL,v_{r,a}} = \sum_{k=1}^{n} (E_{EOL,m_k} \times F_{R,m_k} - E_{EOL,b_k} \times F_{R,b_k}) \quad (6)$$

Where,
$E_{EOL,m}$= environmental profile of the EOL stage of the auto part exhibiting the MC
$E_{EOL,b}$= environmental profile of the EOL stage of the baseline auto part
$F_{R,m}$= the auto part replacement factor of the auto part exhibiting the MC
$F_{R,b}$= the auto part replacement factor of the baseline auto part
n = total number of auto parts exhibiting the MCs (58 auto parts)
The total net change in the EOL stage of "not-changed" auto parts is "null".

Formula 7 is applied to calculate the total net change in the cradle-to-grave environmental profile, with powertrain adaptation ($\Delta E_{Total,v_{r,a}}$), of the Mach-I auto parts:

$$\Delta E_{Total,v_{r,a}} = \Delta E_{P,v_{r,a}} + \Delta E_{Use,v_{r,a}} + \Delta E_{EOL,v_{r,a}} \quad (7)$$

The total net change of the cradle-to-grave environmental profile of "not changed" auto parts is "null".

Allocation Rules

"Mass" was deemed as the most appropriate physical parameter for allocation for the Mach-I LCA study. "Mass" was used as allocation parameter to calculate the cradle-to-grave inventory profiles (inputs/outputs) of all auto parts for both 2013 Ford Fusion and Mach-I. The total net change of production, use, and EOL stage, and cradle-to-grave LCA results were allocated on "mass" basis across all auto parts. The LCA applies the ISO 14044 conformant end-of-life recycling approach (avoided burden) [3]. This approach is also in conformance with CSA Group LCA Guidance and the "Declaration by the Metals Industry on Recycling Principles", endorsed by 17 international industry associations in 2007, which states: "For purposes of environmental modeling, decision-making, and policy discussions involving recycling of metals, the metals industry strongly supports the end-of-life recycling approach over the recycled content approach" [15].
Where applicable, the avoided burden allocation approach is applied for all metals. According to this approach, all scrap outputs during the life cycle prior to end-of-life recycling is balanced out with scrap inputs into manufacturing to avoid double-counting. The appropriate mass of the remaining net scrap is then modeled as being sent to material recycling at end-of-life. If more secondary material is generated by the product system than is used in the manufacturing stage (a positive net amount of EOL scrap), the product system receives a "credit" equal to the burden of primary material production minus the burden of recycling scrap into secondary material. Similarly, if less secondary material is generated by the product system than is used in the manufacturing stage (a negative net amount of EOL scrap), the product system receives a "debit" equal to the burden of primary material production minus the burden of recycling scrap into secondary material. For the purpose of this study, the default value assumed for the collection rate of old vehicles in North America and the shredder yield are assumed to be 0.94 and 1, respectively. This LCA study assumes that none of the EOL auto parts are reused.

LCA Indicators and Results

Table IV shows the total net savings of the cradle-to-grave LCA of Mach-I auto parts. The LCIA results were calculated using the characterization factors of US EPA TRACI v2.1, 2012 methodology with the LCA software SimaPro 8.03, 2014. All energy-related LCI indicators (NRF, NRN, NRB, RH, RSGW, and RB) were calculated using characterization factors of CED v.1.8, SimaPro. Two material-related LCI indicators (NRMR and RMR) were manually calculated based on the sums of input/output LCI flows calculated with SimaPro. Positive (+) values represent an environmental burden related to the cradle-to-grave LCA. Negative (-) values represent a reduction of the environmental burden, in other words "environmental savings", a benefit to the environment. Mach-I design shows significant environmental benefits compared to the 2013 Ford Fusion, both built and driven for 250,000 km in North America, in terms of all selected LCA and LCI impact categories. In regards to the primary energy indicators, Mach-I consumes less fossil fuels based and nuclear energy and more renewable energy, overall resulting to the total primary energy savings of 156,197 MJ.

Table IV: Total Net Change in the Cradle-to-Grave Environmental Profile of the Mach-I Auto Parts (LTDDv= 250,000 km; Total Auto Parts= 67)

LCIA and LCI Indicators	Indicator units	Cradle-to-grave total net change of the Mach-I Auto Parts
Acidification potential, AP	kg SO2-eq	-11.8
Eutrophication potential, EP	kg N water-eq	-0.06
Global warming potential, GWP	kg CO2–eq	-10,817
Photochemical ozone creation potential, POCP	kg O3-eq	-266
Human health particulate potential, HHPP	kg PM2.5-eq	-0.83
Depletion potential of the stratospheric ozone layer, ODP	kg CFC-11-eq	-1.05E-03
Total Primary Energy, TPE (TPE breakdown is as follows)	MJ	-156,197
Non-renewable, fossil, NRF	MJ	-157,345
Non-renewable, nuclear, NRN	MJ	-83
Non-renewable, biomass, NRB)	MJ	0.004
Renewable, hydropower, RH	MJ	1,422
Renewable, solar, geothermal, wind, unspecified, RSGW	MJ	166
Renewable, biomass, RB	MJ	-357
Use of non-renewable material resources,	kg	-933
Use of renewable material resources, RMR (CO_2 in air, N_2 in air, O_2 in air, wood)	kg	7.00[1)]

Note:

1) Since the RMR indicator includes the "CO_2 in air", a positive value is not considered to be an environmental burden.

Table V details the cradle-to-grave total net change of Mach-I auto parts on a life cycle stage basis (in absolute basis). The total net change of production stage LCA results of the Mach-I auto parts are significantly influenced by three important factors:

1) The *high* Mach-I full vehicle mass reduction (total of 364 kg), dominated by the iron and steel based material reduction (see Tables I and II);

2) The *high* and *moderate-to high* amount of input scrap for NA semi-fabricated aluminum and steel products which has contributed to an improved environmental profile of these products (see Table III); and

3) The *75% hydro based (clean)* electricity grid mix used by aluminum smelting facilities in North America, which has contributed to an overall improved environmental profile of NA aluminum products (see Table III).

The total net change of use stage LCA results of the Mach-I auto parts are determined by the overall 364 kg (23%) full vehicle mass reduction, *enabling engine downsizing*, which resulted in a combined fuel economy of 34 mpg (6.9 l/100 km) compared to 28 mpg (8.4 l/100 km) for the 2013 Ford Fusion. The total net change of EOL stage LCA results of the Mach-I auto parts are significantly influenced by the *high* North American EOL recycling rate of both steel and aluminum products and the EOL recycling approach.

Table V: Cradle-to-Grave Total Net Change of Mach-I Auto Parts per Life Cycle Stage – in Absolute Basis (LTDDv= 250,000 km; Total Auto Parts= 67)

Impact category	Unit	Total Net Change- Cradle-to-grave	Total Net Change- Production stage	Total Net Change- Use stage	Total Net Change- EOL stage
AP	kg SO2 eq	-11.8	1.0	-9.3	-3.5
EP	kg N eq	-0.06	2.47	-0.35	-2.18
GWP	kg CO2 eq	-10,817	513	-11,071	-259
POCP	kg O3 eq	-266	0.49	-228	-39
HHPH	kg PM2.5 eq	-0.83	1.28	-1.27	-0.84
ODP	kg CFC-11 eq	-1.05E-03	-1.06E-03	-4.26E-07	9.03E-06
TPE	MJ	-156,197	6,624	-154,618	-8,203
NRF	MJ	-157,345	1,734	-154,253	-4,826
NRN	MJ	-83	943	0	-1,026
NRB	MJ	0.004	0.009	0	-0.004
RH	MJ	1,422	3,693	0	-2,271
RSGW	MJ	166	245	0	-79
RB	MJ	-357	8	-365	-0.3
NRMR	Kg	-933	-2076	0	1143
RMR	Kg	7.0	10.5	0	-3.4

Figure 1 and Figure 2 depict the GWP and TPE break-even points for both vehicle designs. In order to accurately depict the total net change of life cycle GWP and TPE due to the MMLV Mach-I design, life cycle impacts were structured according to the production, use, and EOL stages. Only the differences of the absolute greenhouse gas emissions and total primary energy flows from both vehicle design life cycles are shown, respectively. Since the lightweight Mach-I design contributes to fuel savings, the slope of the graphical representation of the resulting GWP and TPE over the LTDDv of 250,000 km is negative (i.e. a downward slope).
Figure 1 illustrates the break-even point of life cycle GWP is 11,589 km for Mach-I auto parts. The total net savings of life cycle GWP is 10,817 kg CO2 eq.

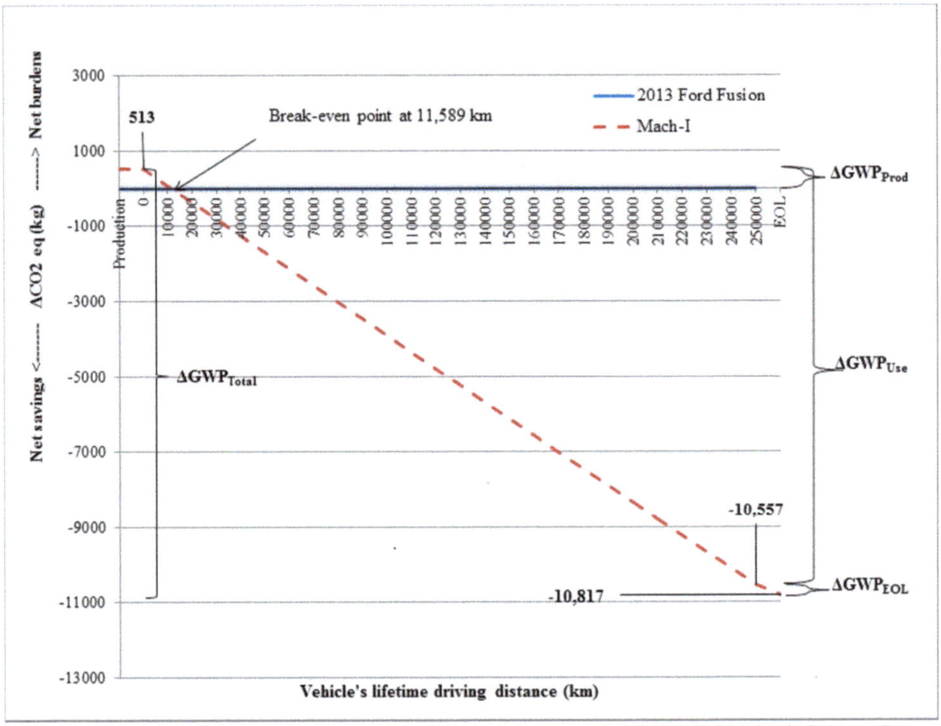

Figure 1: Total Net Change of GWP During the Life Cycle of the MMLV Mach-I Auto Parts and the Break-Even Point (LTDDv= 250,000 km)

Figure 2 illustrates the break-even point of life cycle TPE is 10,710 km for Mach-I auto parts. The total net savings of life cycle TPE is 156,197 MJ.

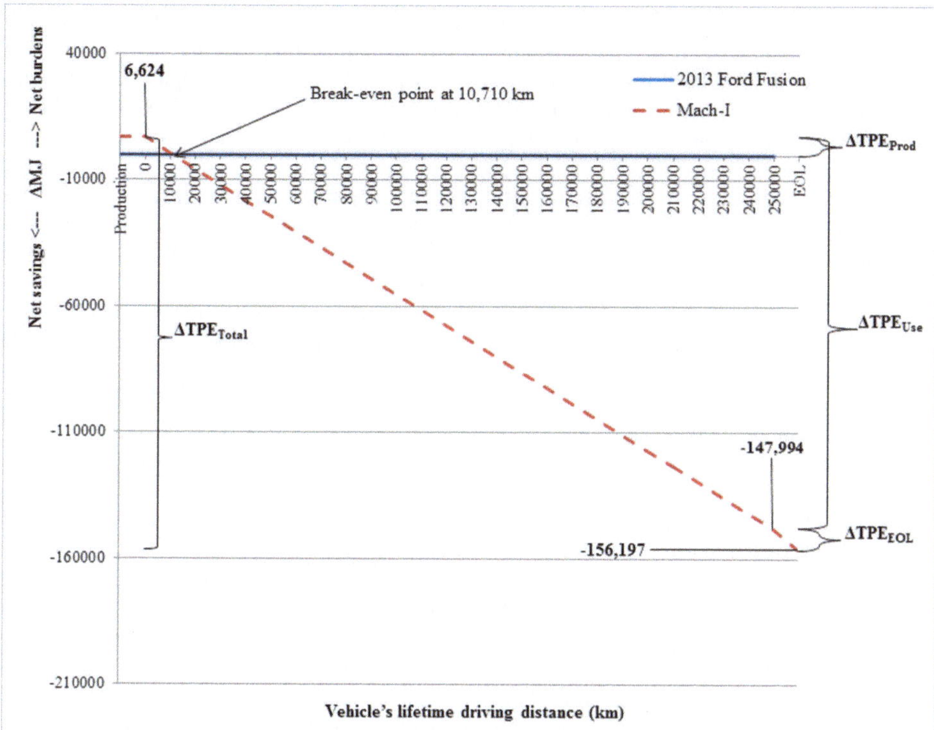

Figure 2: Total Net Change of TPE During the Life Cycle of Mach-I Auto Parts and the Break-Even Point (LTDDv= 250,000 km)

Conclusions

Based on the ISO 14040/44 and CSA Group LCA Guidance conformant life cycle inventory, impact assessment and the sensitivity analysis, the following conclusions can be reached:

- The MMLV Mach-I design is environmentally superior to the 2013 Ford Fusion, both built and driven for 250,000 km in North America, in terms of all selected LCA and LCI impact categories. More specifically, the cradle-to-grave total net savings of the MMLV Mach-I (in percentage basis), relative to the cradle-to-grave LCA of the 2013 Ford Fusion, resulted in significant environmental benefits of AP (14%), EP (0.4%), GWP (16%), POCP (16%), HHPH (8%), ODP (16%), TPE (16%), NRMR (29%) and RMR (9%).
- The MMLV Mach-I design is environmentally superior to the 2013 Ford Fusion in terms of GWP (beyond 11,589 km) and TPE (beyond 10,710 km).
- The use stage, including both fuel production and combustion, accounts for most of the total net savings of the MMLV Mach-I design life cycle AP, GWP, POCP, HHPP and TPE.
- The production stage is the dominant stage for the total net savings of life cycle ODP, NRMR and RMR. The total net savings of life cycle EP are dominated by the EOL stage. Overall, the EOL stage contributes to an increase of the cradle-to-grave total net savings across all impact categories (with the exception of ODP).

- The MMLV Mach-I design achieved an overall 364 kg (23%) mass reduction (curb weight is reduced from 1,559 kg to 1,195 kg), enabling engine downsizing, which resulted in a combined fuel economy of 34 mpg (6.9 l/100 km) compared to 28 mpg (8.4 l/100 km) for the baseline vehicle. Total life cycle mass-induced fuel savings due to the lightweight auto parts resulted in significant fuel economy benefits of 3,642 L (or 962 gal.).
- The body (sub-system 1) is the largest contributor to the cradle-to-grave total net savings of life cycle AP (43%), GWP (35%), POCP (44%) and HHPP (57%). Interior (sub-system 2) and Chassis (sub-system 3) are the main contributors of the cradle-to-grave total net savings of life cycle ODP (100%) and EP (over 100%).
- The body-in-white, panels, seat and wheel auto part sub-assemblies were the most significant contributors to the total life cycle mass-induced fuel savings at 758 L (21%), 292 L (8%), 292 L (8%), and 290 L (8%), respectively. These four auto part sub-assemblies have incorporated the highest weight reduction in the framework of the MMLV Mach-I vehicle design.
- The body-in-white, panels, tires and seat auto part sub-assemblies were the most significant contributors to the cradle-to grave total net savings of life cycle AP, EP, GWP, POCP and HHPP. The HVAC auto part sub-assembly is the main contributor to the cradle-to grave total net savings of life cycle ODP.
- Given that the LCA results show that over 79% of the cradle-to-grave total net savings of most impacts (except ODP and NRM) are related to use stage (total net change of gasoline production and combustion), the total net savings were deemed sensitive to the Mach-I CFE_M and $LTDD_V$ use stage parameters. In addition, EOL recovered scrap rate and allocation rules for recycling ("end-of life" versus "recycled content" approach) were identified as the third and fourth distant significant issues. The rest of seven varied assumptions (such as well-to-wheel LCI profile of gasoline, Al scrap EOL LCI profile, metal stamping process, EOL disposal scenario of carbon fiber, plastic and tire auto parts and transportation data), could only impact the GWP and TPE insignificantly, less than 1% and 3%, respectively.

References

1. L. Bushi, "Comparative LCA Study of Lightweight Auto parts of MMLV Mach-I Vehicle as per ISO 14040/44 LCA Standards and CSA Group 2014 LCA Guidance Document for Auto Parts" (A LCA report prepared for Promatek Research Centre, Canada, 2014, 169 pp.).

2. International Organization for Standardization, *Environmental Management - Life Cycle Assessment - Principles and Framework*, ISO 14040:2006.

3. International Organization for Standardization, *Environmental Management - Life Cycle Assessment - Requirements and guidelines*, ISO 14044:2006.

4. CSA Group, *SPE-14040-14- Life Cycle Assessment of Auto Parts- Guidelines and Requirements for Conducting LCA of Auto Parts Incorporating Weight Changes Due to Material Composition, Manufacturing Technology, or Part Geometry*, 2014.

5. US DOE EERE 2005, Advanced Vehicle Technology Analysis and Evaluation Activities, FY 2005 Annual Progress Report, "*Automotive System Cost Model (ASCM) Development*", Author: Das, S, pp 60-64.

6. PRé Consultants, SimaPro LCA software, Version 8.0.3, 2014.

7. US EPA, ORD/NRMRL/Sustainable Technology Division, Systems Analysis Branch: Tool for the Reduction and Assessment of Chemical and other Environmental Impacts (TRACI), TRACI version 2.1- User's Manual, (Report, 2012, 24 pp).

8. Argonne National Laboratory, GREET.net 2013 software (Version 1.1.0.9745).

9. Argonne National Laboratory, GREET^{TM} 2012 software, GREET2 Model.

10. World Steel Association, "*Life cycle assessment methodology report, Life cycle inventory study for steel products*", (A LCA report, 2011, 95 pp).

11. The Aluminum Association, "*The Environmental Footprint of Semi-Finished Aluminum Products in North America*", (A LCA report, 2013, 124 pp).

12. PE INTERNATIONAL, Inc., "*Life Cycle Assessment of Polymers in an Automotive Assist Step*", (A LCA report prepared for American Chemistry Council, 2012, 56 pp).

13. US EPA fueleconomy.gov website: http://www.fueleconomy.gov/feg/bymodel/2013_Ford_Fusion.shtml.

14. C. Koffler and K. Rohde-Brandenburger, On the Calculation of Fuel Savings through Lightweight Design in Automotive Life Cycle Assessments, International Journal of LCA 15 (1): 128-135, 2010.

15. J. Atherton, Declaration by the Metals Industry on Recycling Principles, International Journal of LCA 12 (1) 59 – 60 (2007).

Acknowledgement

Authors would like to thank Magna International and the Provence of Ontario for funding this LCA project, as well as Magna International, Ford Motor Company and the US DOE for the support and assistance associated with the LCA vehicle-specific input data. Last but not least, authors would like to thank Simone Ehrenberger, third-party critical reviewer, for her valuable comments and recommendations that help improve the LCA report.

Disclaimer

This material is based upon work supported by the Department of Energy National Energy Technology Laboratory under Award Number No. DE-EE0005574.

This report was prepared as an account of work sponsored by an agency of the United States Government. Neither Magna International, Ford Motor Company, the United States Government nor any agency thereof, nor any of their employees, makes any warranty, express or implied, or assumes any legal liability or responsibility for the accuracy, completeness, or

Engineering Solutions for Sustainability: Materials and Resources II
Edited by: Jeffrey W. Fergus, Brajendra Mishra, Dayan Anderson, Emily Allyn Sarver, and Neale R. Neelameggham
TMS (The Minerals, Metals & Materials Society), 2015

IMPROVEMENT OF LOW TEMPERATURE FORMABILITY OF AZ31 MAGNESIUM ALLOY BY HIGH SPEED ROLLING

Jing Su[1], Abu Syed H. Kabir[1], Mehdi Sanjari[1], In-ho Jung[1] and Steve Yue[1]

[1] *Department of Mining and Materials Engineering, McGill University, 3610 Rue University, Montreal, Quebec, Canada H3A 0C5*

Keywords: Magnesium alloy, Twin roll casting, High speed rolling, Formability

Abstract

Magnesium alloys have attracted great attention for wide applications as structural materials for automotive industry due to their low density, high specific strength and stiffness, compared to aluminium and steel. Therefore, Mg alloys have potential to reduce the weight of the vehicles and thus to decrease the fuel consumption and CO_2 emission. However, the limited formability of Mg alloys at low temperatures, which is owing to the insufficient number of slip systems and the strong basal texture, impedes the applications of the wrought Mg products. In the present study, a high reduction of 72% was achieved at a low temperature of 100 °C by employing high speed rolling (1000 m/min) in single pass. As compared, a low rolling speed of 15 m/min was selected to produce the Mg sheets at the same temperature. The sheet fractured at a reduction of 37% during low speed rolling. Deformation mechanisms of low temperature rolled AZ31 Mg alloy sheets were studied by comparing two significantly different rolling speeds.

Introduction

Magnesium and its alloys have low density, high specific strength and high specific stiffness compared to steel and aluminum, which makes them promising candidates for automotive industry where weight reduction of the vehicles is one of the most important concerns. The conventional way to produce Mg alloy sheets is to cast slabs with the thickness of 20 to 30 cm and homogenize for several hours; then hot rolled to the desired thickness [1, 2]. However, by twin-roll casting (TRC), the Mg alloy cast strip can be produced to less than 5 mm in one step [2-4]. Compared to conventional process, the processing costs of TRC are only a third to a half, operating and investment costs are only a quarter to a third, as well as lower space and labour requirements [4]. Therefore, TRC is highly effective, time-saving, and energy saving [1, 4]. In addition, the solidification rates during twin roll casting is much faster than the conventional ingot casting [2], which reduces the segregation and refines the microstructure and thus is beneficial for the mechanical properties [5].

As is well known, magnesium has a hexagonal close-packed (HCP) structure, which shows poor formability at room temperature. This is due to the lack of sufficient independent slip systems and therefore pronounced basal texture [6, 7]. When Mg alloy sheets rolled at low temperatures, edge cracks easily appeared or sheets fractured at low reductions [8-10]. The maximum achievable reductions during cold rolling of pure Mg and AZ31 (Mg–3Al–1Zn) alloy before failure were around 30% and 15%, respectively [9]. Even at a warm temperature of 200°C, a reduction of 30% was only achieved without side-cracks during rolling AZ31 alloy [10].

To improve formability, Mg alloy sheets are manufactured at elevated temperatures by multi-pass rolling at small reductions and intermediate annealing between each pass in the industry [11]. The rolls are often heated to avoid temperature drop during rolling [8, 11]. Low productivity, high energy consumption and high cost limit the application of Mg alloy sheets [8, 11]. Strong basal texture inevitably occurs during hot rolling of Mg alloy sheets [12] and therefore further limits the formability of sheets. In addition, extensive dynamic recrystallization (DRX) in hot rolled Mg sheets reduces the effectiveness of texture weakening by static recrystallization during annealing [13].

By adding rare earth elements or Y in Mg alloys was also proved to randomize the basal texture, which can improve the formability of Mg alloy sheets. Barnett et. al showed that more than 90% reduction can be reached during cold rolling of Mg–0.2Ce alloy [9]. However, addition of rare earth elements largely increases the cost of Mg sheets. Therefore, the alloy chosen for the present study was the most widely developed commercial Mg alloy, AZ31 (Mg-3Al-1Zn).

It has been reported that the formability is greatly improved by high-speed rolling [8, 14, 15]. Koh et. al and Li et. al showed that a reduction of 60% can be achieved in AZ31 alloy sheet by single-pass rolling at a speed of 2000m/min from room temperature to 350°C [8, 14]. Sanjari et. al also showed that the reduction of 60% without edge cracking was attained for AZ31 alloys in single pass at speed of 1000m/min at 300°C, while cracks initiated at the low rolling speed of 15m/min [15]. The better formability is related to the decrease in heat transfer from the sheet to the rolls with shorter touching time [8, 16] and internal heat generated by plastic deformation and friction between the sheet and the rolls. The increase of temperature during high speed rolling leads to activation of additional slip systems and thus improves the formability [8, 14]. In addition, high speed rolling has potential to modify the texture [15, 17-19].

In the present study, twin roll casting was employed to produce the initial Mg plates. Then, a low temperature of 100°C was chosen for rolling Mg alloy sheets at a high rolling speed of 1000m/min. For comparison purposes, rolling was also performed at the low rolling speed of 15m/min using the same initial temperature. The objective of this research was to study the deformation mechanisms during high speed rolling and to account for the improvement in formability in this way.

Experimental Procedure

The alloy used for the present research is an AZ31 magnesium alloy (Mg-3.2%Al-0.77%Zn-0.34%Mn), which was supplied by the Posco, Korea. Twin roll cast was employed to product the Mg alloy sheets. Then warm rolling was conducted to break down the as-cast structure. The as-received materials were annealed at 300°C for 1h to recrystallize and homogenize the structure. The annealed plate was subjected to single pass rolling at 100°C at a high rolling speed of 1000m/min at reductions from 8% to 72% and at a low rolling speed of 15m/min at reductions from 8% to 37%. Prior to rolling, the plates were pre-heated to 100°C for 10min and the rolls were not heated. The as-rolled sheets were quenched as soon as possible after rolling to preserve the microstructure. Microstructure and texture were characterized by optical microscopy and X-ray diffraction. The specimens were mechanically ground and polished. To reveal the optical microstructures, a solution consisting of picric acid (4.2g), acetic acid (10ml), water (10ml) and ethanol (70ml) was used.

Results

As Received Microstructure and Texture

The optical microstructure and (0002) pole figure of the as-received AZ31 Mg alloy plate are shown in Figure 1. It shows a bimodal microstructure with small equiaxed grains and elongated large grains. These small grains could be formed through dynamic recrystallization during warm rolling. The (0002) pole figure reveals a basal texture with basal pole splitting along the RD direction. After annealing on the as-received plate, it is found that static recrystallization took place and the grain growth of the recrystallized grains occurred simultaneously, although the elongated unrecrystallized grains still remained (Figure 2 (a)). A basal texture with a single peak, i.e. [0001]//ND, is seen in the (0002) pole figure (Figure 2 (b)) and the maximum intensity is slightly lower than in the as-received material.

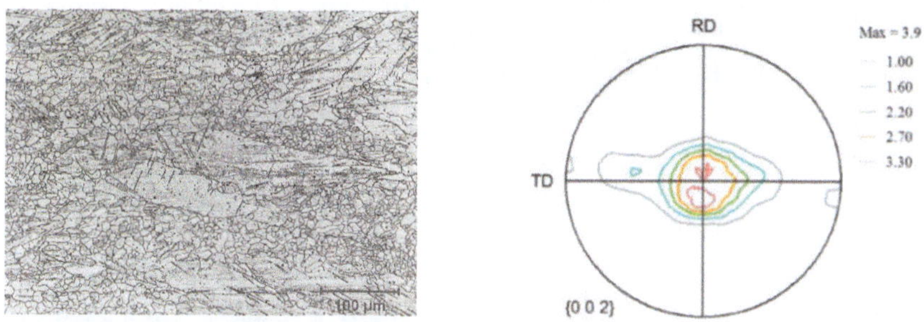

(a) (b)

Figure 1. Optical microstructure and (0002) pole figure of the as-received plate

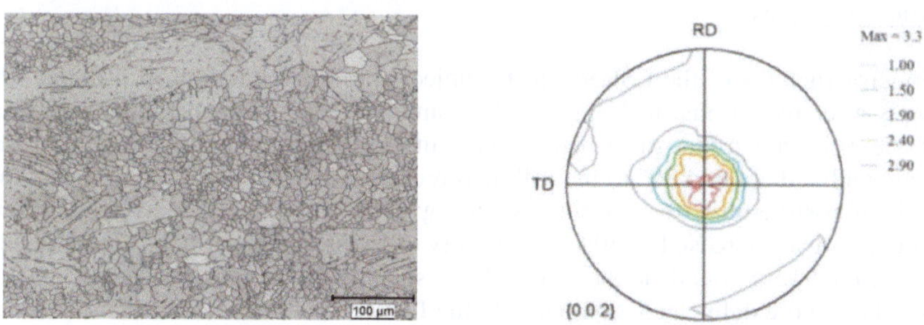

(a) (b)

Figure 2. Optical microstructure and macrotexture of the annealed plate

Formability

The formability of the AZ31 Mg alloy sheets, as characterized by the appearance of cracks, is illustrated in Table 1 for the two rolling speeds. No cracks occur at the reduction of 8% in either HSR or LSR specimens, while the edge cracks appear at the reduction of 30% at both rolling speeds. However, the cracks formed during HSR are less deep than those during LSR. When the reduction was increased to 37%, the sheets fracture into pieces during LSR. By contrast, when HSR was employed, the sheets remain intact and only moderate edge cracking is observed at reductions of 49% and 72%. Therefore, HSR leads to the better formability than LSR. Curiously, the cracks formed during HSR to a reduction of 72% are less severe than those developed at the reduction of 49%.

Table 1. Comparison of formability of under low speed and high speed rolling conditions

	LSR	HSR
8%		
30%		
37%		
49%		
72%		

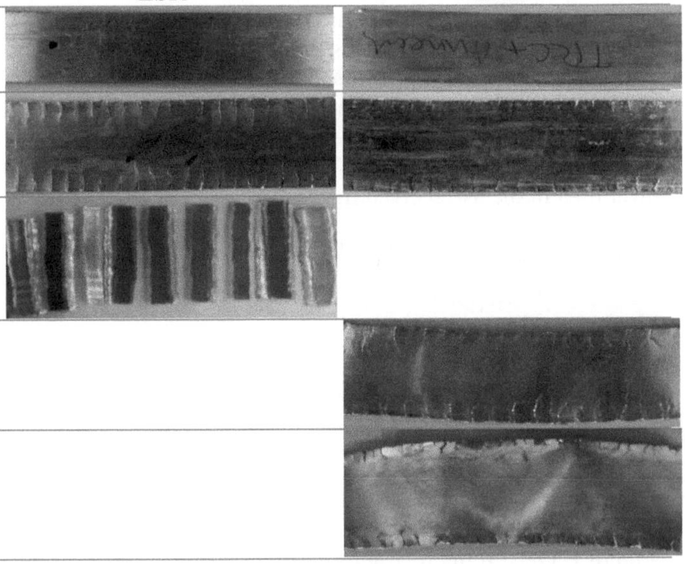

As-Rolled Microstructures

The optical microstructures of the LSRed sheets subjected to the reductions of 8% and 30% and the HSR sheets after reductions of 8%, 30%, 49% and 72% are compared in Figure 3. At the reduction of 8%, it is shown that shear bands and twins formed during both HSR and LSR. The shear bands are inclined about ±30° to the rolling direction and separated by thin twins. In terms of the optical microstructures, the specimens during LSR and HSR display similar features. When the reduction was increased to 30%, the density of twins and shear bands increased and no recrystallized grains were found at the two rolling speeds. However, the shear bands in the HSRed material are more delineated, compared to the LSRed specimen.

With increasing the reduction to 49%, dynamic recrystallization (DRX) can be seen in the vicinity of twins and shear bands in the HSRed specimen. In the cold rolled pure Mg, the nuclei were also found at shear bands where the stored energy is high and is associated with large orientation gradients [9]. At the reduction of 72%, full recrystallization was achieved.

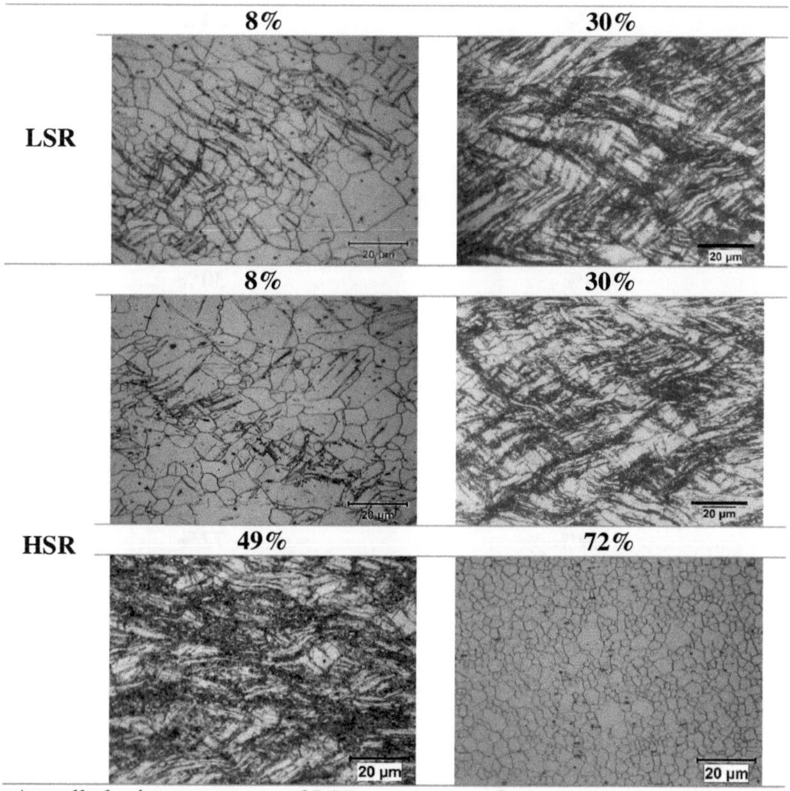

Figure 3. As-rolled microstructures of LSR specimens after reductions of 8% and 30% and HSR specimens after reductions of 8%, 30%, 49% and 72%

Macrotexture of the As-rolled Specimens

The (0002) pole figures of the as-rolled specimens subjected to reductions of 8% and 30% by means of LSR and 8%, 30%, 49% and 72% by means of HSR are illustrated in Figure 4. All the (0002) pole figures reveal the basal textures with double peaks splitting along the RD direction. As the reduction was increased from 8% to 30% at both rolling speeds, the intensity of the basal texture increased and the angle between the two peaks slightly increased as well. During HSR, as the reduction was increased from 30% to 72%, the basal texture continued to intensify, even though dynamic recrystallization was well under way.

Another trend that is evident is that, at the reduction of 30%, the maximum intensities of the basal texture during HSR are lower than those during LSR. This may be due to the occurrence of more secondary and contraction twinning and/or <c+a> slip under HSR conditions.

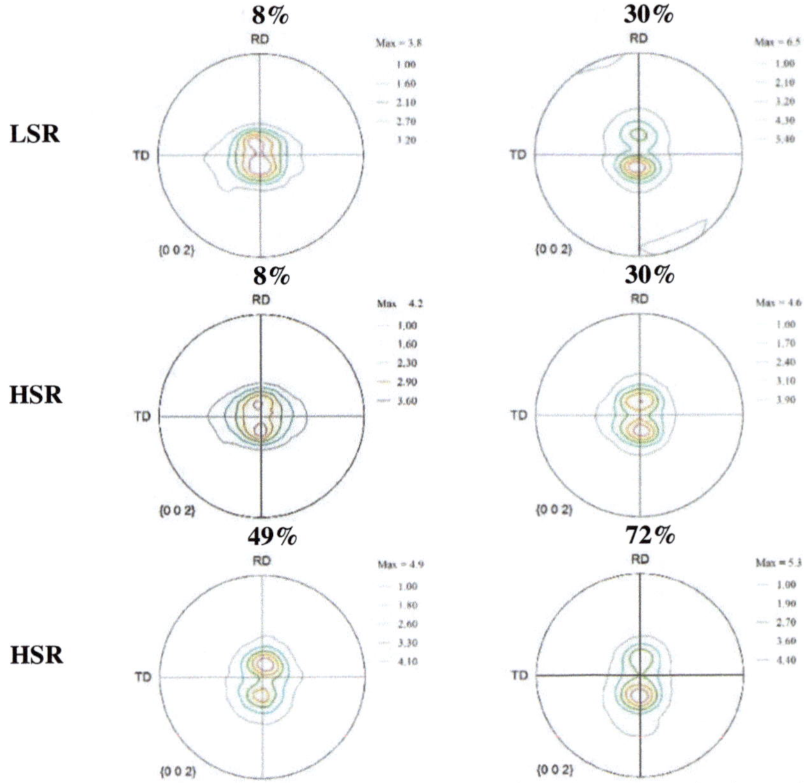

Figure 4. (0002) pole figure of as-rolled specimens subjected to LSR reductions of 8% and 30% and HSR reductions of 8%, 30%, 49% and 72%

Discussion

Heat Transfer during Rolling

The improvement in the formability associated with HSR has been reported to be related to the reduction of heat transfer from the sheet to the rolls due to the shorter contact times [14, 16]. In order to determine the actual deformation temperature *T* in the present study, the amount of deformation and frictional heating must be estimated, together with the effect of roll chilling. The temperature change ΔT during rolling can be evaluated using Equation (1) [20]:

$$\Delta T = \Delta T_p + \Delta T_f - \Delta T_R \quad (1)$$

where ΔT_p is the temperature rise due to the plastic work, ΔT_f is the temperature rise due to friction, and ΔT_R is the temperature drop due to contact with the cold rolls. [20]

As can be seen from Figure 5, the deformation temperature during LSR is much lower than during HSR at the same reductions. This is because that the temperature drop due to roll cooling is higher during the former than the later. Furthermore, the temperature rise due to plastic work and friction during LSR is lower than during HSR. In this way, during LSR, at reductions of 8% and 15%, the actual temperatures are lower than the entry temperature, only reaching 115°C at the reduction of 30%. By contrast, the deformation temperatures during HSR remain higher than the entry temperature for all reductions and increase with the reduction. At 30% reduction, this temperature is about 203°C increasing to 582°C at 72%.

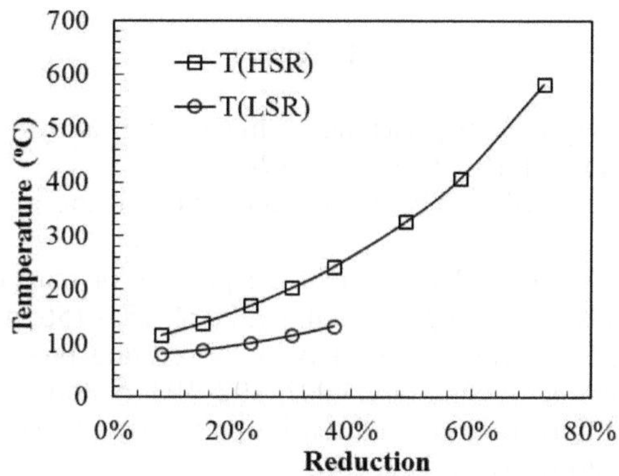

Figure 5. Effect of roll speed and reduction on rolling temperature (°C)

Deformation Mechanisms

During both HSR and LSR, at the low reductions from 8% to 30%, the microstructure was dominated by twins and shear bands which are typical in Mg alloys during rolling at low and intermediate temperatures [9, 18, 21, 22]. With increasing the reduction to 49% during HSR, DRXed grains were found in the vicinity of twins and shear bands and the microstructure was fully recrystallized at the reduction of 72%, which is in turn responsible for the better formability of the HSRed sheets. This is because that the temperatures generated during HSR were much higher than during LSR, which compensated the higher strain rate and thus decreased the critical strain for dynamic recrystallization. In addition, during HSR, the higher temperatures at higher reductions maybe promote the activation of <c+a> slip systems, therefore weakened the basal texture.

Conclusion

1. Reductions of 72% can be readily achieved in one pass at the high rolling speed. Conversely, sheets fracture at 37% reduction when low speed rolling is employed. The greater formability of HSR is related to the higher temperatures generated, which in turn decreases the critical strain for DRX and promotes the activation of <c+a> slip systems, improving the ductility.
2. Basal pole splitting along the RD was observed under all deformation conditions. This can be primarily attributed to the occurrence of contraction and secondary twinning.
3. At reductions of 30%, the maximum intensity of the basal texture was lower during HSR than during LSR. This is again related to the formation of more contraction twins as well as to the activation of <c+a> slip during high speed rolling.

Acknowledgements

The authors are grateful to the Natural Sciences and Engineering Research Council (NSERC) Magnesium Strategic Research Network (MagNET) in Canada for funding this project and Posco Inc. in South Korea for providing the twin roll cast AZ31B alloy plates. They are also thankful to Prof. Utsunomiya for making available the high speed rolling facility at Osaka University.

References

1. Ding, P., et al., Twin-roll strip casting of magnesium alloys in China. Transactions of Nonferrous Metals Society of China, 2008. 18: p. s7-s11.
2. Liang, D. and C.B. Cowley, The Twin-Roll Strip Casting of Magnesium. JOM, 2004. 56(5): p. 26-28.
3. Watari, H., et al., Semi-solid manufacturing process of magnesium alloys by twin-roll casting. Journal of Materials Processing Technology, 2004. 155–156: p. 1662–1667.
4. Watari, H., et al., Feasibility study of twin roll casting process for magnesium alloys. Journal of Materials Processing Technology, 2007. 192–193: p. 300–305.
5. Park, S.S., et al., Microstructure and tensile properties of twin-roll cast Mg–Zn–Mn–Al alloys. Scripta Materialia, 2007. 57: p. 793–796.
6. Agnew, S.R., M.H. Yoo, and C.N. Tomé, Application of texture simulation to understanding mechanical behavior of Mg and solid solution alloys containing Li or Y. Acta. Mater., 2001. 49: p. 4277–4289.
7. Styczynski, A., et al., Cold rolling textures in AZ31 wrought magnesium alloy. Scr. Mater., 2004. 50: p. 943–947.
8. Koh, H., et al., Deformation and Texture Evolution during High-Speed Rolling of AZ31 Magnesium Sheets Materials Transactions, 2007. 48(8): p. 2023-2027.
9. Barnett, M.R., M.D. Nave, and C.J. Bettles, Deformation microstructures and textures of some cold rolled Mg alloys. Mater. Sci. Eng. A, 2004. 386: p. 205–211.
10. Jeong, H.T. and T.K. Ha, Texture development in a warm rolled AZ31 magnesium alloy. J Mater Process Tech, 2007. 187–188: p. 559–561.
11. Zarandi, F. and S. Yue, Magnesium Sheet: Challenges and Opportunities in Magnesium Alloys - Design, Processing and Properties, F. Czerwinski, Editor. 2011, InTech. p. 297-320.
12. Valle, J.A.d., M.T. Pe´rez-Prado, and O.A. Ruano, Texture evolution during large-strain hot rolling of the Mg AZ61 alloy. Mater Sci Eng A, 2003. 355: p. 68-78.
13. Huang, X., K. Suzuki, and Y. Chino, Different annealing behaviours of warm rolled Mg–3Al–1Zn alloy sheets with dynamic recrystallized microstructure and deformation microstructure. Mater. Sci. Eng. A, 2013. 560: p. 232–240.
14. Li, H., et al., Deformation mechanism and texture and microstructure evolution during high-speed rolling of AZ31B Mg sheets. J. Mater. Sci., 2008. 43: p. 7148–7156.
15. Sanjari, M., et al., Promotion of texture weakening in magnesium by alloying and thermomechanical processing-II high speed rolling. J. Mater. Sci., 2014. 49: p. 1426-1436.
16. Koh, H., et al., Deformation and texture evolution during high-speed rolling of AZ31 magnesium sheets. Mater. Trans., 2007. 48(8): p. 2023-2027.
17. Sanjari, M., et al., Promotion of texture weakening in magnesium by alloying and thermomechanical processing: (I) alloying. J. Mater. Sci., 2014. 49: p. 1408–1425.

18. Chun, Y.B. and C.H.J. Davies, Texture effects on development of shear bands in rolled AZ31 alloy. Mater. Sci. Eng. A, 2012. 556: p. 253–259.
19. Sandlöbes, S., et al., The relation between shear banding, microstructure and mechanical properties in Mg and Mg-Y alloys. Mater Sci Forum, 2011. 690: p. 202-205.
20. Muraoka, A., et al., Improvement in rolling workability of Fe3Al by high-speed rolling. Journal of Physics, 2012. 379: p. 1-6.
21. Sandlöbes, S., et al., On the role of non-basal deformation mechanisms for the ductility of Mg and Mg–Y alloys. Acta Mater., 2011. 59: p. 429–439.
22. Barnett, M.R. and N. Stanford, Influence of microstructure on strain distribution in Mg–3Al–1Zn. Scr. Mater., 2007. 57: p. 1125–1128.

Engineering Solutions for Sustainability: Materials and Resources II
Edited by: Jeffrey W. Fergus, Brajendra Mishra, Dayan Anderson, Emily Allyn Sarver, and Neale R. Neelameggham
TMS (The Minerals, Metals & Materials Society), 2015

PREPARATION OF NANO CRYSTALLINE FORSTERITE SYNTHESIZED BY MECHANICAL ACTIVATION TO USE ORTHOPEDIC AND DENTAL APPLICATIONS

Hassan Gheisari[a] and Ebrahim Karamian[b]

Ceramic Group, Department of Technical, Lenjan Branch,

Islamic Azad University, Isfahan, Iran

[b]Assistant Professor, Department of Materials Engineering, Najafabad Branch,

Islamic Azad University, P.O.Box 517, Isfahan, Iran

Corresponding Author: ***Gheisary.iauln.ac.ir**, Tel: **+989135344540**

Abstract:

Recently, forsterite (Mg_2SiO_4) has been introduced as a possible candidate for load bearing applications. In this investigation, forsterite powder was synthesized by mechanical activation method. Specimens were composed of a blend of pure magnesia (MgO) (98% purity, Merck) and silica amorphous (SiO_2) (98% purity, Merck) powder with 57 % wt. and 43%wt, respectively. The powder mixture was milled by high energy ball mill for 5 h and 10 h with ball to powder ratio 10. Then, the mixtures milled have been annealed at 1000 ℃ for 2 h in muffle furnace. X-ray diffraction (XRD), scanning electron microscopy (SEM), BET and PSA performed on the samples. According to XRD results, the sample milled for 10 h just shown the forsterite phase. In fact, our study indicated that forsterite powder was composed of nano-crystalline structure, about 40 nm, can be prepared by mechanical activation to use as a new biomaterials for orthopaedic purposes.

Keywords: Forsterite, Ball milling, Mechanical Activation, Synthesis, Nano Crystalite

Introduction

The forsterite (Mg_2SiO_4) ceramic is a new bioceramic with good biocompatibility [2]. Compared with hydroxyapatite ceramics, forsterite shows a significant improvement in the fracture toughness, superior to the lower limit reported for cortical bone [1,2].However,the sintering temperature of forsterite is too high, which impoverishes its mechanical properties [2,3].

Recently, forsterite (Mg_2SiO_4) has been introduced as a possible candidate for load bearing applications. Although it has a better bending strength and fracture toughness than those of commercially available hydroxyapatite ceramics, the degradation rate of forsterite ceramic is tremendously low, and the apatite formation ability is poor as well [4]. The aim of the present work was to investigate characterization of nanostructure forsterite ceramic as opposed to previous studies. The results of this paper can be used for further researches and it would promote the possibility of usages of nanostructure forsterite ceramic in orthopaedic application.

Materials and Methods

Recently, forsterite (Mg_2SiO_4) has been introduced as a possible candidate for load bearing applications. In this investigation, forsterite powder was synthesized by mechanical activation method as a solid state process. Forsterite was composed of nano-crystalline structure with proximately diameter 30-50 nm. Specimens were composed of a blend of pure magnesia (MgO)(98% purity, Merck) and silica amorphous (SiO_2) (98% purity, Merck) powder with 57 % wt. and 43%wt, respectively. The powder mixture was milled by high energy ball mill, ball-to-powder ratio 10:1 and rotation speed 600 rpm for 5 h and 10 h. Then, the mixture milled has been annealed at 1000 °C for 2 h in muffle furnace at air atmosphere. X-ray diffraction (XRD) and scanning electron microscopy (SEM) on the samples to characterize. According to XRD results, the sample milled for 10 h just shown the forsterite phase, while the sample milled for 5 h shown forsterite along with several phases. In fact, our study indicated that forsteritepowder was composed of nano-crystalline structure, 30-40 nm, can be prepared by mechanical activation to use as a new biomaterials for orthopaedic purposes. SEM micrograph was performed using a LEO 435 VP to investigate the morphology. Samples coated with Au by sputter spraying, low vacuum and 100-120 V accelerating voltage. The specific BET surface area of forsterite powder has been done by Kelvin B100.

The Modified Scherrer

The modified Scherrer equation can provide the advantage of decreasing the sum of absolute values of errors, $\sum(\pm\Delta \ln\beta)$ 2, and producing a single line through the points to give a single value of intercept lnKλ/L. At this sample, Figure 1, the linear regression plot is obtained as $y = 3.355X - 5.540$. This is equivalent to Ln β = ln (1/CosΘ) + ln (kλ/L). From this line, the intercept is -5.646 and $e^{-5.6486}$ = Kλ/L and L=39 nm. So, TCP crystallite size average is 39 nm.

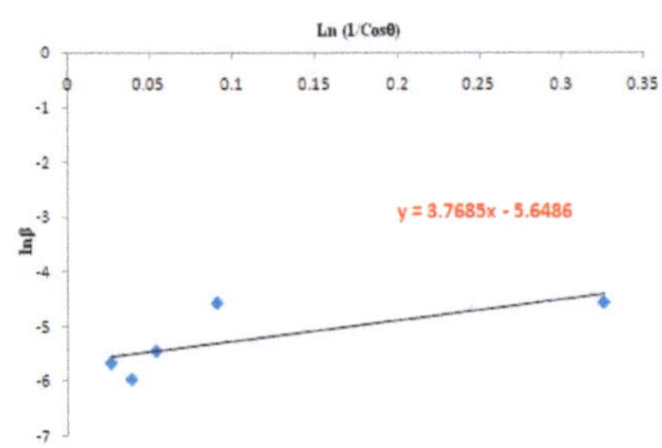

Fig 1. Plot of lnβ vs. Ln (1/Cosθ) of sample heated at 1000 °C for 2 h

Results

SEM Micrographs

Figures 2-4 show the SEM micrograph the materials milled before and after heating.

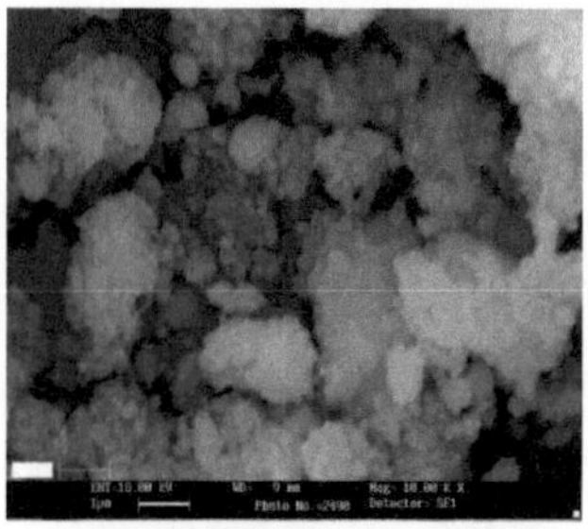

Fig. 2. SEM micrograph of the powder mixture milled before heating

Fig. 3. SEM micrograph of the sample milled for 10 h and then heated at 1000 °C for 2 h

Fig. 4. SEM micrograph of the sample milled for 5 h and then heated at 1000 °C for 2 h

XRD Results

Fig. 5 and 6 show the XRD patterns of the materials mixture milled heated at three temperatures.

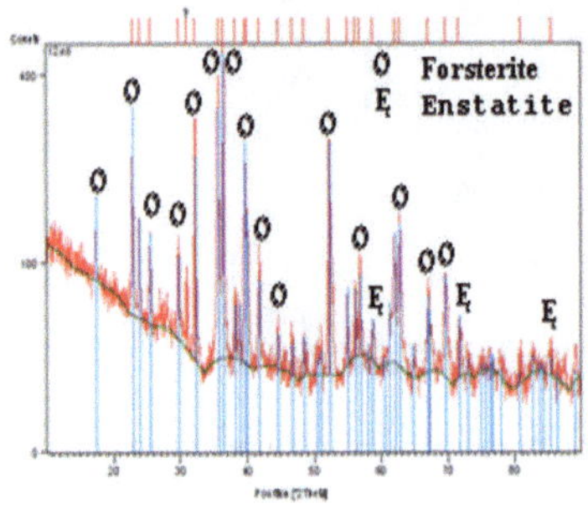

Fig. 5. XRD pattern of the materials mixture milled for 5 h and then heated at 1000 °C for 2 h

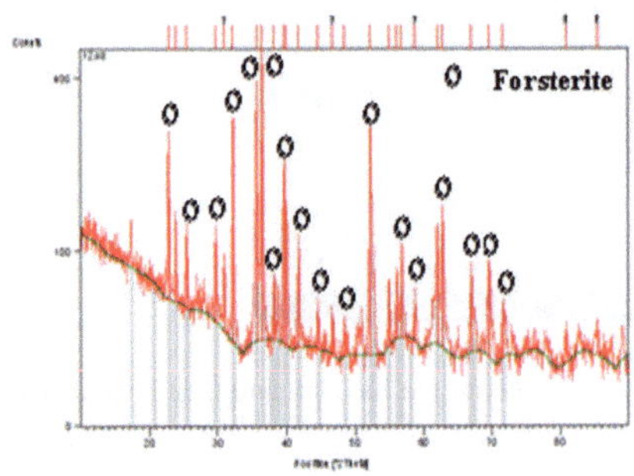

Fig. 6. XRD patterns of the materials milled for 10 h and then heated at 1000 °C for 2 h

Considering SEM micrograph in figure 2, average particle size of powder mixture materials is micron range.

Pure Density

Real density or pure density of the forsterite powder was calculated by pycnometer. Considering to this method pure density was 2.87 g/cm^3.

BET Results

The specific surface area of the prepared powder was calculated from the N_2 gas adsorption isotherms using the multipoint BET technique The average particle size of the prepared powder, assuming that the particles synthesized were spheroid, was calculated as shown in Equation 1.

$D=6000/ (S_{BET} * d)$ Eq. 1

Where, d and D are true density (g/cm^3) and the average particle size (micron) of materials mixture milled, respectively. The specific surface area determined by BET was 4.5 m^2/g. And pure density calculated by Pycnometer method was 2.87 g/cm^3. So, according to Eq. 1 the particles size was estimated about 470 nm. In fact, the powder mixture is micron size. In addition, it is confirmed in SEM micrograph (figure 2).

Discussion

According to the XRD patterns, forsterite phase was synthesized in both the samples (Figures 5 and 6). But, there is single phase, forsterite, just the sample milled for 10 h (Figure 6). Therefore, synthesis of forsterite phase is affected by using mechanical activation. In fact, mechanical activation intense the synthesis of forsterite phase in the sample. In addition, forsterite average crystallite size is almost 40 nm.

Conclusions

In conclusion, forsterite nano crystallite, about 40 nm, have been synthesized at 1000 °C by materials mixture mechanical activated for 10 h. Whereas, based on the previous studies, forsterite phase was synthesized at above 1100 °C.

Acknowledgment

The authors would like to extend their gratitude for the supporting provided by Lenjan Branch, Islamic Azad University, Isfahan, Iran.

References

[1] Hoexter DL. Bone regeneration graft materials. J Oral Implantol, 2002; 28: 290 –294.
[2] Jarcho M. Calcium phosphate ceramics as hard tissue prosthetics, Clin Orthop 1981;157: 259 – 278.
[3].Burchardt, H. The biology of bone graft repair. Clin Orthop 1983;174:28–42.
[4] Blokhuis TJ et al., Resorbable calcium phosphate particles as a carrier material for bone marrow in an ovine segmental defect. J Biomed Mater Res 2000;51:369 –375.

IV

Power Production

Engineering Solutions for Sustainability: Materials and Resources II
Edited by: Jeffrey W. Fergus, Brajendra Mishra, Dayan Anderson, Emily Allyn Sarver, and Neale R. Neelameggham
TMS (The Minerals, Metals & Materials Society), 2015

DETERMINING THE ENERGY VALUE ON DIFFERENT COMPRESSIONS OF SAWDUST BRIQUETTES

I. U.Onyenanu[1], C. E. Ilochonwu[2*], P. N. Atanmo[1]

[1]Department of Mechanical Engineering, Anambra State University, Nigeria
[2]Scientific Equipment Development Institute Enugu, Nigeria
*Corresponding author email: mecury4eva@yahoo.co.uk

Keywords: binding materials, sawdust briquettes, energy values, domestic fuel.

Abstract

Biomass and other industrial wastes is on the increase and causes a lot of problem especially the effect on our environment. In developing countries like Nigeria, there are inadequate measures of disposing these wastes, hence, converting them to other useful products such as briquettes for domestic fuel is desirable. In this research work, the energy values of briquettes made from sawdust using two binders were assessed. The sawdust briquettes of different compressions was produced using sawdust of different grain sizes (fine and coarse) and two different types of binders (clay and starch) in percentage compositions of 75:25, 80:20, 85:15, 90:10 and 95:05, respectively. The briquettes were subjected to energy evaluation test using the Bomb Calorimeter. The results indicated that the higher the compression, the more the energy value for both samples. Therefore sawdust briquette of higher compression is more suitable for starting and maintaining fire for cooking and other domestic heating.

1. Introduction

Biomass waste is a general term given to all organic waste and dry plant materials and is characterized with a variety of conversion and end use. It has found its way as one of the favorable means of creating a cleaner environment, waste recycling to valuable products, renewable energy and inexpensive energy conversion etc. Examples of these include; sawdust, rice husk, palm kernel shell, groundnut shell, sugarcane bagasse etc.

Again, the uses of biomass fuel such as composite sawdust briquette have been found to be a good source of renewable energy for domestic cooking (O. N. Aina, et. al, 2009). In seventeenth century, the rural poor often burn dried cow dung because of the acute shortage of wood fuel and wide spread deforestation. The conversion of Agricultural by products, wood waste and coal dust to high energy value briquettes for cooking and drying have been investigated and found to be feasible.

Furthermore, the importance of sawdust as a source of fuel cannot be over-emphasized as it is readily available and cheap. Most often, it is usually dumped or burnt off at sawmill sites in Nigeria thus a lot of energy is being wasted and uncontrolled heat generated. Over the years, sawdust has been used for the production of heat and power in gasification of plant and for domestic cooking (kurstdet, 1981). This possibly raises the issue of how best to utilize this economical waste. Moreover, the production of biomass briquette contributes to income generation of micro enterprise or any entrepreneur who produces them and sells them. In this ways, more money stays within the community rather than being exported for foreign fuels.

Briquetting can be defined as a way of converting loose biomass residue such as sawdust, straw or rice husk into high density solid blocks that can be used as a fuel. Briquetting can be of two types - high pressure briquetting and low pressure briquetting. High pressure briquetting uses a power-driven press to raise the pressure of dry, powdered biomass to about 1500bar (150Mpa).This will raise the temperature of the biomass and as a result melts the lignin in the woody (sawdust) material. Low pressure briquetting can be used for materials with a low amount of lignin, such as paper and charcoal dust. In this process, the powdered biomass is mixed into a paste with a binder such as starch or clay, and water. A press is used to push the paste into a mould or through an extruder or can simply be shaped by hand. The best materials for high pressure briquetting are sawdust and woody residues, because they contain a high proportion of lignin. However, most dry agricultural residues can be used if they are ground into a coarse powder. Some do not make good briquette on their own (e.g. grain straw and dried grasses) but work well when mixed with woody materials to provide the lignin.

Thus, this work investigate the energy values on different compressions of sawdust briquettes using a bomb calorimeter. But before measuring the energy values of the sawdust briquette, binders for different briquettes will be reviewed.

2. Materials and Methods

This study was developed by using the facilities and materials laboratory at the Scientific Equipment Development Institute, Enugu State – Nigeria.

2.1 Raw Material Preparation

The sawdust used was a mixture of different grain sizes (0.05cm & 0.1cm and 0.3cm & 0.5cm respectively) constituting two major samples for the experiment. The mixture of the different grain sizes for each sample was in the ratio of 50:50 and 70:30. Each of the grain sizes sample was gotten from two different sawmills located in Aba and Enugu (Eastern part of the country, Nigeria). The moisture content (MC) of each sample before briquetting was 15% and 20% respectively. Finally, we designated the finer grain samples as Group A, then the coarse sample as Group B.

2.2 Briquetting Production and Quality Evaluation

The sawdust was poured into a flash dryer which automatically reduce the moisture content to 5% at 250 - 3000C for 50 minutes for both samples.

Group A: The sawdust mixture of finer grains was fed into a bowl and mixed with two different types of binders (clay and starch) in percentage compositions of 75:25, 80:20, 85:15, 90:10 and 95:05, respectively *(See Table 1).* The agitating process was done in a mixer to enhance proper blending prior compaction. A steel cylindrical crucible (die) of dimension 14.3mm height and 49.3mm in diameter was used for this project. The die was freely filled with known amount of weight (charge) of each sample mixture and be positioned in the briquetting press machine for compression. The piston was actuated manually for 10, 12, 14, 16 and 18 respective oscillation counts of hammer head with respect to piston movement to compress the samples. Compacted pressure ranged from 3.0 – 9.0MPa. After pressure was applied at a time to the material in the die, the briquette formed was extruded.

Prior the release of applied pressure, each briquette was replicated five times according to the level of process variables. The briquettes produced were allowed to dry in the sun for two days and then assessed for their energy.

Group B: Again, the sawdust mixture of coarse grains was fed into a bowl and mixed with two different types of binders (clay and starch) in percentage compositions of 75:25, 80:20, 85:15, 90:10 and 95:05, respectively *(See Table 2)*. The same procedure for Group A was utilized for Group B in making the briquettes.

2.3 Data Collection

The length, mass, diameters of the briquettes were determined. These measurements were used to compute the volume and density of each of the One Hundred and Fifty samples of briquettes produced *(See Table 3)*.

NB: We used the value we obtained from the average of the five samples for each sample in the table. This was done to get a more accurate value for the research

Data were also collected on the physical properties of briquettes produced from the sawdust. The methods and formulas used are given below.

Density: The weights of briquettes were determined on the balanced in the laboratory. Then, the volumes of briquettes were determined by a simple calculation based on the direct measurement of length and diameter of the briquettes.

The Formula:

D (kg/m3 or g/cm3) = M/V;

Where; D = Density

M = Mass

V = Volume.

2.4 Measurement of the Energy Content of the Briquettes Produced

To measure the energy content of the briquettes produced from different compressions, a bomb calorimeter was used.

Heat Value: The heat value of both samples was determined using a Bomb calorimeter. Each samples of the briquette from each Group was weighed. And placed in the crucible before covering it tightly. The bomb was closed and charged in with oxygen up to 30 atm. The bomb was fired up by depressing the ignite switch to burn the sample in an excess of oxygen. The maximum temperature rise in the bomb was measured with the thermocouple and galvanometer system

Formula:

G.E. (Kcal/g) = (G.meter deflection x Calibration)/ (Weight of sample.)

3. Results and Discussions

3.1 Physical Characteristics of the Briquettes

The briquettes were produced using a screw press briquetting machine. The outer surface of the briquettes was carbonized and solid with no holes at the center. The average weight of the briquettes is approximately 651g, while the average length and diameter are 15.5cm and 5.1cm respectively. The average volume is $609m^3$ while the briquettes from the both samples of the sawdust assumed a brown coloration.

Table 1: Showing the % constituent for Group A samples

S/N	Sample	Starch %	Clay %	Sawdust % *(Finer Mixture)*
1.	A1	25	-	75
2.	A2	-	25	75
3.	B1	20	-	80
4.	B2	-	20	80
5.	C1	15	-	85
6.	C2	-	15	85
7.	DI	10	-	90
8	D2	-	10	90
9.	E1	5	-	95
10.	E2	-	5	95

Table 2: Showing the % constituent for Group B samples

S/N	Sample	Starch %	Clay %	Sawdust % *(Coarse Mixture)*
1.	A1	25	-	75
2.	A2	-	25	75
3.	B1	20	-	80
4.	B3	-	20	80
5.	C1	15	-	85
6.	C2	-	15	85
7.	DI	10	-	90
8	D2	-	10	90
9.	E1	5	-	95
10.	E2	-	5	95

Table 3: Showing the physical properties of the briquettes

S/N	Sample	Vol. before Briquetting (cm^3)	Bulk Density Before briquetting (g/cm^3)	Vol. after Briquetting (cm^3)	Bulk Density after briquetting (g/cm^3)
1.	A1	52.314	0.956	96.599	0.518
2.	A2	52.314	0.956	90.849	0.550
3.	B1	52.314	0.956	119.599	0.418
4.	B2	52.314	0.956	97.749	0.512
5.	C1	52.314	0.956	100.433	0.498
6.	C2	52.314	0.956	96.312	0.519
7.	D1	52.314	0.956	119.216	0.419
8.	D2	52.314	0.956	116.149	0.430
9.	E1	52.314	0.956	115.766	0.432
10.	E2	52.314	0.956	123.816	0.404

Table 4: Comparative assessment of energy values of the briquettes.

Sample	Starch %	Clay %	Sawdust %	Vol. before Briquetting (cm^3)	Bulk Density Before briquetting (g/cm^3)	Vol. after Briquetting (cm^3)	Bulk Density after briquetting (g/cm^3)	Mean
A1	25	-	75	52.314	0.956	96.599	0.518	0.534
A2	-	25	75	52.314	0.956	90.849	0.550	
B1	20	-	80	52.314	0.956	119.599	0.418	0.465
B2	-	20	80	52.314	0.956	97.749	0.512	
C1	15	-	85	52.314	0.956	100.433	0.498	0.509
C2	-	15	85	52.314	0.956	96.312	0.519	
DI	10	-	90	52.314	0.956	119.216	0.419	0.425
D2	-	10	90	52.314	0.956	116.149	0.430	
E1	5	-	95	52.314	0.956	115.766	0.432	0.418
E2	-	5	95	52.314	0.956	123.816	0.404	

Table 5: Mean calorific values of briquettes

Sample	Briquette	Calibration Constant	Galvanometer deflection	GE (Kcal/g)
A1	Sawdust and Starch	0.7872	2.39	4.421
A2	Sawdust and Clay	0.7872	2.20	3.976
B1	Sawdust and Starch	0.7872	2.40	4.553
B2	Sawdust and Clay	0.7872	2.30	3.911
C1	Sawdust and Starch	0.7872	2.60	4.652
C2	Sawdust and Clay	0.7872	2.44	4.009
D1	Sawdust and Starch	0.7872	2.85	4.701
D2	Sawdust and Clay	0.7872	2.50	4.011
E1	Sawdust and Starch	0.7872	3.0	4.723
E2	Sawdust and Clay	0.7872	2.55	4.014

3.1.2 Moisture Content

One of the main parameters determining briquette quality is moisture content of the sawdust used as the input material. The most durable briquettes of sawdust are of the moisture content of 5%. The moisture content of the sawdust studied is 5% which ensured production of good quality briquettes. This lower moisture content of briquettes implies higher calorific value.

Moisture content in excess of 20% would result in considerable loss of energy required for water evaporation during combustion at the expense of the calorific value of the fuel. Such a fuel may not also be stable in storage.

3.1.3 Heat Value

Heat value or calorific value determines the energy content of a fuel. It is the property of biomass fuel that depends on its chemical composition and moisture content.

The most important fuel property is its calorific or heat value. Table 5; revealed that, briquettes of higher compression had more energy values.

3.2 Discussion

The lignocellulose waste (sawdust) used for the production of briquettes was randomly collected from sawmills located in Aba and Enugu (Eastern part of the country, Nigeria). The moisture content (MC) of each sample before briquetting was 15% and 20% respectively. They were flash dried to reduce the moisture content automatically to 5% at a temperature of 90 ^{0}C – 100 ^{0}C for 50 minutes.

The quality of briquette is determined by the moisture content of the sawdust used as the input material. The higher the moisture content of the sawdust, the higher the loss of energy required for water evaporation during combustion at the expense of the calorific value of the briquette. The dried sawdust was freely fed into the cylindrical crucible and positioned in the briquetting press for compression. The piston was actuated manually for 10, 12, 14, 16 and 18 respective oscillation counts of hammer head with respect to piston movement to compress the samples. Compacted pressure ranged from 3.0 – 9.0MPa. After pressure was applied at a time to the material in the die, the briquette formed was extruded.

In determining the calorific value of the briquettes, the Bomb Calorimeter was used. First, the samples of the briquettes produced was allowed to dry naturally in the sun for a period of 2 days. After which the average mass of the group sample was gotten. The bomb's body was placed and tightly screwed in position. The thermocouple wire was plugged into the bomb until the pressure rose to 25 bars. The light spot index was set to zero using the galvanometer zero knob ensuring a stable temperature before the firing knob was depressed and released to fire the bomb. Heat is released and the maximum deflection of the galvanometer scale was recorded after which the burnt gases were released from the apparatus with the aid of the pressure release valve.

The maximum deflection obtained in the galvanometer was converted to energy value of the briquette material by comparing the rise in galvanometer deflection with that obtained when a sample of known calorific 5.1cm 15.5cm value of benzoic acid is combusted. The process is repeated to determine the energy value of the different compressions. The energy values of all the samples burnt were recorded.

4. Conclusions

In this study, a comparative assessment of the calorific values of briquettes produced from different compressions of sawdust biomass materials was carried out in terms of the grain sizes and different binders (starch and clay). It is concluded that briquettes made from sawdust using starch as binder gave the highest energy value during combustion while the least energy was produced by briquettes produced from sawdust using clay as binder. Again, it was observed that the higher the compression, the higher the energy values of the briquettes. Therefore sawdust briquette of higher compression is more suitable for starting and maintaining fire for cooking and other domestic heating.

Furthermore, the experiment test performed have also revealed that production of agricultural by-products (e.g. sawdust) briquettes can greatly provide alternative sources of energy for domestic cooking in Nigeria and also serve as a great measure in reducing the environmental hazard posed by poor methods of agricultural waste disposal, in addition to curbing the popular use of charcoal which has an adverse effect on our environment (deforestation).

5. Acknowledgments

The authors would like to thank Anambra State University (Department of Mechanical Engineering). The efforts of technical staff of Scientific Equipment Development Institute, Enugu State – Nigeria is also acknowledged.

6. References

Adegbulugbe, A. O. (1994). Energy - environmental issues in Nigeria. International Journal of Global Energy Issues6 (12): 7 – 18.

Akinbami, J. F. K. (2001). Renewable energy resources and technologies in Nigeria. Present situation, future prospects and policy framework. Mitigation and Adaptation strategies for Global Change6: 155 – 181.

C.O. Adegoke (2001, April) Waste to wealth: sawdust briquette as a case study, Paper presented at the Mechanical Division of NSE conference, held at Lafia Hotel, Ibadan 20–21.

D. Ayhan and S. Ayse (1998) Evaluation of biomass residue1. Briquetting waste paper and wheat straw mixtures. Fuel Processing Technology, 55, 175-183.

Engineering International the CIGR E-Journal of Scientific Research and Development Vol., VII Manuscript E.E 05 003: 10 pp.

Engineering International: the CIGR E Journal of Scientific Research and Development Vol., 111, 19 pp.

Jekayinfa, S.O. and O.S. Omisakin (2005). The energy potentials of some agricultural wastes as local fuel materials in Nigeria. Agricultural

K.K. Jacob, N.M. Simon, M. Jonathan, N.S. Canon and E. Wanjongo, (2005) Recycling waste into fuel briquettes, www.formatkenya.org/ormbook/ chapters/chapter15htm.

O. M. Aina, A. C. Adetogun, and K.A. Iyola (2009) Heat energy from value-added sawdust briquettes of Abizia zygia. Ethiopian Journal of Environmental Studies and Management, 2 (1), 42-49.

Engineering Solutions for Sustainability: Materials and Resources II
Edited by: Jeffrey W. Fergus, Brajendra Mishra, Dayan Anderson, Emily Allyn Sarver, and Neale R. Neelameggham
TMS (The Minerals, Metals & Materials Society), 2015

HIGH TEMPERATURE FUEL CELLS FOR EFFICIENT CONVERSION OF FOSSIL FUEL ENERGY

Jeffrey W. Fergus[1]

[1]Auburn University; 1301 Shelby Center, Auburn, AL 36849

Keywords: Solid oxide fuel cells, molten carbonate fuel cells

Abstract

The amount of the world's energy supply from renewable sources is increasing rapidly but still constitutes only a small proportion of the total supply. Most of our energy is derived from fossil fuels, so improving the efficiency of the conversion of fossil fuels is critical and can be achieved with electrochemical devices. In particular, high-temperature fuel cells, such as molten carbonate and solid oxide fuel cells, have the fuel flexibility needed to allow for the electrochemical conversion of fossil fuels. This fuel flexibility is a result of high operating temperatures, which also create challenges in the development of materials used in the devices. In this paper, the opportunities for using high-temperature fuel cells as part of a sustainable energy strategy and the materials challenges that need to be overcome to take advantage of these opportunities are discussed

Introduction

Reducing the impact of the meeting the world's energy demands on climate change and other environmental effects requires a range of approaches. The development of renewable energy technologies, such as wind and solar, are needed. However, the amount of energy from renewable sources is small at this time, so these developments should be complemented with other approaches such as reducing consumption and improving the efficiency of the conversion of energy from non-renewable sources.

One of the ways to more efficiently convert chemical energy to electrical energy is to use electrochemical conversion rather than combusting the fuel. Electrochemical conversion is not subject to the Carnot cycle in combustion of fuels from chemical energy to mechanical energy and subsequent efficiencies in converting the mechanical energy to electrical energy.

Fuel Cell Operating Principles

Electrochemical energy conversion occurs by the oxidation of a fuel at the anode of the electrochemical cell and the reduction of oxygen at the cathode. For example, in an oxygen ion conducting electrolyte, such as a the yttria stabilized zirconia used in solid oxide fuel cells (SOFCs), hydrogen can be oxidized by oxygen ions according to Reaction 1

$$H_2 + 4O^{2-} \rightarrow 4H_2O + 8e^- \quad (1)$$

which produces water and electrons. The electrons cannot pass through the electrolyte and thus pass through the external circuit, which provides electric power and allows for the reduction of oxygen at the cathode according to Reaction 2,

$$2O_2 + 8e^- \rightarrow 4O^{2-} \quad (2)$$

as shown schematically in Figure 1. Figure 1 also illustrates that the fuel for oxidation at the anode can be provided by reforming methane according to Reaction 3,

$$CH_4 + 2H_2O \rightarrow 4H_2 + CO_2 \quad (3)$$

which can occur in an external reformer, or, because of the high operating temperature, can occur internally with in the fuel cell.

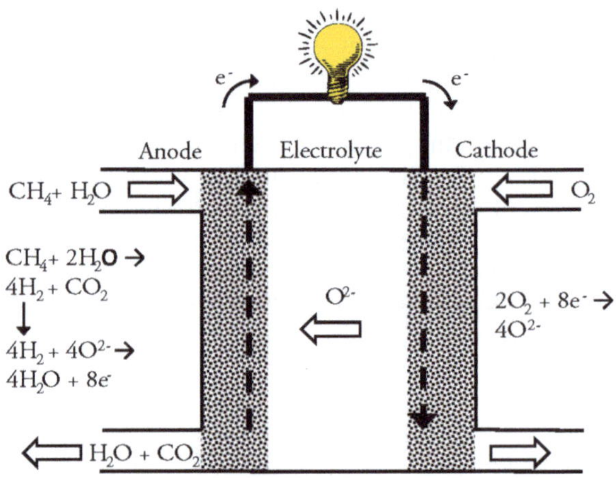

Figure 1. Schematic operation of solid oxide fuel cell.

Another high temperature fuel cell is a molten carbonate fuel cell (MCFC) which uses a molten electrolyte that conducts carbonate (CO_3^{2-}) ion, so that the oxidation of the hydrogen would occur according by Reaction 4.

$$H_2 + 4CO_3^{2-} \rightarrow 4H_2O + 4CO_2 + 8e^- \quad (4)$$

as shown in Figure 2. The reduction reaction occurs by Reaction 5,

$$2O_2 + 4CO_2 + 8e^- \rightarrow 4CO_3^{2-} \quad (5)$$

and CO_2 produced at the anode is fed back to the anode.

Fuel Cell Applications

Fuel cells have been promoted as a clean source of electrical power using hydrogen fuel, since the only byproducts are electricity, heat and water. The challenges associated with development of the infrastructure to supply and distribute hydrogen fuel have led to decreased interest in fuel cells. Unfortunately, this decreased interest has included both fuel cells that require hydrogen as the fuel as well as fuel cells that can be applied even before a hydrogen infrastructure is established [1]. In particular, high temperature fuel cells like SOFCs and MCFCs can operate using hydrogen as the fuel, but fortunately can also operate using other fuels such as natural gas, gasified coal or biomass-derived hydrocarbon fuels. This fuel flexibility creates opportunities for the expanded use of electrochemical energy conversion.

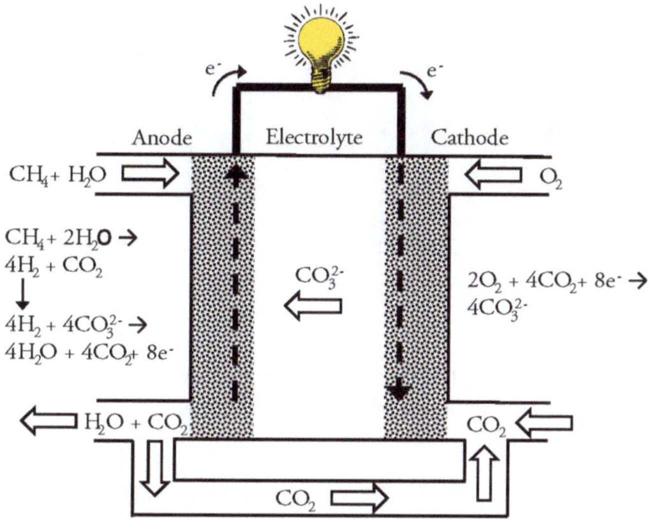

Figure 2. Schematic operation of molten carbonate fuel cell.

In addition to this fuel flexibility, the energy conversion of fuel cells does not depend on the size of the power plant [2], which also expands the range of applications to include from Watt- to Megawatt-range plants. Fuel cell power plants typically consist of modules which are combined to build plants of the desired size. Figure 3 shows 60-kW SOFC module that consists of four 15-kW SOFC stacks and can be combined using multiple modules to meet a greater power demand. Figure 4 shows a larger scale 1.4-MW MCFC module which consists of four 350-kW fuel cell stacks. These modules can be combined in larger plants, such as the 59-MW fuel cell park shown in Figure 5.

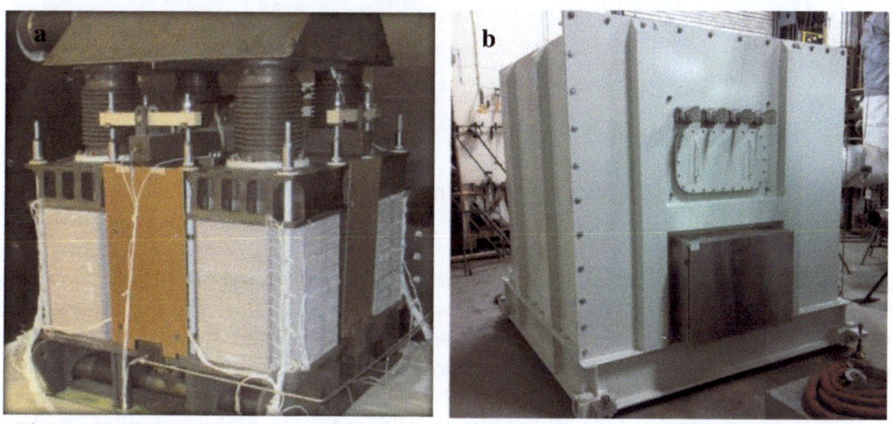

Figure 3. 60-kW SOFC module: a. Four 15-kW module, b. Completed module. (Photographs courtesy of FuelCell Energy)

Figure 4. 1.4-MW MCFC plant: a. Single cell and 350-kW module, b. Four 350-kW modules, c. One-module 1.4-MW plant. (Photographs courtesy of FuelCell Energy)

Figure 5. 59-MW, 5.2-acre fuel cell park (Photograph courtesy of FuelCell Energy)

In addition to the large-scale applications shown above, there are opportunities for smaller-scale distributed power sources [3]. The ability to use natural gas enables the use of combined heat and power (CHP) units for residential applications. In addition, the use of reformed diesel fuels allows for the use of SOFCs in potential transportation applications, such as in auxiliary power units (APU) for trucks. Similarly, the use of JP-8 fuel allows for the potential use in distributed power for military applications. In addition to these fossil-fuels, high temperature fuel cells can operation using fuel derived from renewable biomass sources.

Although, the fuel flexibility of SOFCs and MCFCs is derived from their high operating temperature accelerating the electrode reactions, the high temperature also increases the rates of undesired reactions and processes, such as the oxidation of interconnect alloys and coarsening of electrode microstructures, and thus can accelerate degradation processes. In addition, the need for a high operating temperature increases start-up times. Because of these challenges, one area of research in SOFCs is in decreasing the operating temperature to a range that maintains the desired fuel flexibility with decreases degradation rates and start-up times [4].

Conclusions

High temperature fuel cells can be used with a variety of fuels (e.g. hydrogen, natural gas, reformed diesel / JP-8, biomass-derived hydrocarbons) and in a wide range of sizes, which enables the use of efficient electrochemical energy conversion in a variety of applications. The implementation of high temperature fuel cells can increase the conversion efficiency of fossil fuels in the near time, while also enable development of the technology in preparation for the increased availability of other fuels, such as biofuels and hydrogen, in the future.

Acknowledgment

Input and photographs from Dr. Hossein Ghezel-Ayagh of FuelCell Energy are gratefully acknowledged.

References

1. E.D. Wachsman, C.A. Marlowe and K.T. Lee, "Role of Solid Oxide Fuel Cells in a Balanced Energy Strategy," *Energy & Environ. Sci.* 5 (2012), 5498-5509.

2.. J.D. Nicholas, "Highlights from the 2013 National Science Foundation Solid Oxide Fuel Cell Promise, Progress and Priorities (PPP) Workshop," *ECS Interface* 13(4) (2013), 49-54.

3. J.W. Fergus, "Solid Oxide Fuel Cells," in *Electrochemical Technologies for Energy Storage and Conversion*, ed. R.-S. Liu, X. Sun, H. Liu, L. Zhang and J. Zhang (Wiley-VCH, Weinheim, Germany, 2012), 671-700.

4. E.D. Wachsman and K.T. Lee, "Lowering the Temperature of Solid Oxide Fuel Cells," *Science* 334 (2011), 935-939.

Engineering Solutions for Sustainability: Materials and Resources II
Edited by: Jeffrey W. Fergus, Brajendra Mishra, Dayan Anderson, Emily Allyn Sarver, and Neale R. Neelameggham
TMS (The Minerals, Metals & Materials Society), 2015

A PATHWAY TO NEAR ZERO EMISSION ELECTRIC ENERGY THROUGH ADDITIVE MANUFACTURE OF SUPERCONDUCTING ELECTRIC TRANSMISSION CONDUITS

Ian Sheehy[1]

[1]RMIT University Bundoora East Campus, Plenty Road, Bundoora, Victoria, 3083, Australia

Keywords: additive manufacture, superconducting transmission, climate change, near zero emissions, directed pioneering, political feasibility, vacuum insulation, corrugated shells

Abstract

Moderately priced electricity for industry and commerce, with minimal climate changing emissions, is integral to the continued development of a stable world society. Economies of scale applied to cost-effective clean energy resources can achieve appropriate prices, but suitable resources are often far from users, requiring safe, reliable, low-cost, low loss, long-distance, environmentally acceptable transmission lines.

Superconducting lines can be used, but they are currently precluded by excessive cost and inadequate reliability. Advances beyond the present state of the art of additive manufacture, and lower operating temperatures with helium coolant can enable construction of practical superconducting transmission systems.

As a first step, short corrugated cylindrical shell conduit sections have been built in titanium, steel, and aluminium alloys by Selective Laser Melting.

Development of an additive manufacturing technology for substantial build lengths is thus on a pathway to the goal of world-wide electric energy supply with near zero emissions.

Introduction

A coherent, detailed engineering approach to worldwide cost-effective atmosphere safe electricity supply is outlined. Coherence can help re-assure the public that action towards the production of near zero emission electricity is both cost-effective and practical. Politicians and investors can then take confident action to rapidly build suitable supply systems.

Coherent presentation is best achieved starting from the global and local necessity for decarbonisation, along with reminders of the importance of adequate electricity supply in accordance with demand.

A concept of directed pioneering development zones, designated for safe electricity generation, to supplement the current opportunistic approach to decarbonisation is introduced. This focused approach can hasten the construction of new high-power transmission lines and generation facilities, and lower costs.

To further assist in the rapid implementation of development zones, an additively manufactured proof of concept transmission cable core, and a potential new additive manufacturing technology, 3D dot by dot additive manufacture are presented. These are intended to lead to development of a special-purpose manufacturing system to economically produce robust, highly reliable superconducting cable. It is expected that this technology will allow construction of underground transmission lines at lower cost than equivalent overhead lines. Lower transmission line cost widens the choice of pioneering zone locations, and the wider choice can increase cost-effectiveness through access to resources producing more year-round power.

The need for near zero emission electricity

Emissions from the production and burning of fossil fuels accumulate in Earth's atmosphere and impede the radiation of heat into space. This increased retention of energy results in persistent changes to weather patterns. Electric energy supply emits a significant part of these 'greenhouse gas' emissions, as fossil fuels are burned for energy[1].
Complex human systems fail catastrophically when conditions change beyond limits. Although these limits are not predictable in advance, unmitigated climate change, with its consequent ecological changes will sooner or later become catastrophic[2]. In the limiting case of human greenhouse gas emissions ceasing immediately, the temperature will still rise by 1° C.

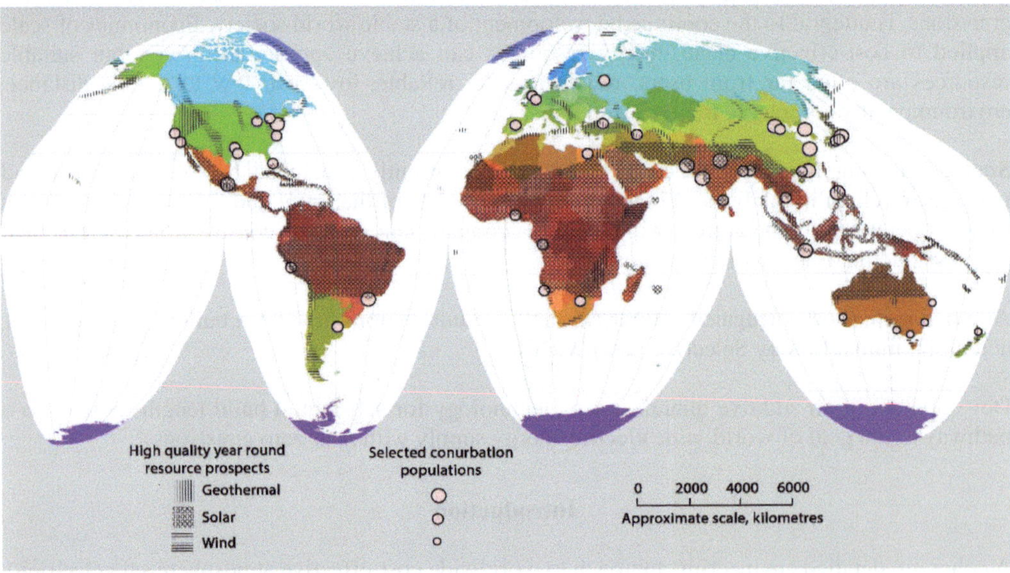

Figure 1: High quality year-round electricity resource prospects. Wind[3] and solar[4] resources which have an economical energy flow in both January and July are indicated. This approximation to seasonal data is needed because electricity is used year round, and long-term storage is impractically expensive. Storage of energy for electrical use can be cost-effective when the expensive storage is charged and discharged frequently, but it is not economical for holding energy from season to season.
Selected conurbation population sizes[5] and geothermal[6] energy data are also shown on this sketch map (drawn by hand using imaging software). Maps of this kind, on larger scales and with more detail, can show areas which can be aggregated into directed pioneering safe electricity zones. They can also couple resource availability with presentation of the social, environmental, and financial effects of resource development. Such maps can be used to create popular optimism, provoke resolution of objections, and influence politicians and investors to action in building near-zero electric system capabilities.

So the risk of catastrophe already exists. The minimum response is to reduce greenhouse gas emissions to near zero as rapidly as possible, with removal of atmospheric carbon dioxide commencing as soon as possible.
Proposed schemes for reducing greenhouse gas emissions on a large scale[7] typically consist of multiple scenarios but little decisive evaluation. The technology assemblies proposed *are* capable of achieving their goals, but they are not usually subject to the political feasibility[8] requirement of cost-effectiveness combined with social and environmental acceptability.

This paper suggests a pathway for the engineering and political efforts necessary to achieve near zero emissions from the electricity industry worldwide in a politically feasible way. Political feasibility is addressed by an unusually stringent 'triple bottom line' evaluation, taking financial, social, and environmental benefits and costs into consideration.

A stringent technology review

The electricity supply industry employs systems which can adjust their output to match electricity demand. Society has adapted to and now depends on this capability. For social acceptance, any new developments must meet this precedent, with marginal variations allowed by smart grids and enthusiasts using only their own locally generated variable power.

- Gas fired power with local carbon capture and storage (CCS) of its combustion products in depleted wells in adjacent depleted gas wells is planned for Peterhead and Captain, UK, and in a project at Taweelah to power Masdah, a 'zero-carbon sustainable' city in the UAE. Gas wells have a small footprint, can be remediated to a high standards, and so the triple bottom line evaluation may be acceptable, provided attention is given to fugitive emissions
- Hot rock geothermal electricity has been demonstrated several times, has small footprints, but remains expensive[9],[10],[11]: further development of this technology is appropriate
- Hydroelectric resources[12] can provide electric energy on demand, but typically spectacular sites for dams mean few acceptable locations are available for new facilities
- Wind power, using turbines prominent in the landscape, can be used with sensitive siting
- Electric energy from biomass is feasible, may be cost-effective, and in some cases suits seasonal energy storage, but is always area intensive. The use of biomass for energy production is contested for other productive uses, such as food, paper, and sightseeing
- Carbon capture and sequestration from coal exists at Boundary Dam Power Station in Saskatchewan[13]. Retrofitting CCS systems can be more expensive than many alternative means of electricity production[14]. Coal mines fail on environmental and social grounds
- Atomic fusion electricity is too little developed to be relevant to this discussion
- Electricity from nuclear fission reactors is claimed to have small gaseous emissions from mining and fuel processing over reactor life cycle.[15] Occasional very serious incidents, underdeveloped radioactive waste emissions disposal, and a long time scale necessary to develop safer alternatives, leaves nuclear power with a poor bottom line.

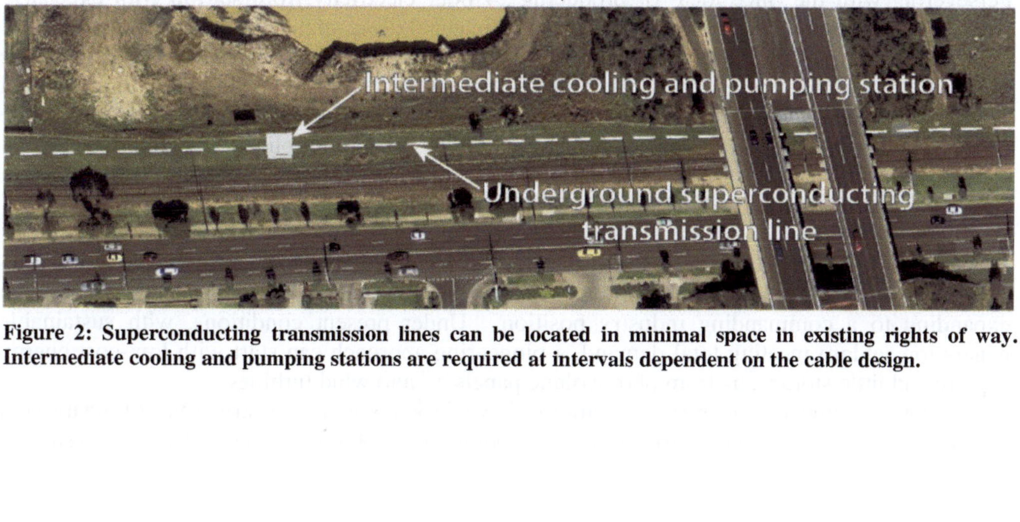

Figure 2: Superconducting transmission lines can be located in minimal space in existing rights of way. Intermediate cooling and pumping stations are required at intervals dependent on the cable design.

A directed pioneering approach to safe electricity generation

The central concept of the directed pioneering approach to safe electricity generation is creation of an appropriate authority to designate and promote development of generation zones for near zero emission power. Each zone will have new high-power transmission lines to large aggregations of electricity users.

The concept is called 'pioneering' because it encourages a small start with large prospects. It is 'directed' because areas of substantial generating potential are identified, mapped, and their development promoted. The approach is not novel, but is an extension of typical large-scale prospecting practice.

Directed pioneering begins with mapping of prospects for large scale energy resources with reasonably large and constant year round availability. Figure 1 is a simple example of such a map. This map confirms Eckroad and Marian's comment that 'Areas with abundant green energy sources are typically located far (up to several thousand kilometres) from the major consumption centers.'[16]

The mapping combines planning information about generation and transmission technologies, existing land use, ecological and social land value, and conditions and hazards such as temperature, flood, fire, and relevant politics. To justify pioneering transmission line construction, regions are aggregated from areas which meet these conditions and are capable of producing large amounts (>5 GW) of electricity year round, in order to ensure adequate power for the transmission lines to carry, once generation is fully developed.

Selection of transmission routes and cost estimates are made, and means sought to construct pioneering transmission lines and initial electricity generation facilities in the zones.

Superconducting transmission lines produced by the low-cost technology described below can cost less than overhead lines, and easements following existing railways and roads also cost less.

Disappointments and misunderstandings

However, this procedure can prove disappointing, as found by some 'Desertec' projects. Some Desertec concept developments failed to take account of political instability in the nations hosting generators[17], long investment timelines, and the assumption that many European governments would accept the project, despite powerful local interests.

Persevering with the 'nice idea' of producing 24 hour electricity from startup with expensive concentrating thermal generation has also been a financial disadvantage to the Desertec efforts.

Misunderstanding occur with the sometimes recommended use of geographic diversity to reduce variability of variable energy resources. But this approach is extravagant, implying duplication of generating capacity. With a focus on creating power sources in regions with year round safe energy availability, and the future deployment of storage systems, duplication is minimised.

Achieving commanding industry positions

For large-scale facility builders, a pioneering strategy means starting to develop transmission lines and low cost initial generating facilities in electricity zones with substantial potential, and then expanding to a commanding industry position. Under present conditions, with sustainable generation seldom meeting total demand, a pioneering approach can start with low-cost energy capture and little storage, as from photovoltaic panels [18] and wind turbines.

The important achievement from the point of view of society is the construction of transmission lines to zones of year-round electricity. More generation and storage can be added later to smooth

the variable energy, especially as the cost of electricity storage can be expected to continue falling as intensive engineering development and production proceeds.[19]
Incremental development then becomes possible. Additional capacity can be added later to support storage and extend productive periods.
Finally, in strategic planning, allowance is required for eventual replacement of natural gas heating in winter by electric power, and the adoption of electrically powered transportation.

Long distance electric transmission and superconducting transmission lines

Electric transmission can be by overhead, near ground, or underground lines. The triple bottom line evaluations differ greatly between these options. All long lines are of overhead construction, carrying hydropower from remote locations to large conurbations. The longest *conductive* transmission line is 2385 kilometres in length. These high-power lines involve large overhead structures, and so are prominent in their landscapes. Rights of way are usually maintained with artificially shortened vegetation.

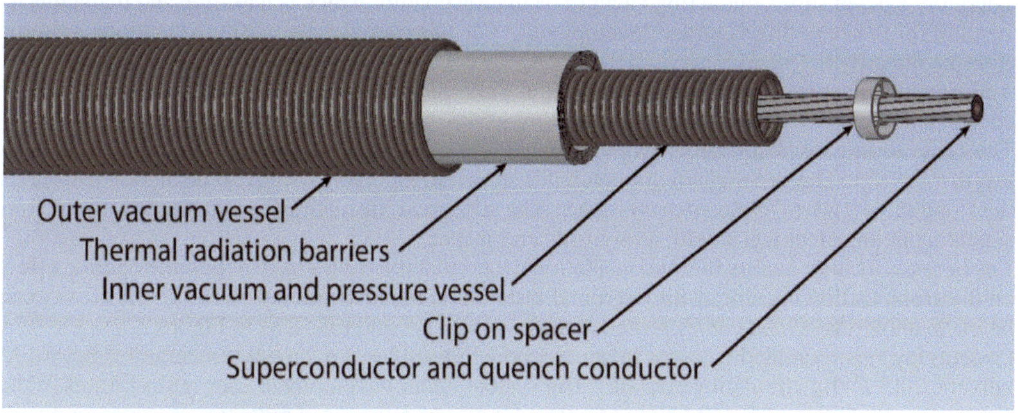

Figure 3: Computer model of the cryogenic conduit concept. When a length of conduit is complete, near standard cable-making equipment will add electrical insulation, shield, armoring and jacket.

Underground transmission cables are sometimes used, but the heat generated by the electric current is dissipated in the ground, further increasing the temperature of the conductor and restricting power transmission. Multiple cables, side by side, can increase power transmission capacity, but require wide rights of way with controlled vegetation height, which is socially objectionable.
Eckroad and Marian[16] also discuss the possibility that superconducting transmission lines could facilitate very low emission electricity generation, given enough engineering development. As adopted in the pioneering approach, superconducting electric transmission lines could connect 'large green-power resources' to 'load centers', and offer the advantages of compact underground installation for the landscape, environment, right of way acquisition, and reduced opposition to the line from neighbors concerned with lowered property values. Reduced vulnerability to wind, snow, and ice compared with overhead lines are further advantages.
Superconducting power lines have been constructed and used under actual grid conditions, but the longest superconducting lines built so far are less than one kilometre long, and all have had lives less than ten years.
For long cables, heat inflow is a major consideration, so highly effective thermal insulation is required. An extensive survey by the author concludes that only vacuum insulation is sufficiently effective for practical long-distance transmission lines.

Vacuum failure and the consequent increase in heat inflow is the main reason for disappointment in superconducting transmission projects. The principal reason is that the vacuum insulated conduit tubes are formed and welded from stainless steel sheet. Although high quality automated welding is used, weld failure rates are excessive, given the length of weld required for long transmission lines.[20, 21]
A design for superconducting transmission cable from the Electric Power Research Institute, EPRI,[22] which attempts to avoid the problem of vacuum leaks, is the starting point for the work presented here. In this concept, a superconducting cable is enclosed in a large diameter welded 'gas pipe' envelope with continuous vacuum pumping.
A weakness in this proposal is the acceptance of vacuum leaks into the vacuum vessel, which necessitates the vacuum pumping. Water carrying solids from the pipe environment will be sucked in through leaks, with the water evaporating and leaving solid deposits which accumulate and block the pipe, eventually rendering the pumping ineffectual.
Therefore a novel superconducting vacuum insulation conduit which is leak-free and maintains its vacuum for the cable life is under development. For this purpose, the vacuum envelope requires extreme integrity.
Additive manufacture, with its ability to produce complex shapes and tight processing parameter control is therefore selected for the manufacture of the thermal insulation vacuum envelope.
The cable form chosen for initial work is the simple structure used in a developmental cable by Pirelli[22] pages 4-8. The superconductor conduit is shown in Figure 3. Electrical insulation surrounds the conduit: a 'warm dielectric' system. The electrical insulation is surrounded by a non-superconducting electrical shield, armouring, and jacket.
The necessarily high quality thermal insulation means that the cables have negligible cooling effect on the ground, allowing simple underground installation in narrow trenches. So rights of way can be made unobtrusive, and cables routed alongside railways and roads, as in Figure 2.
Present-day superconducting wires in superconducting cables are kept at temperatures below 73 kelvin (-200° C) by circulating coolant. The system under development uses temperatures in the range 40 to 80 kelvin. The 40 kelvin temperature allows a four-fold reduction in superconductor quantity and cost compared with liquid nitrogen cooled systems, while the system can be operated economically at higher temperatures when loading is low.
However, the commonly used circulating coolant, liquid nitrogen, freezes at these temperatures. Hydrogen has been suggested as a cheap alternative coolant, supercritical (gaseous) in this temperature range, but the combination of large amounts of electrical energy with hydrogen risks initiation of fierce fires when cables are breached. Therefore, helium is selected, with the cost of helium being outweighed by the saving in superconductor.

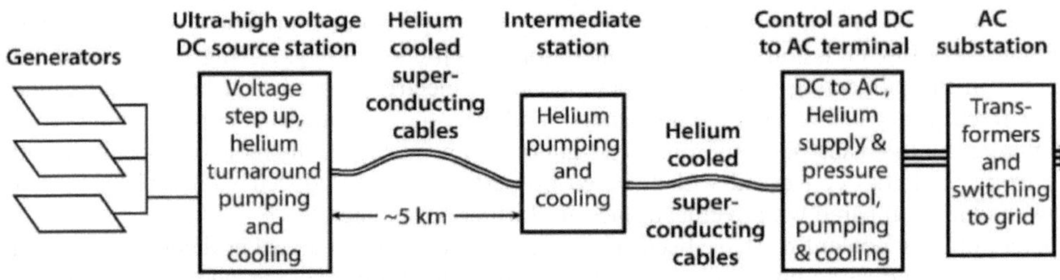

Figure 4: Block diagram of a high power near zero emission electricity generation system using DC superconducting transmission. Maximum power transmission is achieved at the lowest temperatures, with helium storage tanks accommodating helium expansion when the coolant temperature is raised to reduce system power consumption at light loads.

Figure 4 shows a block diagram of an atmosphere safe generation and superconducting transmission system. The power sources for this are joined by relatively short non-superconducting (conductive) lines to an ultra-high voltage transmission source station. This station launches DC power, and pumps re-cooled recirculated helium into the superconducting cables. The coolant is re-cooled at intermediate stations, continuing on to a control and terminal station, where it is converted to three phase alternating current and connected to the existing distribution grid.

Experimental

Initial Proof of Feasibility of the additive manufacture of superconducting cable conduit shells is in metal, because of the need for impervious vacuum and pressure envelopes. A number of models for developmental specimens were prepared using Solidworks Computer Aided Engineering software. The same software was used to simulate the properties of the developmental specimens in compression, tension, bending, and under pressure. Results showed that the shells satisfy the specification that they could be wrapped around a 3.0 metre diameter cable drum core, contain vacuum, and the inner conduit tube safely sustain pressures to 5 MPa.

All additive manufacture specimens were built in an SLM Solutions GmbH SLM250 powder-bed selective laser melting system with an argon atmosphere. Exploratory specimens were built in titanium alloy, maraging steel, and aluminium, with shells 0.25 and 0.5 mm thick, and corrugation profiles from ripple to U shape. It was soon discovered that specimens made with 0.25 mm thick walls are not vacuum or pressure tight. Wall thicknesses of 0.5 and 1.0 mm were built in aluminium and titanium, culminating in a decision to use 1.0 mm thick walls throughout.

Following the developmental builds, a Proof of Concept conduit for a vacuum insulated superconducting cable, Figure 5, was built.

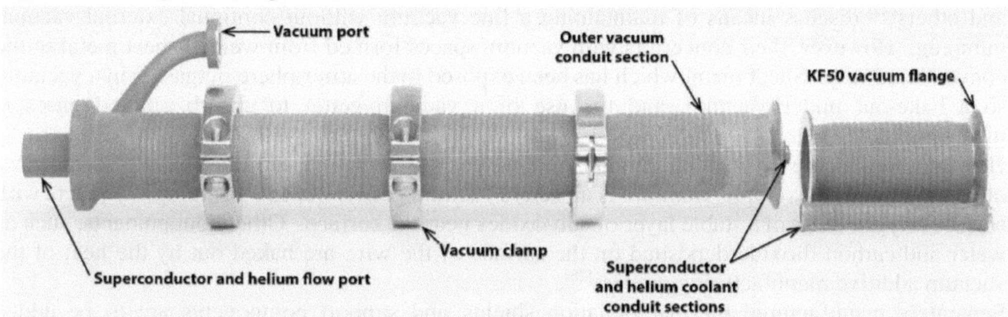

Figure 5: Partially assembled Proof of Concept additively manufactured titanium superconducting cable conduit. Approximately half of the built sections are shown.

These parts were made in titanium alloy Ti-6Al-4V, because its low thermal conductivity reduces heat inflow through the inner tube supports.

The Proof of Concept provides for thermal insulation with multiple closely spaced reflective concentric shim cylinders supported by the inner tube support structures. Simulations show that closely spaced reflectors have lower heat inflow than the values reported for the superinsulation used in other projects.

Because of the limited build length capacity of the laser powder-bed additive manufacturing system, these tubes are flanged for joining. The outer tube sections clamp together with standard KF50 vacuum clamps. The inner tubes are built with custom flanges for joining with indium wire, and a central rod is provided for stability.

The shim cylinders are sufficiently short, and space is allowed between the shims and the corrugated inner and outer tubes for the conduit to be bent without contact when wound onto a cable drum. The dimensions of the sections built were all within 1.5% of the model used in their production, and are proving vacuum tight.

3D dot by dot additive manufacture

It is obvious that the length of sections that could be made by any powder-bed system is inadequate for long superconducting cables. In addition, it would be impractical to empty residual powder from long hollow cylindrical sections.

To achieve low costs, the goal was set of making complete superconducting cable cores, already evacuated, in a single pass additive manufacturing system in a vacuum atmosphere. A survey of available additive manufacturing methods showed that wire-fed systems are best suited to this application.

However, it was noted that the continuous linear melting commonly used in wire fed additive manufacture has only been used to produce walls many millimetres thick. Since thinner walls than these additively manufactured walls are required, the technique of 3D micro (arc) welding[23] has been taken as a starting point. This technique deposits metal from a fine wire feed one dot at a time. Each deposit is formed as a molten meniscus, which can be less than one millimetre in diameter. The deposit is allowed to solidify before the any adjacent dot is deposited. The authors show that high quality walls can be built with this technique.

The concept is now to transfer the concept of 3D micro welding technique to additive manufacture in a vacuum. The low deposition rate can be compensated for by operating multiple wire fed deposition systems concurrently, as shown in Figure 6.

The ability of this system to produce shells with enduring vacuum must be considered. Waynert and others[24] discuss means of maintaining a fine vacuum without continual external vacuum pumping. However, their concern is with vacuum spaces formed from welded sheet metal in the conventional way. Sheet metal which has been exposed to the atmosphere outgasses in a vacuum, so a bake-out under vacuum, and the use of a vacuum getter to absorb emitted gases is incorporated.

However, producing corrugated conduit in a vacuum by wire-fed additive manufacture allows other options. For example, titanium is an active vacuum getter. At 450° C, it rapidly reacts with surface oxygen to form a stable layer of sub-oxides near the surface. Other contaminants, such as water and carbon dioxide deposited on the surface of the wire are baked out by the heat of the vacuum additive manufacturing process[25].

Separately manufactured thermal radiation shields and support components are to be added robotically after bake-out while in the vacuum before assembly.

A final issue is the durability of the cable core structure as a vacuum and pressure envelope. Could re-crystallisation over time lead to failures similar to those of welded corrugated conduits? Process and product design must provide for benign re-crystallization of the cable core.

Because of the smaller amount of materials used, simpler installation in the field, and lowered cost of right of way acquisition, the installed cost of such superconducting transmission lines is expected to be about half that of overhead conducting lines. As an example of present-day long transmission lines, a non-superconducting line with a power capacity of 7.2 gigawatts operating at 500 000 volts DC, has a reported cost of 1.67 million US dollars per kilometre[26]. Costings by the author have all been less than USD 1.0 million per kilometre for superconducting lines (USD 3.0 billion for a 3000 kilometre line). Generating systems for this amount of power cost at least 14 billion, so the cost of a long power-line to allow a given installation to be located for maximum output is easily compensated by increased output.

Conclusion

Coherence of worldwide electricity decarbonisation efforts can be enhanced by official designation of safe electricity generation zones in which pioneering development of large-scale generation and long transmission lines is encouraged. The zones are to be chosen after stringent evaluation of the 'triple bottom line' of financial, social, and environmental benefits and costs, and the availability of year-round resources for minimum duplication of plant.

In support of this pioneering approach, development of 3D dot by dot additive manufacture in a vacuum, and development of special purpose systems for producing cable cores is proposed. The pathway through lower cost additively manufactured transmission systems and the safe electricity zone pioneering system leads towards world-wide electric energy supply with near zero emissions.

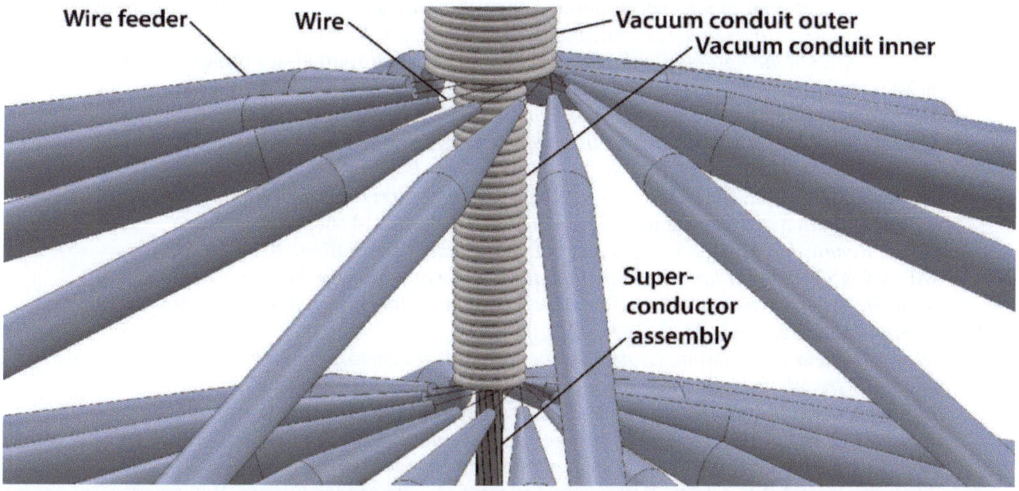

Figure 6: Model of multiple wire-feeders building corrugated vacuum containment tubes around a superconductor assembly. Laser beams melt the surface of the build and the end of a wire together one point at a time into a tiny raised meniscus of molten metal to achieve fine definition. The metal is allowed to solidify before an adjacent menisci are created. Elements connecting the inner and outer tubes are built where necessary by appropriate movement of the wire feeder radial positions. Other supports and radiation barriers are added robotically. The flexible finished core moves upwards and is reeled onto a cable drum.

Acknowledgements

Thanks to Professors John Andrews and Grahame Holmes for support and guidance in the research, and to Professor Milan Brandt and Aaron Patteras of the RMIT University Advanced Manufacturing Precinct for generous support in the manufacture of the additively manufactured specimens. The sustained work of Don, Patrick, Ian, Mark, David, Eli, and all others of the RMIT University Bundoora East workshops in the production and use of the apparatus used in this project is gratefully acknowledged, as is the RMIT University for a supporting Scholarship.

References

1. IPCC, *IPCC Fifth Assessment Synthesis Report-Climate Change 2014 Synthesis Report.* 2014.
2. Adger, W.N., et al., *Are there social limits to adaptation to climate change?* Climatic Change, 2009. **93**(3-4): p. 335-354.
3. NASA. *NCEP/NCAR Reanalysis 1000mb Vector Wind (/s) Composite Mean*. ND; [Copies of January and July winds from NASA, original not found.]. Available from: http://www.climate-charts.com/World-Climate-Maps.html#vector-wind.
4. Colombia University. *Earth's Environmental Systems - Absorbed Solar Radiation (January and July)*. ND; Available from: http://eesc.columbia.edu/courses/ees/slides/climate/swrad.gif.
5. Brinkhoff, T. *Major Agglomerations of the World*. 2013; Available from: http://citypopulation.de/world/Agglomerations.html.
6. GENI. *Global Renewable Energy Resources - Geothermal Energy* 2007; Available from: http://www.geni.org/globalenergy/library/renewable-energy-resources/geothermal.shtml.
7. DDPP *Pathways to deep decarbonization 2014 report*. 2014.
8. Majone, G., *On the notion of political feasibility.* European Journal of Political Research, 1975. **3**(3): p. 268.
9. Humphreys, B. *Habanero Geothermal Project Field Development Plan*. 2014.
10. Cummings, R.G., et al., *Mining earth's heat-Hot dry rock geothermal energy.* Technology Review, 1979. **81**: p. 58-74.
11. Beckers, K.F., et al., *Levelized costs of electricity and direct-use heat from Enhanced Geothermal Systems.* Journal of Renewable and Sustainable Energy, 2014. **6**(1).
12. IPCC, *Renewable Energy Sources and Climate Change Mitigation.* 2011: p. 167.
13. MIT CC&ST *Power Plant Carbon Dioxide Capture and Storage Projects*. 2014.
14. Voll, D., et al., *Cost estimation of fossil power plants with carbon dioxide capture and storage.* Energy Procedia, 2012. **23**: p. 333-342.
15. World Nuclear Association. *Global number of nuclear reactors | Electricity supplied by nuclear energy | Greenhouse gas emissions avoided.* 2014 [cited 2014; Available from: http://www.world-nuclear.org/Nuclear-Basics/Greenhouse-gas-emissions-avoided/.
16. Eckroad, S. and A. Marian, *Long-distance dc transmission of green power, in Superconductivity and the environment: a Roadmap.* Superconductor Science and Technology, 2013. **26**(11): p. 7-9.
17. Massetti, E. and R.E. Claire, *Rethinking African Solar Power for Europe.* Review of Environment, Energy and Economics, 2014.
18. Devabhaktuni, V., et al., *Solar energy: Trends and enabling technologies.* Renewable and Sustainable Energy Reviews, 2013. **19**(0): p. 555-564.
19. Medina, P., et al. *Electrical Energy Storage Systems: Technologies' State-of-the-Art, Techno-economic Benefits and Applications Analysis*. in *System Sciences (HICSS), 2014 47th Hawaii International Conference on*. 2014. IEEE.
20. Gouge, M.e.a., *Flexible Cryostats for Superconducting Cables: Reliability and Lifetime Issues*, in *DOE 2006 Wire Development Workshop*. 2006.
21. Gouge, M.J., et al., *Vacuum-insulated, Flexible Cryostats for Long HTS Cables: Requirements, Status and Prospects.* AIP Conference Proceedings, 2008. **985**(1): p. 1343-1350.
22. EPRI *Program on Technology Innovation: a Superconducting DC Cable*. 2009.
23. Horii, T., S. Kirihara, and Y. Miyamoto, *Freeform fabrication of superalloy objects by 3D micro welding.* Materials & Design, 2009. **30**(4): p. 1093-1097.
24. Waynert, J., et al. *Long Term Vacuum Maintenance in HTS Equipment without External Pumping*. in *Advances in Cryogenic Engineering: Transactions of the Cryogenic Engineering Conference-CEC*. 2004. AIP Publishing.
25. Mizuno, Y., et al., *Temperature dependence of oxide decomposition on titanium surfaces in ultrahigh vacuum.* Journal of Vacuum Science & Technology A, 2002. **20**(5): p. 1716-1721.
26. Power-technology.com *The world's longest power transmission lines*. 2014.

Engineering Solutions for Sustainability: Materials and Resources II
Edited by: Jeffrey W. Fergus, Brajendra Mishra, Dayan Anderson, Emily Allyn Sarver, and Neale R. Neelameggham
TMS (The Minerals, Metals & Materials Society), 2015

THE INGRID PROJECT: DEVELOPMENT OF SOLUTIONS FOR SUSTAINABLE AND HIGHLY INTERCONNECTED GRIDS

Fabrizio D'Errico[1], Adamo Screnci[2], Massimo Bertoncini[3]

[1]Politecnico di Milano, Department of Mechanical Engineering,
Via La Masa 34, 20156 Milan (Italy)
[2]Mc Phy Energy, 40 rue de Berges; 38000 Grenoble (France)
[3]Engineering Ingegneria Informatica S.p.A., Via San Martino della Battaglia, 56
00185 Rome (Italy)

Keywords: Sustainable electricity, smart grids, energy career, hydrogen, energy storage

Abstract

One of the current main challenges in green-power storage and smart grids is the lack of effective solutions for accommodating the unbalance between renewable energy sources - offering intermittent electricity supply - and a variable electricity demand. Integrating intermittent renewable energy sources by safe and cost-effective energy storage systems is today achievable. Coupled with electrolizers, high-capacity solid-state storage of green-hydrogen is practicable to sustain integration, monitoring and control of large quantity of GWh from renewable generation. The 23.9 MLN Euros INGRID European large demonstrative project started in July 2012 combines magnesium-based material solid-state hydrogen storage systems with advanced ICT technologies to intelligently interconnect miscellaneous energy networks (i.e. electricity and gas) and safely delivering green-hydrogen to various existing or forthcoming markets. One solution INGRID project addresses is an off-grid utility to store renewable electricity captured from wind sources to refill full-battery electric cars.

Introduction

The raw materials needed for energy supplies are taken from deposits which are not evenly distributed around the world and are, in many cases, located in regions affected by political instability. Globalization is leading to higher economic growth in successful emerging countries but also to more competition for fossil fuels. Relevant environmental and economic consequences of such a completion cannot be negligible today. Increasing of pollutant emissions as byproducts of power generation is high environmental price to sustain economic growth when largely based on combustion processes of fossil fuels. Over last decade, energy companies have been investing billions of euros in the expansion of power supply from renewable sources such as wind, water, solar, biomass and geothermal energy with some big challenges to face to compete with fossil fuels. It is known power generation plants using fossils as energy resources are normally built at sites which are conveniently located. Similarly to fossil fuels sources, renewable energy sources (RES) are available in certain areas. However, unlike fossil fuels, these energy sources are initially not linked to specific materials and cannot be simply transported and stored until they are required. Additionally, renewable energy sources are available when and where the wind blows or the sun shines; thus, power generation from RES subjected to variability and fluctuation. As a result, one major challenge power generation companies have been facing is to govern renewable energy peaks that cannot be accepted by grid for distribution (grid must be stabilized for their safe functioning). Despite wind farms would be economically competitive over their lifespan – considering both capital and operating

expenditures at a whole - they suffer of highly fluctuant raw energy source, the wind. Peak loads cannot be accepted by grid, as above stated, thus wind farms are usually uneconomically managed as they are often totally or partially shut down. This fact results in a lot of power generation wasted, with negative impact onto unitary energy selling price.
Attempts are being made to avoid these problems by expanding power grids, at high investment costs and government centralized decisions on investments. On the other side, macro-trends moving toward new paradigm in the structure of the future energy system: energy management shall be more decentralized, more integrated on a local level with broadened application focus. Being respectful of operational security of the grid, the location of the new RES plants, mainly wind farms, would be determined by the owners and local authorities. On the other hand, situation for distribution is greatly complicated by small generator units widely distributed, since they would be often completely out of control (and sometimes even the observability) of the system operator. With large generator units, the system operator would be capable to manage multiple sources production. Unfortunately, in European system, Distributed Generation (DG) cannot dispatch energy as it likes, as the current system is driven by Transmission System Operators (TSO) which largely function within national boundaries. To promote change to distributed renewable energy generation, energy storage facilities would add more flexibility to the system. Especially in case of wind source, power generation is highly fluctuating with the time and often difficult to reliably predict. These facts make the planning and operation of the power system very difficult. In addition, DG is connected to the Low and Medium Voltage (LV and MV) grids. These systems have been planned to deliver electrical energy in a single direction, thus they have been designed to optimize the power flows from High Voltage (HV) stations to the loads. In this scenario, the introduction of minor amount of DG results in a reduced net loading of the transmission grid and it can be considered as a benefit. Nevertheless, a large amount of DG produces relevant effects on both distribution and transmission networks and it could dramatically impact on several technical frontlines. Over such scenario, an energy storage solution for intermittent highly available wind energy sources is a key for future energy securing in several EU Countries. Hydrogen batteries we discuss in the following here are one possible ready-to-market solution for large-capacity RES storage.

The storage opportunity of the hydrogen energy carrier

The production of hydrogen from water electrolysis fed by renewable intermittent sources represents a major long-term renewable energy storage opportunity. A number of benefits are provided to existing and new market applications. Potentially unlimited capacity of RES storage via-hydrogen conversion is key to unlock wind and solar distributed generation. An example of possible highly integrated supply-chain of RES power is shown in the scheme of Fig.1. The flow chart in Fig.1 considers numerous hydrogen advantages. Moreover in this scheme emerges how hydrogen can serve multiple energy uses with high integration. As Fig.1 illustrates, hydrogen produced by water electrolysis sourced by RES can be stored - in small or terawatt hour-scale quantities and in a variety of ways – can:

- *Be used as centralized power generation*, to be exploited by two possible pathways: a) use large scale hydrogen-compatible combined cycle gas turbines (CCGT), now under development[1]; b) use large-scale stationary fuel cell plants. Today the latter pathway is not a cost-effective solution, because of high cost of fuel cell technology

[1] By 2010 in Fusina (close to Venice, Italy), started demonstration plant consisting of a combined cycle in which a turbogas is fuelled with hydrogen in order to produce electricity and heat. The turbogas is equipped with a new conception burner specifically designed to be fuelled with pure hydrogen, and at the same time generate extremely low emissions of nitrogen oxide. The thermal energy produced by the combustion is converted into electricity in the gas turbine, reaching power which amounts to approximately 12 MW, whilst the flue gas are constituted exclusively of hot air and water vapour.

and limited efficiency of reconversion of hydrogen into electricity. However much progresses are being in cost reduction and efficiency increase driven by rising up interest of some nations in this stationary power generation solution[2].

- *Be used as green-hydrogen* for fuelling Fuel Cell Electric Vehicle (FCEV) fleet. Hydrogen produced from water electrolizers can be dispatched to the hydrogen refueling station close to the RES plant, piped to refueling station or transported to stations where pipeline network is not available[3];
- *Be supplied to industry as a "green" commodity* or feedstock for current industrial uses, mainly petroleum and chemical industries. Currently 95% of hydrogen is produced by reforming fossil fuel, thus employing processes that release lots of CO2;
- *Be chemically combined with carbon* to produce synthetic hydrocarbon fuels[4] for transportation. Depending on process, the synthesis of hydrogen into hydrocarbon can produce synthetic natural gas (methane) or liquid synthetic fuel as methanol. As power is required to produce the synthetic products, either gas or liquid, and it is provided by the stored energy in hydrogen that is used for the specific synthesizing process, these concepts are referred in recent literature as *power-to-gas* and *power-to-liquid* respectively. The most advantaged is the transportation and early exploitability issues: as synthetic methane is produced, it can be injected and stored in the existing gas grid as it is not distinguishable from fossil methane; furthermore it can be burned in conventional equipment to produce power and heat. Obviously, as the overall supply-chain is thought starting from the RES source, this option would allow to feed large quantity of power stored as renewable synthetic green-methane for sustainable mobility immediately available on market. Also the synthetic renewable methanol is ready for early exploitation as several fuel-flexible cars are taking advantages from methanol as clean-burning and easy-to-handle fuel.
- *Be injected directly in the natural gas network*, within the range 6-20% renewably upgrade the carbon dioxide (CO2) fraction of biogas.

---

[2] In 2012 it was announced a series of large fuel cell installation 60 MW fuel cell park is under development by POSCO Energy along with partners in Hwaseong.

[3] Around the World there are a total of 80 hydrogen refueling stations (33 in Europe) supplying hydrogen produced by water electrolysis.

[4] Synthetic gas production specifically refers to methane following the reaction (1) CO2 + 4 H2 –> CH4 + 2H2O and the CO from sequestered CO2 following the reaction (2) CO2 + H2–> CO + H2O. The hydrogen gas, together with CO produced by (2), is component of syngas (SNG) a valuable feedstock for the production of liquid fuel and other commodities. The production of SNG through hydrogen methanation also makes the recovery of CO2 for material use (e.g. in the chemicals sector) possible. The EU Emission Trading System (EU ETS), CO2 reduction goals, raw material scarcity, and price considerations are the major drivers for these markets.

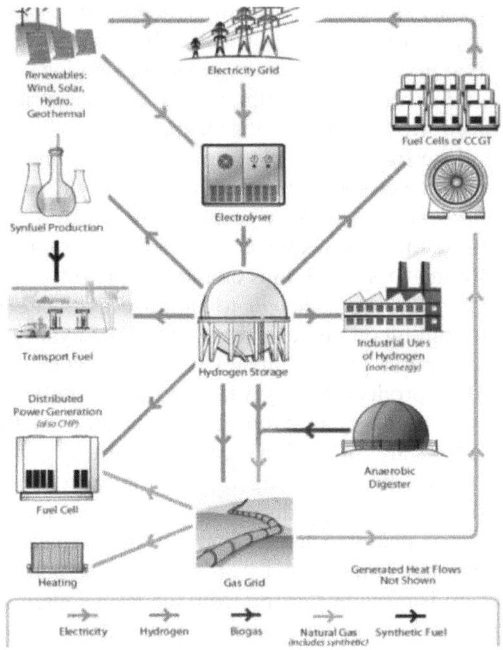

Figure 1 - The highly integrated energy network by water electrolysis RES-sourced (source: "Water electrolysis and Renewable Energy System "released on May 2013 (latest check on June 2013 at website www.fuelcelltoday.com).

The INGRID concept design and the real scenario

The main scope of INGRID project is to realize in South of Italy a demonstrating plant capable to capture intermittent renewable generation produced by a wind farm located in such a highly congested local grid. Italy has witnessed a strong intermittency decentralized penetration over the past few years, as well as recurrent curtailments and congestion issues[5]. Decentralized generation causes more recurrent power flow inversions in High Voltage/Medium Voltage transformers (150/20 kV substations) as well as local high voltage. As negative consequence, high distributed RES generations causes large energy curtailment, a critical issue in Italy especially in such areas with weak transmission infrastructure and large supplying offer of renewable sources, such as Southern Italy. In such regions, as shown in the map of Fig.2, wind generation is highly developed. Wind curtailment represented 5.6% of the total wind generation in 2010 (470 GWh). In 2009, 10,7% of the total wind generation was curtailed (around 700 GWh). In some regions, for example in the Southern Apennine Mountains, wind curtailment can reach up to 23% of total wind generation[6].

[5] ENEL Distribuzione, partner of INGRID project and the main distributor system operator (DSO) in Italy, claims for 315,000 connections of decentralized generation in 2011, representing 19.5 GW, 11.8 GW of which of solar and 3.2 GW wind (Source, TERNA Report 2012).

[6] In 2013, the net generation was 289 TWh whereas the demand was 332 TWh: this deficit of 43 TWh (12.9% of the demand) was covered through imports. Accordingly curtailment minimization is the major field requirement for INGRID energy storage service/application.

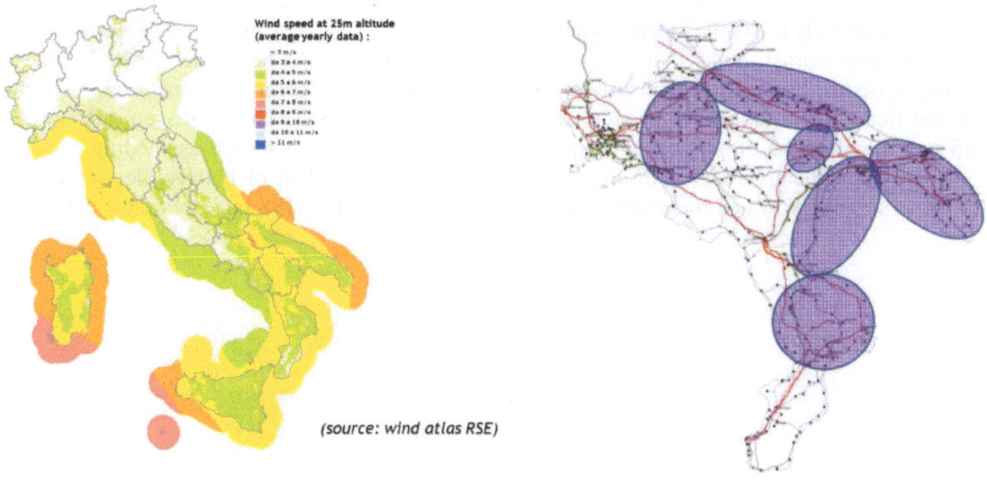

(a) (b)

Fig. 2 – a) Italian wind source situation (RSE Wind Atlas); b) main critical patterns for evacuation of wind energy produced in the south of Italy (source: TERNA 2013)

The INGRID Project aims to design, construct and to operate a large energy storage demonstrator. A novel fast responding 1.2 MW electrolysis hydrogen generator coupled with safe hydrogen solid storage (HSS) units will provide smart balancing support for the local distribution in power grid. As main INGRID expected outcome, intermittent renewable energy sources available in wind farm will captured continuously, surpassing the problem of frequent shutting down due to distribution network congestion.

The INGRID hardware architecture

The INGRID plant rationale is depicted in Fig.3. As it is conceived, the plant aims to integrate intelligent monitoring, control and communication system with affordable and reliable technologies. About 1 GWh is the order of stored electricity in INGRID plant. The plant is directly linked to medium voltage (MV) node close to local wind farm in city of Troia, in Apulia Region, South of Italy. The plant is designed to serve a local grid node by accepting power peaks generated by close wind farm. The Energy Management System, or EMS in the scheme of Fig.3, is the ICT intelligent device that can analyze multiple level information in order to manage (in real-time) distribution of electricity to INGRID generation plant, to by-pass partly or totally the RES power to the grid (i.e. weather forecast and statistical analysis for predictability of peak loads, real time market prices for hydrogen or electricity; etc.)[7]. Water electrolizers (WE) in

[7] The Energy Management System (EMS) is the software stack which manages the INGRID node, and is in charge of decoupling intermittent RESs from power and gas distribution networks. It is the intelligent component of the INGRID system node, which will be in charge of managing and optimizing energy production, storage and control determining, by supervising and optimizing how much hydrogen should be produced with a view to flow out an optimized hydrogen/power amount. EMS main roles are to supervise and control power production from nearby RESs, manage power surplus stored as hydrogen and optimize energy delivery as hydrogen surplus through either a

INGRID plant have been designed specifically for an instantaneous reply to such peak loads. Downstream to WE unit, the Green Energy Storage (GES) module realized by hydrogen solid storage modules (HSS) is the storing equipment of the plant. The hydrogen gas generated by WE unit flows into storage tanks that realize the solid-state storing. As better described in the next section, gaseous hydrogen is absorbed by magnesium metal composite material, the basic material that constitutes the Hydrogen Solid Storage tanks (HSS). The hydrogen gas stored in HSS can therefore be employed when requested in several manners such as feeding fuel cells (FC) that equip the INGRID demonstrator for fueling a small fleet of battery electric vehicles (BEV), to supply hydrogen merchants directly by transporting the HSS tank modules by trucks.

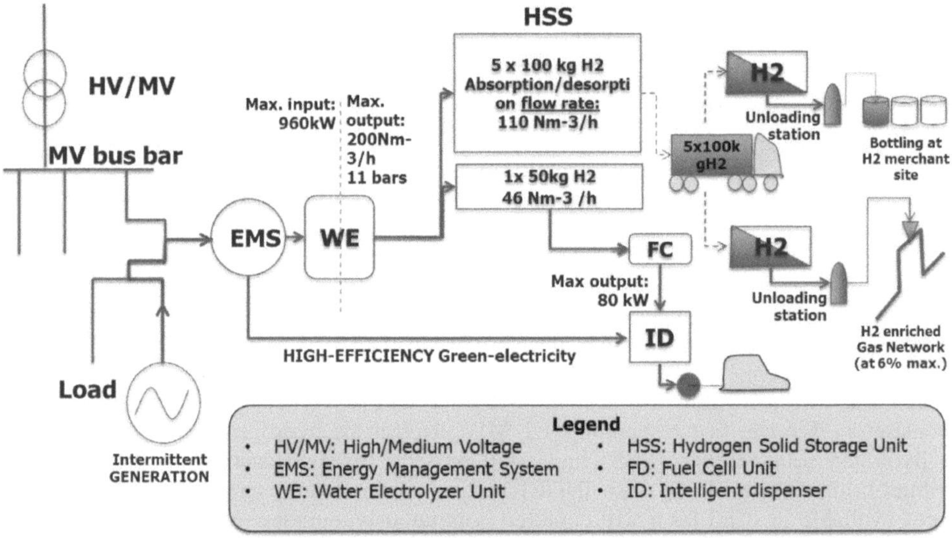

Fig. 3 – The INGRID demonstration site architecture as an example of the INGRID system solution.

Instantaneous response to peak load for the grid stabilizing

Water electrolizers and fuel cell stacks here employed have the major advantages of an instantaneous response to peak loads produced by wind farm. Highly fluctuant electricity produced by wind turbines flows into the water electrolyzer that uses it to quite instantaneously produce hydrogen. Any request of electricity by downstream users (i.e. demand) is therefore decoupled from production (supplying) side thanks to water electrolizers and storage unit. More specifically, as shown in Fig.3, the INGRID demonstrative plant is directly connected to the Medium Voltage (MV) bar, where wind farm is connected. In such a real scenario, RES production cannot increase indefinitely at the MV access node of the grid, as such a high electricity variability cannot be managed by local grid. The INGRID plant is therefore structured to connect work with constant absorption of RES electricity in order to either reduce or increase power absorption – to produce less or more hydrogen – to smooth power peaks and stabilizes the downstream grid's users. This approach allows keeping a high quality power curve also in case of large quantity of RES power plants connected. The Energy Management System (EMS) can decide the multiple-supply side basing on real-time information gathered by a multiple-demand

fuel cell serving local electric cars charging, either injecting hydrogen into a close gas network, or in a biomass plants, or in a gas turbine.

requests (such as how much hydrogen shall be produced for the next hours? who are the multiple customers among hydrogen merchants, refilling stations, chemical industry to serve?) and profitability analysis (at which price could be bought by multiple customers hydrogen that we produce in next hours?) to produce less quantity of hydrogen (by reducing the power to electrolizers) and maximize the RES sourced electricity to the Intelligent Dispenser (ID) - refer to Fig.3 - that links the EMS directly with the dispatching point of green-electricity.

Magnesium metal as key-solution for safe and viable hydrogen solid storage

Production, storage and dispatching of hydrogen as carbon-free green-energy carrier is technically allowed in INGRID plant by development of safe transportability of solid state hydrogen. Solid storage of gaseous hydrogen produced in WE is realized by use of high-capacity module made of composite material mainly based on magnesium metal. Magnesium metal is capable to store large quantity of hydrogen by absorbing atomic hydrogen to form solid compounds, namely the magnesium hydrides (MgH2), that are thermodynamically stable up to around 350°C. Such a solid-state hydrogen storage method was selected as storage technology in INGRID project since it is intrinsically safe (no pressurized reactive hydrogen gas is present) and it is a very long-term storage. Regarding metallurgical phenomena involved, high diffusivity of hydrogen is recognized for several metals (hydrogen has a small atomic radius specie with high diffusivity in metal crystal lattice. Stability of solid compounds hydrogen can form with atomic metals as well as the reversibility of hydrogen atoms absorption (namely, the desorption reaction) are keys for deploying metals in hydrogen storing purposes. However, the kinetics of absorption and desorption phase is most relevant key issues to allow metal base material to be employed industrially as materials for hydrogen solid state storage. Among several metals to be employed for industrial applications, magnesium metal is one of most interesting materials, due to its low density and relative high storing capacity (when transportability of solid tanks is an issue)[8], rapid kinetics for absorption and desorption and relatively competitive costs. Specific researches have been conducted onto metallurgical issues to optimize kinetics for absorption and desorption to be fully compatible with industrial applications. To this scope, it is worth noticing hydrogen atoms penetrate and diffuse into metal by preferential routes; such routes are mainly grain boundaries because of their higher concentration of free spaces (vacations) left by magnesium atoms. Thus the "storing" phase develops preferentially through formation of intermetallic compounds at interstitial sites present in crystal lattice[9]. In last two decades further breakthroughs in hydrogen sorption kinetics – namely, solution to speed up the absorption and desorption phase - have been achieved by pre-treating magnesium by high-energy ball milling operations [1-5]. Ball milling mechanical treatment consists in applying severe plastic deformation to material to produce a highly refined microstructure, nanostructure, with a dramatic increase of grain boundaries, most preferred patterns for hydrogen diffusion. The importance of hydrogen kinetics sorption and lightness of hosting material (magnesium has very low density) are two key-issues to take into account for commercially exploiting solid-state hydrogen. Regarding to HSS unit specifically designed for INGRID plant, a minimum time for

[8] Magnesium has a good storage capacity, achieving theoretically the 7.6 wt.% and it has density of 1.74 g/dm^3.

[9] Further constraints that work against absorption phase are: a) the presence of oxide layers that spontaneously formed in air on metals and, b) such a "barrier effect" that impedes hydrogen atoms to more deeply penetrate inside bulk because top layers saturate. These represents actually physical (oxide layers) and chemical (saturated layers) barriers that can prevent a fully hydrogen absorption inside bulk material; these barriers are generally responsible of lower sorption kinetics and practical absorption rate (weight percentage of hydrogen absorbed into host metal) far from theoretical values.

complete loading/unloading operations of a 100kg capacity hydrogen module takes about 8 hours.
Additionally, one further drawback is due to the fact gaseous hydrogen can be absorbed by magnesium with an exothermic reaction, meanwhile the reversible desorption is an endothermic reaction. When hydrogen is absorbed and reacts with Mg to form MgH_2, the reaction $Mg + H_2 \rightarrow MgH_2$ releases heat. This heat shall be managed and thus evacuated. Conversely, the reaction of hydrogen desorption by magnesium $MgH_2 \rightarrow Mg + H_2$ is endothermic, and this fact implies to provide to HSS module sufficient heat to start reaction and proceed. This heat energy has to be provided by external auxiliaries like electrical heating[10]. This facts lead to proper HSS module design to be safely handled, easily transported by truck and to allow rapid connection to auxiliaries for providing heating in the desorption stage. To satisfy all such technical constraints, the HSS module has been designed as show in Figure 4. Particularly, the single moveable standard tank container is realized with multiple of 100kg hydrogen capacity HSS modules (Fig.4). Internally the HSS module are realized as detailed in Fig. 5, while in Fig.6 it is shown a schematic top view of the storing and charging area in INGRID plant. As it is visible, the HSS tanks are connected to hydrogen generation units (WE units) and when the HSS module is charged and ready for be transported to customer, it is linked to truck and moved. In the left-bottom part of Fig.6 it is visible the fuel cell units that is directly connected to stationary solid storage module; in INGRID plant the FC unit serves for recharging BEV car locally available.

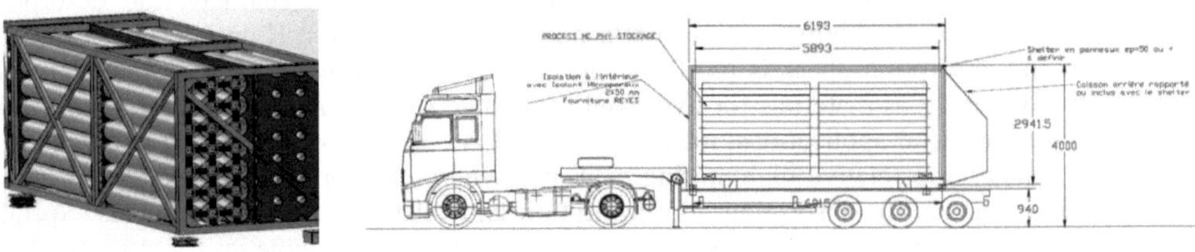

(a) (b)

Figure 4 - Scheme of the transportable storage: a) the internal structure of HSS module, vessels for hydrogen storage are visible across metal frame; b) the HSS tank moveable container on truck.

[10] The exothermic and endothermic reactions of hydrogen absorption and desorption respectively in magnesium lattice occurs around 300 °C. One possible route to produce Mg hydride reactions with lowered reaction enthalpy (i.e. low temperatures for uptake and release phases) is to add elements which exhibit lower enthalpies of formation. Mg_2Ni is an example of this class of materials, as the enthalpies of formation for the ternary hydride Mg_2NiH_4 and the Mg_2Ni is of the order of -67 $kJmol^{-1}$ per H_2, therefore lower than the -78 $kJmol^{-1}$ per H_2 that is enthalpy for the MgH_2 formation. Unfortunately this route reduces drastically the storage capacity and increases the density of the modified Mg-based material.

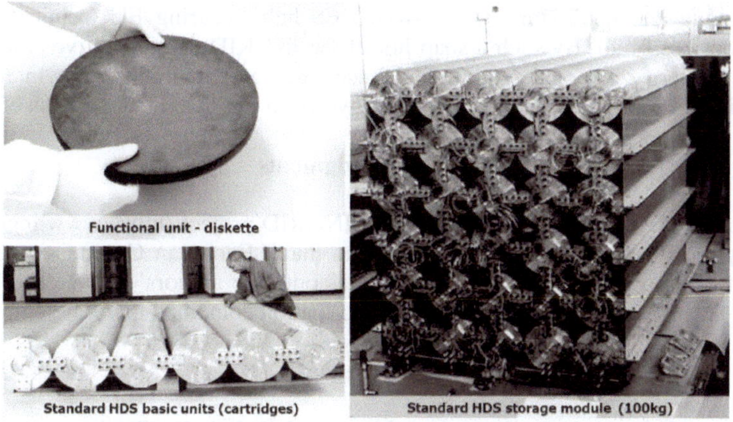

Figure 5 - Hydrogen solid storage module; anticlockwise view from up left: the single diskette module made of nanostructured magnesium; the single cartridge where several diskettes are placed; the 100kg module with several cartridges.

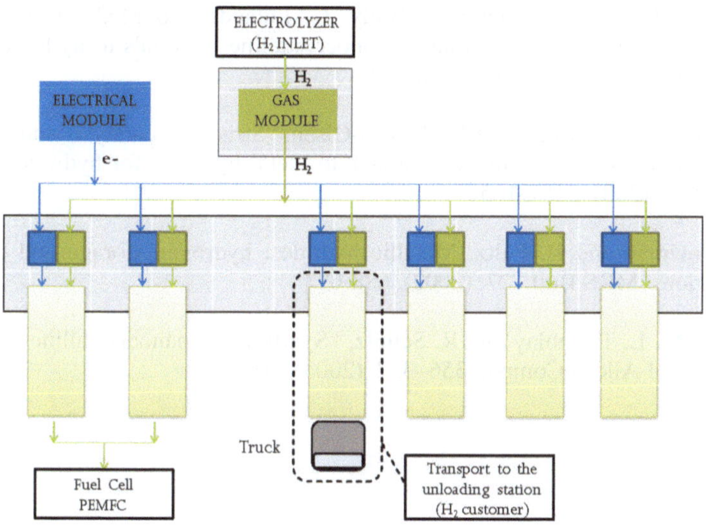

Figure 6 – Scheme of charging area in INGRID Plant.

Conclusions

The EU Commission Roadmap 2050 – Practical guide to a prosperous, low-carbon Europe (April 2010) states that the energy transition towards a decarbonized economy has benefits. This Roadmap has developed a reference projection of the energy transition to test different scenarios for technology and climate policies in the next half-century. Among those ones, lower energy costs per unit of output, more stable and predictable energy prices, an increased security of energy supply for much more economic stability, a more stable sustainable energy. In such a context, the INGRID concept can contribute in a concrete and fundamental portion in sustaining the European vision for this pathway. It is thought in order to building the pre-requisite conditions for the setting up of the EU "green energy economy from RES" in the near term period. Managing and deploying energy from RES help securing EU Countries from energy dependency on fossil fuel. By such a snapshot of the INGRID demonstrative case supply-chain, it is possible to verify in practice some key-issue, such as potentialities for interconnection of electricity grid and gas grid by exploiting green-hydrogen produced from excess of RES power.

Acknowledgments

This work makes use of results produced by the INGRID project (http://www.ingridproject.eu), co-funded by the European Commission. Authors thank European Commission for supporting project dissemination activities. This work reflects only the authors' views. The Community is not liable for any use that may be made of the information contained therein.

References

1. G. Liang, J. Huot, S. Boily, A. van Neste and R. Schulz, Catalytic effect of transition metals on hydrogen sorption in nanocrystalline ball milled MgH2–Tm(Tm = Ti, V,Mn,Fe and Ni) systems. J Alloys Compd, 292 (1999), p. 247-252

2. R. Shulz, J. Huot, G. Liang, S. Boily, G. Lalande and M.C. Denis, et al. "Recent developments in the applications of nanocrystalline materials to hydrogen technologies. Mater Sci Eng, A267 (1999), 240-245

3. A. Zaluska, L. Zaluski and J.O. Stroem-Olsen, "Structure, catalysis and atomic reactions on the nanoscale: a systematic approach to metal hydrides for hydrogen storage", Appl Phys, A72 (2001), 157-165

4. R.C. Bowman and B. Fultz, Metallic hydrides: hydrogen storage and other gas-phase applications. MRS Bull, 27 (2002), 688-693

5. J. Huot, M. L. Tremblay and R. Schulz, "Synthesis of nanocrystalline hydrogen storage materials", J Alloys Compd, 356–357 (2003), 603-607

Engineering Solutions for Sustainability: Materials and Resources II
Edited by: Jeffrey W. Fergus, Brajendra Mishra, Dayan Anderson, Emily Allyn Sarver, and Neale R. Neelameggham
TMS (The Minerals, Metals & Materials Society), 2015

RECHARGEABLE MAGNESIUM BATTERIES WITH NOVEL PVdF–PAN GRAFT COPOLYMER ELECTROLYTE MEMBRANES

M. Latha[b], Ch. SandeepReddy[b], V. Kanakaiah[b], G. Venkateshwarlu[b], K. Srinivasulu[b] Ch. KotiReddy[a], D. Shailaja[a]*& J. Vatsala rani[b]*

[a]Polymers & Functional Materials Division, [b]FluoroOrganic Division
CSIR-Indian Institute of Chemical Technology, Hyderabad– 500007 India

Abstract

Rechargeable magnesium cells will be an important storage medium when the sustainable economy of the future is looking towards an alternative to lithium ion batteries as the energy storage medium. Rechargeable magnesium cells with Mg metal anode and graphite fluoride cathode were assembled with graft copolymer (PVDF-g-PAN50) soaked in imidazolium based ionic liquid and basic electrochemical studies, cyclic voltammetry and impedance were carried out to analyze the performance of the copolymer. Cyclic voltammetric studies show good mobility of Mg ions in the copolymer and also reversibility of Mg ions in the cathode material. Impedance studies reveal reduced resistance of the cell which may be due to formation of thinner or better conducting solid electrolyte interface on the electrode. Charge-discharge results of the test cells with grafted copolymer revealed negligible capacity loss.

Keywords: PVDF, PAN, graft copolymer, rechargeable Mg battery, graphite fluoride

Corresponding Authors: D. Shailaja & J. Vatsala Rani

J.Vatsala rani, email : vatsala@iict.res.in

D.Shailaja, email : sdonempudi@iict.res.in

Tel. No: +91-40-27193208, Fax No: +91-40-27193991

1.0 Introduction

Magnesium based energy systems, inherit advantages of low cost, low toxicity, high safety and high negative potential and charge density. Thus magnesium is a natural choice for potential use as anode material in metal-ion batteries [1]. Natural graphite is extensively used in metal–ion batteries as electrode material due to its layered structure. The large π-electronic network in graphite aids in formation of intercalation compounds [2]. The carbon fluorides (CF_x) formed at low temperatures (LT) have ionic character and retain very good electronic conductivity [3]. Fluorination or functionalization is an useful way to improve the specific energy of carbon materials, the functional groups or hetero-atoms incorporated can help in adsorption of ions and then improve the hydrophilicity/lipophilicity of the carbon materials. The enhanced wettability assists in rapid electrolyte ion transport within the micropores. Yazami and Hamwi [4] synthesized graphite fluoride at room temperature and used as cathode material in solid state PEO-based lithium batteries, cyclic voltammetric studies of the cathode material displayed reversibility with two reversible peaks at approximately 3V. Reversible Mg deposition in aprotic solvents such as alkyl carbonates, esters and acetonitrile is difficult, due to surface passivation of magnesium metal by the reduction products of the electrolyte [5]. Room temperature molten salts or ionic liquids (ILs) are the next generation electrolytes, because of their wide liquid –phase range, good ionic conductivity, lack of volatility, non-flammability, very low vapour pressure and great thermal and electrochemical stability [6]. Since ILs are solvent-free, there is no solvation shell in ILs, and thus ILs can offer a well identified ion size. Reversible intercalation and de-intercalation of Al-ions was observed in ionic liquid (1,3-di-n-butyl imidazolium Bromide [bim] [Br]) [7].

Organic polymer membranes loaded with inorganic electrolytes are generally used as solid polymer electrolytes for rechargeable battery applications, because it avoids direct solvent interaction with the electrode and helps minimize surface passivation of the metallic anode to large extent. It is known from literature that polyethylene (PE) [7,8], polypropylene (PP) [9], polyethylene oxide (PEO) [10-12], polyacrylonitrile (PAN) [13-15], polymethylmethacrylate (PMMA) [16], polyvinylidenefluoride (PVDF) and its copolymer [17-21], were used as separator membranes in batteries. Amongst them, PVDF is found to be most suitable due to its outstanding mechanical properties, high chemical resistance, good thermal stability as well as high piezoelectric and pyroelectric coefficients. PVDF is often used by modification with other polymers to overcome the limitations of higher crystallinity that gives poor stability to the loaded liquid electrolytes resulting in low ionic conductivity [22]. Composites of PVdF and PAN are known for their efficient performance as gel polymer electrolytes for offering synergistic benefits of individual polymers [23]. While PAN provides superior mechanical stability and interfacial characteristics, PVdF gives good electrochemical stability and affinity for metal ions [13].

In the present study, we have reported the use of (PVDF-g-PAN50) as polymer electrolyte membrane. The membranes were loaded with magnesium salts and soaked in imidazolium based ionic liquid. Test cells with Mg metal anode and graphite fluoride cathode were assembled using the graft copolymer membrane and was subjected to basic electrochemical studies, cyclic voltammetry and impedance to analyze the performance of the copolymer as separator in magnesium batteries.

2.0 Experimental

2.1 *Chemicals:*

Natural graphite foil 0.5mm thick (Alfa-Aesar), pyridinium poly(hydrogen fluoride) (PPHF), 1-n-butyl imidazole, n-butyl bromide, magnesium powder, magnesium-perchlorate, PVdF with a molecular weight of 180 kg/mol, Acrylonitrile (AN), 2, 2'-azobisisobutyronitrile, N,N-Dimethylformamide (DMF), Ethanol were purchased from Sigma-Aldrich.

2.2 *Synthesis of ionic liquid:*

Bromine based ionic liquid 1,3-di-n-butylimidazolium bromide ([bim] [Br]) is prepared by mixing 1-n-butyl imidazole and n-butyl bromide in 1:1.2 molar ratio, at 90°C for 12h, the excess n-butyl bromide is removed under reduced pressure [12] 0.5 M $Mg(ClO_4)_2$ is added to ionic liquid.

2.3 *Synthesis of the PVDF-g-PAN50 graft copolymers:*

Before graft polymerization, the PVDF powder was treated with potassium hydroxide (KOH). In brief, PVDF was immersed in a 10 wt % KOH solution containing 0.05 wt % ethanol, and then the solution was stirred for 30 min at 60^0C. After the solution cooled, the precipitate was collected by filtration and then washed four times with distilled water to remove the alkali solution. The 1g of alkaline-treated PVDF powder was dissolved in DMF at 60^0C with a concentration of 10 wt %, and then 1g of AN and 0.04g of 2,2'-azobisisobutyronitrile were added to the system under an N_2 atmosphere with continuous stirring. The reaction was allowed to proceed for 24 h under constant agitation at 60^0C. The graft copolymer was precipitated by methanol and collected by filtration. To remove the residue of the AN monomer, the graft copolymer was washed with a mass of distilled water and methanol several times and then was dried in vacuum.

2.4 *Preparation of the membrane:*

PVDF and PVDF-g-PAN50 membranes were prepared by phase inversion from a DMF solution with a 15 wt % concentration of the polymer or copolymer. The solution of PVDF or PVDF-g-PAN50 was cast onto a glass plate and subsequently immersed in a coagulation bath. Distilled water at room temperature was used as the coagulation bath. Each membrane separated from the glass plate by itself after immersion in water for about 1 min. The formed and dried membranes were soaked in ionic liquid for 48h under nitrogen in glove box for further characterization.

2.5 *Electrochemical fluorination of graphite:*

Natural graphite foil, 0.5mm thick is cut into 10mm x 20mm strip and used as working electrode in the fluorination cell. Teflon container was used to electrochemically fluorinate graphite with three electrode assembly, graphite plate (10mm x 20mm) as working electrode, Pt plate counter electrode, Pd wire reference electrode, electrolyte was pyridinium poly (hydrogen fluoride). Fluorination of graphite was done at room temperature in inert atmosphere by potential sweep technique, the potential was scanned from -1.0V to +1.0 V at 20 mVs^{-1}, for 30 mins.

2.6 *Mg-ionl CF_x cell assembly*:

Ionic conductivity of the grafted copolymer membranes were analysed by sandwiching the membranes between two SS plates. Prior to placing the membranes between the SS plates it was soaked in ionic liquid for 48 hrs. Test cell 'C' for electrochemical studies was assembled with copolymer membrane which showed maximum conductivity, the membrane was sandwiched between two electrodes Mg metal ($1cm^2$) and graphite fluoride sheet ($1cm^2$). Electroanalytic studies, cyclic-voltammetry (CV), and charge-discharge were performed with test cell. Ionic conductivity of the membrane was measured by electrochemical impedance spectroscopy (EIS). The frequency range used is 10 mHz to 1000 kHz, and the amplitude is 5mV after complete equilibration of the electrode. Intercalation, de-intercalation and reversibility of magnesium - ions in graphite fluoride was analysed by cyclic voltammetry. Cycle life of the test cells were analysed by charge-discharge studies at constant current discharge of $0.1mAcm^{-2}$.

2.7 *Characterization and instrumentation*:

The graphite fluoride was characterized by X-ray spectroscopy (Hitachi with field emission gun), X-ray photoemission spectra (KRATOS AXIS 165 ESCALAB, MgKα anode), and surface area (BET, Micromeritics). Electrochemical measurements cyclic voltammetry, Electrochemical impedance spectroscopy and charge –discharge studies of the cell was performed on (Bio-logic SAS, model 1- VSP, France make), electrochemical work station. Proton NMR spectra were recorded in deuterated DMF solution using a AVANCE 500 MHz spectrometer. Chemical shifts are given in ppm with tetramethylsilane (TMS) as a standard.

3.0 Results and Discussion:

3.1 *^{1}H NMR studies of grafted copolymer membranes*:

Figure:1 shows ^{1}H NMR spectra of PVDF and PVDF-g- PAN50 copolymers. The peaks at 3.32 and 2.90 ppm are due to head-to-tail (ht, $CH_2CF_2CH_2CF_2$) and head-to-head (hh, $CF_2CH_2CH_2CF_2$) bonding arrangements of PVDF. PVDF-g- PAN50 copolymer shows additional peaks in the region of 2.1 ppm and 3.4 ppm, which can be attributed to the methylene and methyne protons of acrylonitrile. This clearly indicates that PAN has been successfully grafted onto the main chain of the alkali-treated PVDF [24, 25].

3.2 *Electrolyte uptake*:

Polymer membranes were transformed into polymer electrolytes by soaking the polymer membranes in ionic liquid solution at room temperature inside the glove box. The % of electrolyte uptake was determined by soaking the membranes in ionic liquid solution at different intervals and knowing the changes in mass of the membranes.

$$\% \text{ of electrolyte uptake} = W_a - W_b / W_b \quad x100$$

Where W_a and W_b are weight of membranes after and before immersion in the ionic liquid electrolyte, respectively. A remarkable increase in the electrolyte uptake capacity of the PVDF-g- PAN50 copolymer membrane with >300% was noticed in comparison to PVDF membrane with 160.8% and this would contribute to enhanced ionic conductivity for the graft copolymer [14].

3.3 *Ionic conductivity of the copolymer membranes:*

The ionic conductivity values of pure PAN, PVDF 1.42×10^{-11} Scm^{-1} and 0.74×10^{-3} Scm^{-1} respectively. Three different PAN grafted PVDF were prepared by varying the PAN composition, the ionic conductivity of the grafted copolymer membranes was better, when compared with the ionic conductivity of PAN, PVDF taken individually. The ionic conductivity of the composition PVDF-g-PAN(50) exhibited maximum ionic conductivity value of 3.71×10^{-3} Scm^{-1}.

3.4 *Electrochemical preparation and XPS characterization of graphite fluoride:*

Cyclic voltammogram of natural graphite in PPHF at -1.0V to 1.0V potential scan is shown in Figure:2 indicates a single oxidation peak at 0.15V and corresponding reduction peak at -0.17V. Repeated CV scans (20 scans) at this potential window show negligible decrease in peak currents. The large π –electronic network in graphite can either be oxidized or reduced quite easily. An electronegative species (F) accepts an electron and forms an ionic bond with the π –electronic network (represented as C_xF below) and an acceptor type graphite intercalation compound (GIC) is produced [2].

$$C_x + A \leftrightarrow C_x^+ \cdot A^- \; (C_xF)$$

C1s binding energy values of graphite fluoride, calculated from XPS spectra was 284.6, 286.1 & 287.6 eV, the binding energy values attribute to ionic C-F bond formation, the values also indicate the formation of three different compositions of C_xF compounds, $C_{2.5}F$, $C_{2.2}F$ & C_2F [26]. The binding energy value of 684.5 eV observed for F1s in graphite fluoride was nearly equal to the ionic F in LiF (F1s: 684.9eV) rather than covalent F in Teflon (F1s: 689.1eV) [27,28], which further specifies formation of ionic GIC. The resistivity of fluorinated graphite was almost halved ($1.2 \times 10^{-3}\Omega cm$) when compared with natural graphite. BET analysis indicated threefold increase ($0.9086 m^2g^{-1}$) in surface area of the natural graphite foil after fluorination. Electrically conducting graphite fluoride C_xF ($5 > x > 2$) was prepared by N. Bartlett et al [29] in the presence of liquid hydrogen fluoride at room temperature. The superior conductivity of graphite fluoride formed was attributed to the formation of bifluoride salt ($C_{12}^+HF_2^-$) as an intermediate in the fluorination process.

3.5 *Cyclic voltammetric studies of test cell 'C':*

Mobility and reversibility of Mg ions in copolymer membrane were analysed by cyclic voltammetric studies. The copolymer membrane exhibiting maximum conductivity was chosen, it was sandwiched between the anode and cathode (cell 'C') and CV studies were performed. The cyclic voltammogram of cell 'C' at $5mVs^{-1}$ scan rate shown in Figure.3 indicates an oxidation hump at 2.1V and corresponding reduction hump was observed at 1.25V. The CV also indicated good reversibility since the total anodic charge was almost equal to the cathodic charge. Reversibility of the cell was further confirmed by subjecting the cell to charge-discharge studies.

3.6 *Charge-discharge studies of the cells:*

The test cell was charge-discharge at constant current of $0.1mA\ cm^{-2}$, charging was continued till the open circuit potential (OCV) of the cell reached 2.5V, discharge was stopped when the cell OCV declined to 0.3V. Figure.4 shows the voltage vs capacity plot of the test cell 'C'. The discharge of the cell was restricted to 0.3V, discharge lower to this potential leads to oxidation of electrolyte and also formation of compounds which cannot be recharged [30]. Test cell demonstrates a well defined and very stable Mg^{2+} intercalation plateau at 2.2 V, the specific capacity of the cell was 68mAhg^{-1}. The discharge curve presented a flat plateau at 2.2V, which estimate to 50% of total discharge capacity. The flat curves while charge and discharge indicate Mg^{2+} de-intercalation and intercalation in the cathode. The galvanostatic charge-discharge cycling (50 cycles) performance of gelatinized Mg cell shown in Figure.5 shows a high degree of reversibility, good capacity retention.

3.7 __XRD studies of the cathode material of cell before and after cycling__:

X-ray diffraction spectra of the natural graphite, fluorinated graphite and charge-discharged fluorinated graphite from cell 'C' is presented in Figure.6. Very small shift was observed in the diffraction peak positions after fluorination of natural graphite, the 002 peak of graphite was predominant in all the three types of graphite, the intensity of other three peaks 100,101,004 was low. The data indicates increase in d-spacing of graphite after electrochemical fluorination, intercalation of fluorine between the graphite layers exfoliates graphite and aids in facilitating Mg-ions in the graphite sheets while discharge of the cell. Increase in d-spacing in the discharged graphite electrode (after Mg intercalation) was also observed but it was lower when compared with the d-spacing increase observed in graphite fluoride. The diffraction spectrum specifies that the basic hexagonal symmetry of the graphite crystal is not disturbed, it is retained after repeated insertion and deposition of Mg–ions in the discharged fluorinated graphite cathode.

The high degree of reversibility (50 cycles), flat discharge plateau at 2.2 V and good columbic competency (97%), in Mg cell with grafted copolymer is quite encouraging. The good reversibility of Mg cell with fluorinated graphite cathode and ILs soaked grafted copolymer electrolyte may be attributed to three important features observed in the system, i) exfoliation of natural graphite on electrochemical fluorination, which facilitates repeated intercalation of Mg ions in the cathode material, while charge –discharge of the cell. ii) the electrolyte is held in the grafted copolymer membrane, which results in indirect contact of the electrode with the electrolyte, which in turn leads to formation of thin conducting solid electrolyte interface (SEI) with lower resistance on the electrode surface, iii) stability or detainment of the basic hexagonal structure of fluorinated graphite after repeated Mg-ion intercalation and deposition extends the cycle life of the system.

Extensive studies are in progress in the direction of enhancing the specific discharge capacity of the cell by modifying the composition of the electrodes and electrolyte. We foresee considerable openings for the chemical and design engineering at nanoscale level of cathode and anode composite materials. Simultaneously we expect noteworthy efforts to initiate Mg-ion conducting electrolytes to augment cell performance at elevated current drains.

4.0 __Conclusion__:

Electrochemical studies of the Mg metal-ion battery assembled with ionic liquid loaded grafted copolymer membrane was studied. It was observed that (PVDF-g-PAN50) had shown maximum electrolyte uptake capacity of >300% thereby eliminating the problem of electrolyte spillage and also improvement in the electrochemical performance of the Mg-ion power source.

As the electrolyte is held in the copolymer membrane, there is no direct contact of the electrolyte with the electrode hence a solid electrolyte interface (SEI) with lower resistance is formed on the electrode surface.

Acknowledgements:
The authors are thankful to CSIR, New Delhi under the TAPSUN program (NWP-0056) for funding. Authors also thank Dr. S. Gopukumar, CECRI, Karaikudi.

References:

[1] T.D.Gregory, R.J. Hoffman, R.C.Winterton, *J. Electrochem.Soc.*,**137,** 775 (1990).
[2] M.Noel, R.Santhanam, *J. of Power Sources.*,**72,** 53 (1998).
[3] Glenn G. Amatucci, Nathalie Pereira, *J.of Fluorine Chem.*,**128**, 243 (2007).
[4] R. Yazami, A. Hamwi, *Solid State Ionics.,* **28**, 1756 (1988).
[5] Z. Lu, A. Schechter, M. Moshkovich, D. Aurbach, *J. Electroanal.Chem.* **466,** 203 (1999).
[6] Yanna NuLi, Jun Yang, Rong Wu, *Electrochem.Comm.*, **7**, 1105 (2005).
[7] Y. S. Chung, S. H. Yoo, C. K. Kim. In Eng Chem Res 48, 2009, 4346.
[8] K. Gao, X. Hu, T. Yi, C. Dai. Electrochim Acta 52, 2006, 443.
[9] G. Venugopal, J. Moore, J. Howard, S. Pendalwar. J Power Sources 77, 1999, 34.
[10] B. K. Choi, K. H. Shin, Y. W. Kim. Solid State Ionics 113–115, 1998, 123.
[11] Y. Kang, H. J. Kim, E. Kim, B. Oh, J. H. Cho. J Power Sources 92, 2001, 255.
[12] B. Laik, L. Legrand, A. Chausse, R. Messina. Electrochim Acta 44, 1998, 773.
[13] A. I. Gopalan, P. Santhosh, K. M. Manesh, J. H. Nho, S. H. Kim, C.G. Hwang, K. P. Lee. J Membr Sci 325, 2008, 683.
[14] B. Huang, Z. Wang, G. Li, H. Huang, R. Xue, L. Chen, F. Wang. Solid State Ionics 85, 1996,79.
[15] S. S. Sekhon, N. Arora, S. A. Agnihotry. Solid State Ionics 136–137, 2000, 1201.
[16] O. Bohke, C. Frand, M. Razrazi, C. Rousselt, C. Truche. Solid State Ionics 66, 1993, 97.
[17] C. M. Costa, L. C. Rodrigues, V. Sencadas, M. M. Silva, G. Rocha, S. Lanceros-Mendez. Journal of Membrane Science 8, 2012, 407.
[18] A. I. Gopalan, K. P. Lee, K. M. Manesh, P. Santhosh, J Membr Sci 318, 2008, 422.
[19] J. R. Kim, S. W. Choi, S. M. Jo, W. S. Lee, B. C. Kim. Electrochim Acta 50, 2004, 69.
[20] A. Manuel Stephan, Y. Saito. Solid State Ionics 148, 2002, 475.
[21] H. Kataoka, Y. Saito, T. Sakai, E. Quartarone, P. Mustarelli. J Phys Chem B 104, 2000, 11460.
[22] X. Yin, H. Cheng, X. Wang, Y. Yao. J Membr Sci 146, 1998, 179.
[23] T.Y. Liu, W.C. Lin, L.Y. Huang, S.Y. Chen, M.C. Yang. Adv Technol 16, 2005, 413.
[24] M. Lazzari, O. Chiantore, R. Mendichi, M. A. L. Quintela. Macromol Chem Phys 206, 2005, 1382.
[25] P. Bajaj, T. V. Sreekumar, K. Sen, R. Kumar, A. S. Brar. Journal of Applied Polymer Science 88, 2003, 1211.
[26] Thomas Mallouk and Neil Bartlett, *J. of Chemical Soc.Chem. Commun.*(1983)103
[27] Y.Kita, N.Watanabe, and Y.Fujii. *J. of American Chem. Soc.*,**101**, 3832 (1979).
[28] Hand book of X-ray Phototelectron Spectroscopy, ed. G,E, Muilenberg, Perkin- Elmer Corporation, Eden Prairie, Minnesota 55344,1978.
[29] Tomohiro Yoshida, Tomomi Hirakawa, Toru Nakamura, Yasuhiro Yamada, Hiroko Tatsuno, Yuki Morita, and Hiroaki Okamoto, *ECS Transactions*. **50** (48),

95 (2013).
[30] J. Vatsala Rani, S. Bhavana Rushi, V. Kanakaiah, S. Palaniappan *J. of Electrochem. Soc.*, **158,** A1031 (2011).

Figure Captions

Figure: 1. ^{1}H NMR spectra of PVDF, PAN and PVDF-g- PAN50 copolymer.

Figure: 2. Cyclic Voltammogram of natural graphite in pyridinium poly (hydrogen fluoride) at $10mVs^{-1}$ sweep rate.

Figure: 3. Cyclic Voltammogram of the cell 'C' at $5mVs^{-1}$.

Figure: 4. Voltage *vs*. capacity of cell C.

Figure: 5 Cycle life *vs*. capacity of Cell C.

Figure: 6 XRD spectra of natural graphite, fluorinated graphite and charge-discharged fluorinated graphite from cell 'C'.

Fig.1

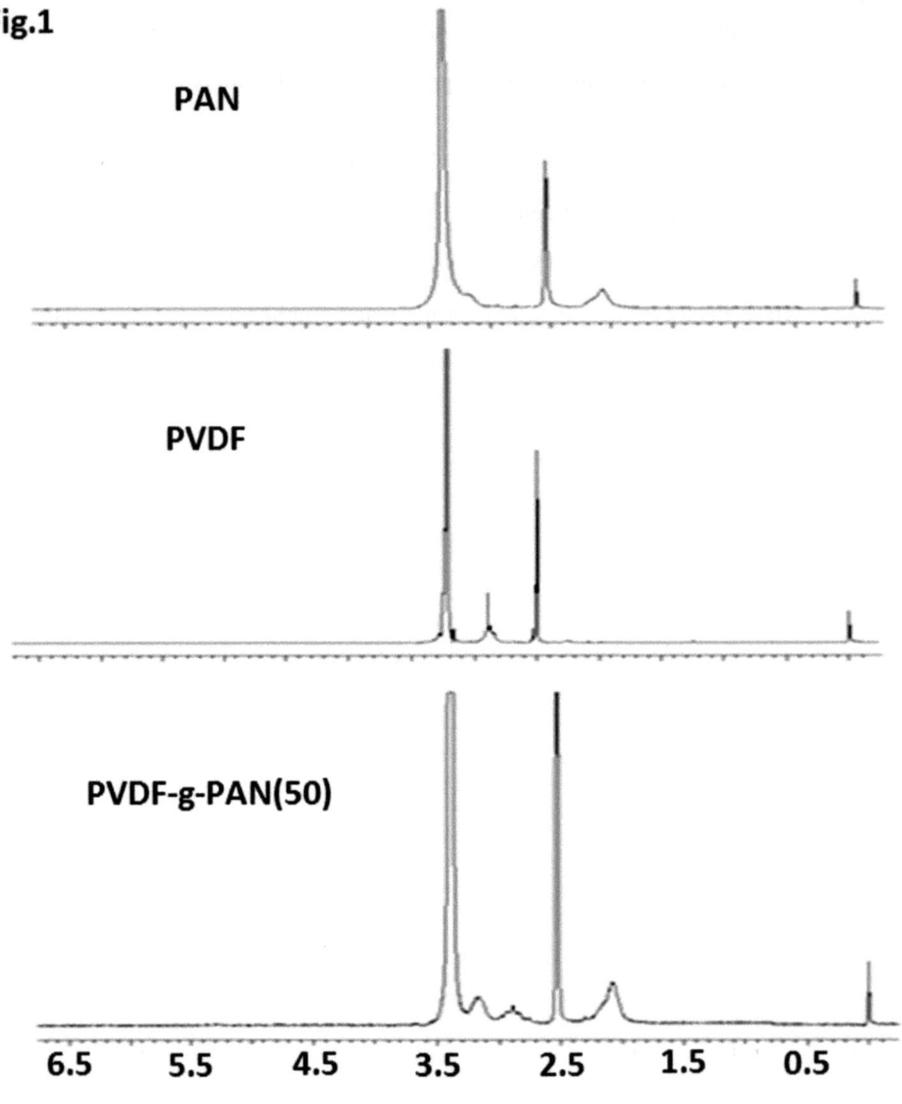

Fig 2

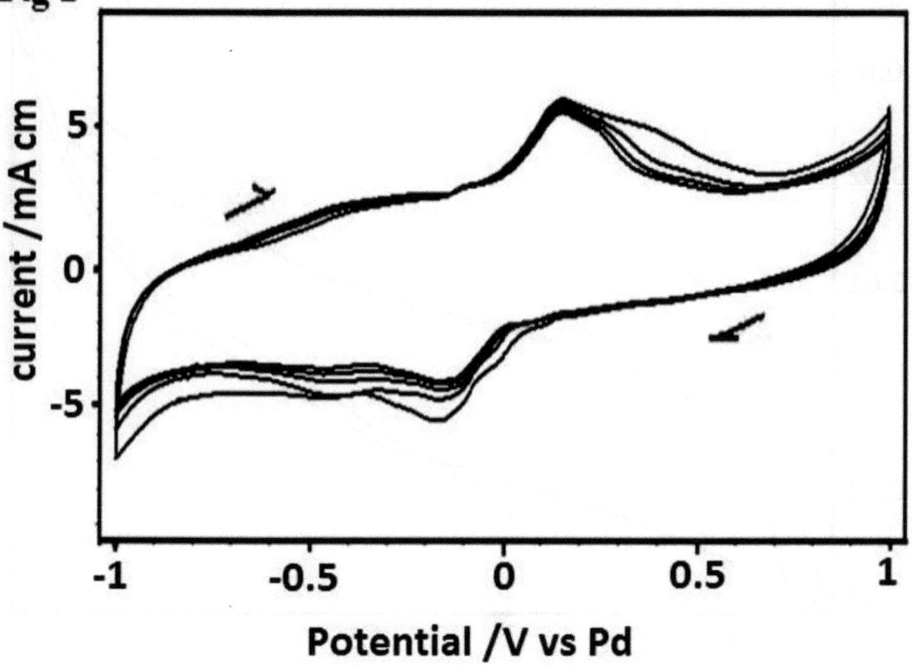

Fig.3

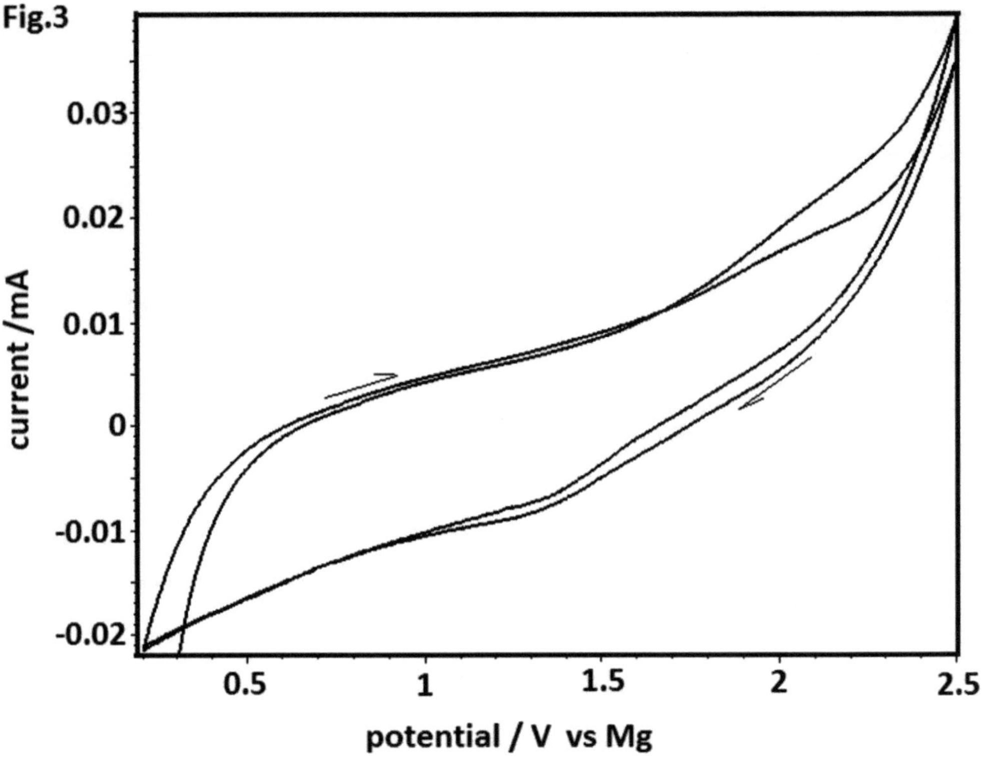

Fig.4

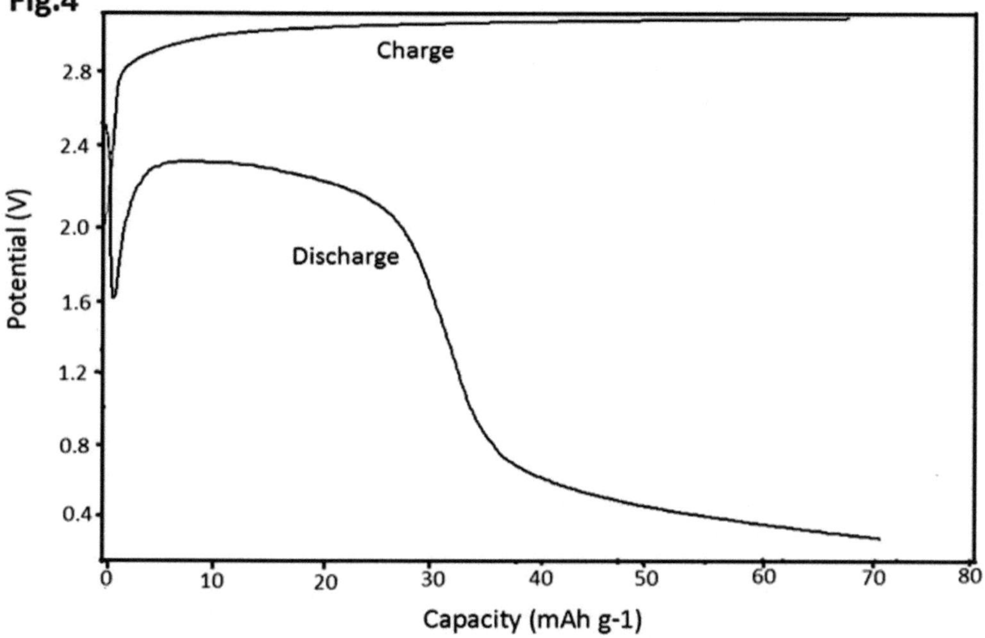

Charge
Discharge
Potential (V)
2.8
2.4
2.0
1.6
1.2
0.8
0.4
0
10
20
30
40
50
60
70
80
Capacity (mAh g-1)

Fig.5

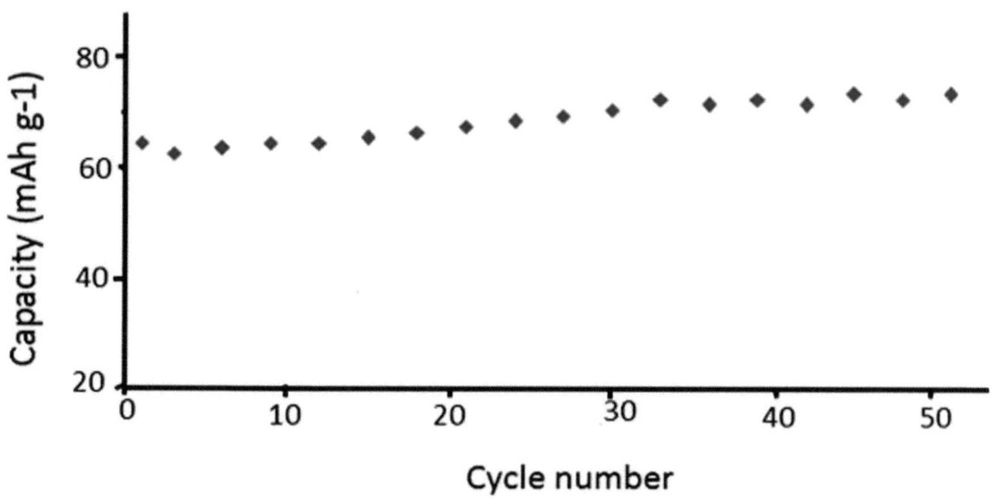

Capacity (mAh g-1)
80
60
40
20
0
10
20
30
40
50
Cycle number

Fig.6

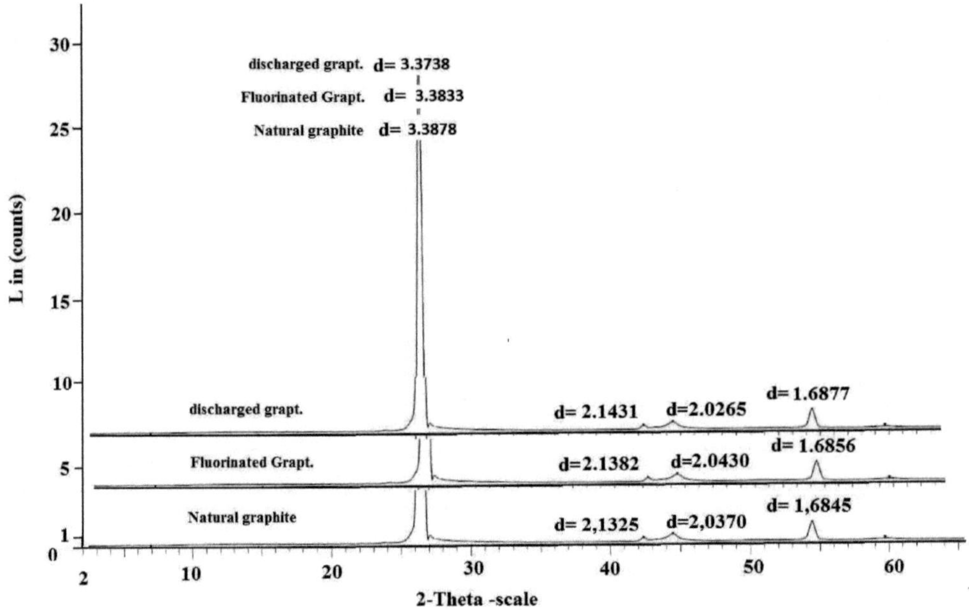

discharged grapt. d= 3.3738
Fluorinated Grapt. d= 3.3833
Natural graphite d= 3.3878
L in (counts)
discharged grapt.
d= 2.1431
d=2.0265
d= 1.6877
Fluorinated Grapt.
d=2.1382
d=2.0430
d= 1.6856
Natural graphite
d= 2,1325
d=2,0370
d= 1,6845
2-Theta -scale

AUTHOR INDEX
Engineering Solutions for Sustainability

SUBJECT INDEX
Engineering Solutions for Sustainability

MIX
Papier aus verantwortungsvollen Quellen
Paper from responsible sources
FSC® C105338

If you have any concerns about our products,
you can contact us on
ProductSafety@springernature.com

In case Publisher is established outside the EU,
the EU authorized representative is:
Springer Nature Customer Service Center GmbH
Europaplatz 3, 69115 Heidelberg, Germany

Printed by Libri Plureos GmbH
in Hamburg, Germany